Mid-Latitude Atmospheric Dynamics

Mid-Latitude Atmospheric Dynamics

A First Course

Jonathan E. Martin
The University of Wisconsin–Madison

John Wiley & Sons, Ltd

Telephone (+44) 1243 779777

Email (for orders and customer service enquiries): cs-books@wiley.co.uk
Visit our Home Page on www.wileyeurope.com or www.wiley.com

Other Wiley Editorial Offices

John Wiley & Sons Inc., 111 River Street, Hoboken, NJ 07030, USA

Jossey-Bass, 989 Market Street, San Francisco, CA 94103-1741, USA

Wiley-VCH Verlag GmbH, Boschstr. 12, D-69469 Weinheim, Germany

John Wiley & Sons Australia Ltd, 33 Park Road, Milton, Queensland 4064, Australia

John Wiley & Sons (Asia) Pte Ltd, 2 Clementi Loop #02-01, Jin Xing Distripark, Singapore 129809

John Wiley & Sons Canada Ltd, 22 Worcester Road, Etobicoke, Ontario, Canada M9W 1L1

Wiley also publishes its books in a variety of electronic formats. Some content that appears in print may not be available in electronic books.

Library of Congress Cataloging-in-Publication Data

Martin, Jonathan E.
Mid-latitude atmospheric dynamics : a first cource / Jonathan E. Martin.
p. cm.
Includes bibliographical references and index.
ISBN 13: 978-0-470-86464-7 (acid-free paper)
ISBN 10: 0-470-86464-8 (acid-free paper)
ISBN 13: 978-0-470-86465-4 (pbk. : acid-free paper)
ISBN 10: 0-470-86465-6 (pbk. : acid-free paper)
1. Dynamic meteorology. 2. Middle atmosphere. I. Title.
QC880.M36 2006
551.5—dc22 2005036659

British Library Cataloguing in Publication Data

A catalogue record for this book is available from the British Library

ISBN 13 978-0-470-86464-7 (HB)
ISBN 10 0-470-86464-8 (HB)

ISBN 13 978-0-470-86465-4 (PB)
ISBN 10 0-470-86465-6 (PB)

Typeset in 10.5/13pt Minion by TechBooks, New Delhi, India

Contents

Preface

Almost no one bears the ceaseless variability of the mid-latitude atmosphere without a firm opinion and at least some degree of interest. The parade of weather systems that are continuously developed and extinguished over this part of the globe ensures that its denizens never need to wait long for unmistakable, and sometimes dramatic, changes in the local weather. For the physical scientist with an interest in (or, as is most often the case for us, the captivated, a *fascination with*) the weather, the unsurprising, yet still remarkable, fact is that this variability is governed by the basic laws of physics first articulated by Newton centuries ago. The exact manner by which those laws are brought to bear upon an analysis of the dynamics of the atmospheric fluid has, especially in the last 100 years, become a separate branch of physics. This book is dedicated to providing an introduction to the physical and mathematical description of mid-latitude atmospheric dynamics accessible to any student possessing a solid background in classical physics and a working knowledge of calculus.

When one begins to wade through the average textbook, one often gets the sense that the author has poured everything he/she knows into the text without regard for whether it is all necessary to accomplish the educational goals of the book. My many years of teaching this material to hundreds of students have provided me with two main motivations for writing this textbook. First, students have invariably complained that the available textbooks are difficult to employ as study tools, often skipping steps in mathematical derivations and thus, on occasion, contributing more to frustration than to edification. They often wonder how the subject matter can seem so clear in lectures and then so confusing that night in the library. Second, there is no other currently available text that serves as a concise primer in the application of elementary dynamics to the central problems of modern synoptic–dynamic meteorology: the diagnosis of vertical motion, fronts and frontogenesis, and the dynamics of the cyclone life cycle from both the ω-centric and potential vorticity perspectives.

In this book I have attempted to remedy both of these shortcomings by presenting an introduction to atmospheric dynamics and its application to the understanding of mid-latitude weather systems in a penetrating conceptual and detailed mathematical fashion. The conversational tone of the book is meant to render its reading akin to attending a lecture given by someone who is profoundly excited by the subject matter.

It is hoped that this tone will increase the likelihood that the book will serve as a genuine study guide for students as they navigate through a first course in this subject.

The first five chapters of the book are specifically targeted at junior-level undergraduates who are taking a first course in atmospheric dynamics. Chapter 1 provides a review of relevant mathematical tools while Chapter 2 considers the fundamental and apparent forces at work on a rotating Earth. Chapter 3 examines the fundamental conservation laws of mass, momentum, and energy producing, along the way, the continuity equation, the equations of motion, and the energy equation. Once developed, the equations of motion are simplified in Chapter 4 through a variety of approximations thus lending insight into basic flow characteristics of the mid-latitude atmosphere. The relationship between circulation, vorticity, and divergence in fluids is examined in Chapter 5 where the quasi-geostrophic system of equations is also introduced.

The last four chapters are targeted toward those students who might subsequently take a course in synoptic–dynamic meteorology in which a significant laboratory component would be a necessary complement. The diagnosis of vertical motions is undertaken in Chapter 6. The meso-synoptic dynamics of the frontal zones that characterize mid-latitude cyclones are considered in Chapter 7 where the examination of frontogenesis and its relationship to transverse vertical circulations is presented in both the quasi- and semi-geostrophic frameworks. Chapter 8 explores the dynamics of the life cycle of mid-latitude cyclones, thus providing a particularly relevant focus for synthesis of the prior chapters. Finally, Chapter 9 provides an introduction to the use of potential vorticity diagnostics for examining the life cycle of mid-latitude cyclones. Much of the material comprising the text comes from years of lecture notes from three distinct courses in the Department of Atmospheric and Oceanic Sciences at the University of Wisconsin–Madison. Both components of the text would be suitably challenging to first-year graduate students with little prior background in meteorology or atmospheric dynamics.

Throughout the text, the emphasis is on conceptual understanding, the development of which for any given topic always precedes the application of mathematical formalism. I recognize that a level of intimacy with the mathematics is necessary but I am certain that it is not sufficient to produce a penetrating understanding of mid-latitude dynamics. Such understanding is, instead, the offspring of a marriage between a conceptual, intuitive sense of the physics of the phenomenon and the corresponding mathematical description of it. At the end of each chapter several problems, characterized by varying degrees of difficulty, are included to assist the student in reinforcing knowledge of the subject matter and in developing solid problem-solving skills. Solutions to selected problems are included at the end of the chapters as well. Complete solutions to all problems are included in a separate *Solution Manual* available from the publisher. Also included at the end of each chapter is an annotated bibliography designed to point the interested student toward seminal or other sources. A more complete, though by no means exhaustive, bibliography can be found at the end of the book.

Acknowledgments

The completion of any significant project in one's life is cause for celebration and reflection. For more than two and a half years the writing, illustrating, refining, and proofreading of this book has occupied me, at odd hours of the day, as a solo endeavor. For this reason, I bear the responsibility for any errors of fact or interpretation that might be found in the text. Of course, in reality, there is nothing "solo" about such an undertaking as a great number of people, some directly and some remotely, have contributed to this effort.

My parents, Leo and Joyce Martin, have provided me with love and constant support throughout my lifetime, affording me numerous opportunities for which I am profoundly grateful. My father's infinite curiosity and creativity exposed me early and consistently to the joys of learning and exploring – especially on excursions to Nahant Beach on stormy autumn days and nighttime walks through the snowy woods of northeastern Massachusetts. These adventures instilled within me an enduring fascination with the atmosphere and the sensible weather it delivers.

I have had the good fortune of encountering a number of excellent teachers during my education whose dedication to clear explanation and deep understanding inspired me to pursue an academic life. Dr. Robert J. Sullivan, C. F. X., whose holistic approach to education provided all those with whom he came in contact an enduring model of excellence, is especially thanked. Dr. James T. Moore provided clear, dissected interpretations of complicated mathematical expressions in my first exposure to dynamic meteorology and thereby made a lasting impression, as did his colleague Dr. Albert Pallmann through his unbridled enthusiasm for inquiry. My dissertation advisor, Dr. Peter V. Hobbs, who passed away the very day the manuscript for this book was completed, exerted a profound influence on my scholarly development through his unwavering insistence on quality, and careful attention to precision in the spoken and written word. Many other teachers, colleagues and the scores of students I have known throughout the years are also gratefully acknowledged here.

Transformation of an illustrated manuscript into a book is not a trivial matter, and demands the efforts of experts. I am grateful to Ms. Lyn Roberts, Ms. Keily Larkins, Dr. Andrew Slade, Ms. Julie Ward, Ms. Lizzy Kingston, and Mr. Jon Peacock at Wiley who guided the book through to production, allowing me the freedom to write it, illustrate it and cover it my way. Thanks also to Mr. Neville Hankins who provided experienced and insightful copyediting to the project. Ms. Jean Phillips provided invaluable assistance in constructing the index.

Special mention is reserved for my family, who have faithfully supported both this enterprise and my sometimes flagging spirits. My daughter, Charlotte, and my son, Niall, have feigned excitement and interest at exactly the necessary moments and have, in other ways as well, provided me with the inspiration to soldier on. A father could not be prouder than I am of them. And finally, to my lovely wife Minh, a rock solid source of encouragement and support, who has lit my soul since the day we met; thank you for the magic and enduring warmth of your love in my life. When one feels so profound a gratitude as I do to these people, language fails to convey even its smallest fraction.

Jonathan E. Martin
Madison, Wisconsin
January 10, 2006

1

Introduction and Review of Mathematical Tools

Objectives

The Earth's atmosphere is majestic in its beauty, awesome in its power, and complex in its behavior. From the smallest drops of dew or the tiniest snowflakes to the enormous circulation systems known as mid-latitude cyclones, all atmospheric phenomena are governed by physical laws. These laws can be written in the language of mathematics and, indeed, must be explored in that vernacular in order to develop a penetrating understanding of the behavior of the atmosphere. However, it is equally vital that a physical understanding accompany the mathematical formalism in this comprehensive development of insight. In principle, if one had a complete understanding of the behavior of seven basic variables describing the current state of the atmosphere (these will be called **basic state variables** in this book), namely u, v, and w (the components of the 3-D wind), T (the temperature), P (the pressure), ϕ (the geopotential), and q (the humidity), then one could describe the future state of the atmosphere by considering the equations that govern the evolution of each variable. It is not, however, immediately apparent what form these equations might take. In this book we will develop those equations in order to develop an understanding of the basic dynamics that govern the behavior of the atmosphere at middle latitudes on Earth.

In this chapter we lay the foundation for that development by reviewing a number of basic conceptual and mathematical tools that will prove invaluable in this task. We begin by assessing the troubling but useful notion that the air surrounding us can be considered a continuous fluid. We then proceed to a review of useful mathematical tools including vector calculus, the Taylor series expansion of a function, centered difference approximations, and the relationship between the Lagrangian and Eulerian derivatives. We then examine the notion of estimating using scale analysis and conclude the chapter by considering the basic kinematics of fluid flows.

Mid-Latitude Atmospheric Dynamics Jonathan E. Martin

1.1 Fluids and the Nature of Fluid Dynamics

Our experience with the natural world makes clear that physical objects manifest themselves in a variety of forms. Most of these physical objects (and every one of them with which we will concern ourselves in this book) have **mass**. The mass of an object can be thought of as a measure of its substance. The Earth's atmosphere is one such object. It certainly has mass[1] but differs from, say, a rock in that it is not solid. In fact, the Earth's atmosphere is an example of a general category of substances known as fluids. A fluid can be colloquially defined as any substance that takes the shape of its container. Aside from the air around us, another fluid with which we are all familiar is water. A given mass of liquid water clearly adopts the shape of any container into which it is poured. The given mass of liquid water just mentioned, like the air around us, is actually composed of discrete molecules. In our subsequent discussions of the behavior of the atmospheric fluid, however, we need not concern ourselves with the details of the molecular structure of the air. We can instead treat the atmosphere as a continuous fluid entity, or **continuum**. Though the assumption of a continuous fluid seems to fly in the face of what we recognize as the underlying, discrete molecular reality, it is nonetheless an insightful concept. For instance, it is much more tenable to consider the flow of air we refer to as the wind to be a manifestation of the motion of such a continuous fluid. Any 'point' or 'parcel' to which we refer will be properly considered as a very small volume element that contains large numbers of molecules. The various basic state variables mentioned above will be assumed to have unique values at each such 'point' in the continuum and we will confidently assume that the variables and their derivatives are continuous functions of physical space and time. This means, of course, that the fundamental physical laws governing the motions of the atmospheric fluid can be expressed in terms of a set of partial differential equations in which the basic state variables are the dependent variables and space and time are the independent variables. In order to construct these equations, we will rely on some mathematical tools that you may have seen before. The following section will offer a review of a number of the more important ones.

1.2 Review of Useful Mathematical Tools

We have already considered, in a conceptual sense only, the rather unique nature of fluids. A variety of mathematical tools must be brought to bear in order to construct rigorous descriptions of the behavior of these fascinating fluids. In the following section we will review a number of these tools in some detail. The reader familiar with any of these topics may skip the treatments offered here and run no risk of confusion later. We will begin our review by considering elements of vector analysis.

---

[1] The Earth's atmosphere has a mass of 5.265×10^{18} kg!

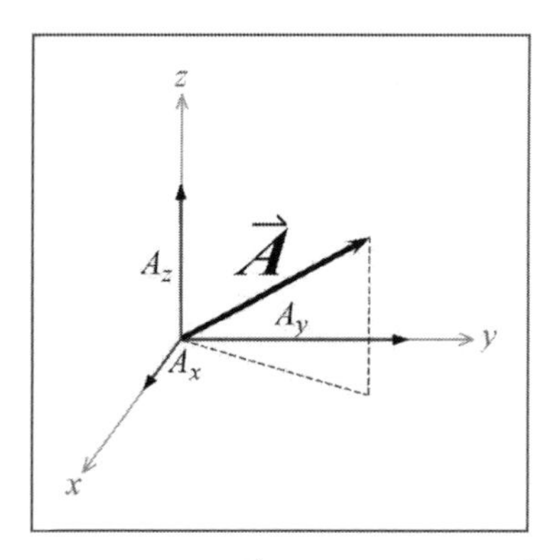

Figure 1.1 The 3-D representation of a vector, $\vec{A}$. The components of $\vec{A}$ are shown along the coordinate axes

1.2.1 Elements of vector calculus

Many physical quantities with which we are concerned in our experience of the universe are described entirely in terms of *magnitude*. Examples of these types of quantities, known as **scalars**, are area, volume, money, and snowfall total. There are other physical quantities such as velocity, the force of gravity, and slopes to topography which are characterized by both magnitude and direction. Such quantities are known as **vectors** and, as you might guess, any description of the fluid atmosphere necessarily contains reference to both scalars and vectors. Thus, it is important that we familiarize ourselves with the mathematical descriptions of these quantities, a formalism known as vector analysis.[2]

Employing a Cartesian coordinate system in which the three directions (x, y, and z) are mutually orthogonal (i.e. perpendicular to one another), an arbitrary vector, $\vec{A}$, has components in the x, y, and z directions labeled A_x, A_y, and A_z, respectively. These components themselves are scalars since they describe the magnitude of vectors whose directions are given by the coordinate axes (as shown in Figure 1.1). If we denote the direction vectors in the x, y, and z directions as $\hat{\imath}$, $\hat{\jmath}$, and $\hat{k}$, respectively (where the ˆ symbol indicates the fact that they are vectors with magnitude 1 in the respective directions – so-called **unit vectors**), then

$$\vec{A} = A_x\hat{\imath} + A_y\hat{\jmath} + A_z\hat{k} \tag{1.1a}$$

is the component form of the vector, $\vec{A}$. In a similar manner, the component form of an arbitrary vector $\vec{B}$ is given by

$$\vec{B} = B_x\hat{\imath} + B_y\hat{\jmath} + B_z\hat{k}. \tag{1.1b}$$

[2] Vector analysis is generally considered to have been invented by the Irish mathematician Sir William Rowan Hamilton in 1843. Despite its enormous value in the physical sciences, vector analysis was met with skepticism in the nineteenth century. In fact, Lord Kelvin wrote, in the 1890s, that vectors were 'an unmixed evil to those who have touched them in any way . . vectors . . have never been of the slightest use to any creature'. Remember, no matter how great a thinker one may be, one cannot always be right!

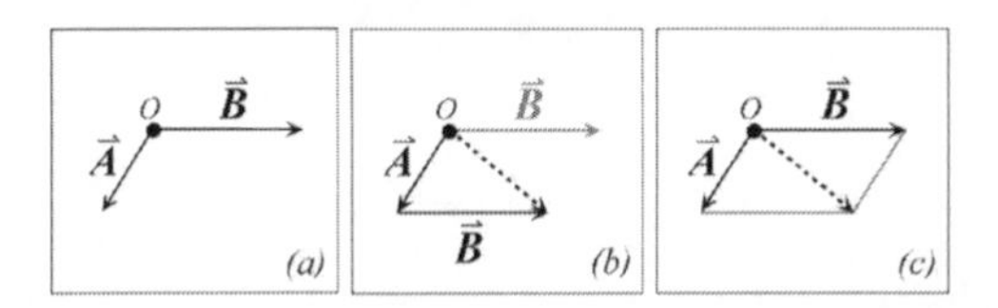

Figure 1.2 (a) Vectors $\vec{A}$ and $\vec{B}$ acting upon a point O. (b) Illustration of the tail-to-head method for adding vectors $\vec{A}$ and $\vec{B}$. (c) Illustration of the parallelogram method for adding vectors $\vec{A}$ and $\vec{B}$

The vectors $\vec{A}$ and $\vec{B}$ are equal if $A_x = B_x$, $A_y = B_y$, and $A_z = B_z$. Furthermore, the magnitude of a vector $\vec{A}$ is given by

$$\left|\vec{A}\right| = \left(A_x^2 + A_y^2 + A_z^2\right)^{1/2} \tag{1.2}$$

which is simply the 3-D Pythagorean theorem and can be visually verified with the aid of Figure 1.1.

Vectors can be added to and subtracted from one another both by graphical methods as well as by components. Graphical addition is illustrated with the aid of Figure 1.2. Imagine that the force vectors $\vec{A}$ and $\vec{B}$ are acting at point O as shown in Figure 1.2(a). The total force acting at O is equal to the sum of $\vec{A}$ and $\vec{B}$. Graphical construction of the vector sum $\vec{A} + \vec{B}$ can be accomplished either by using the tail-to-head method or the parallelogram method. The tail-to-head method involves drawing $\vec{B}$ at the head of $\vec{A}$ and then connecting the tail of $\vec{A}$ to the head of the re-drawn $\vec{B}$ (Figure 1.2b). Alternatively, upon constructing a parallelogram with sides $\vec{A}$ and $\vec{B}$, the diagonal of the parallelogram between $\vec{A}$ and $\vec{B}$ represents the vector sum, $\vec{A} + \vec{B}$ (Figure 1.2c).

If we know the component forms of both $\vec{A}$ and $\vec{B}$, then their sum is given by

$$\vec{A} + \vec{B} = (A_x + B_x)\hat{i} + (A_y + B_y)\hat{j} + (A_z + B_z)\hat{k}. \tag{1.3a}$$

Thus, the sum of $\vec{A}$ and $\vec{B}$ is found by simply adding like components together. It is clear from considering the component form of vector addition that addition of vectors is commutative ($\vec{A} + \vec{B} = \vec{B} + \vec{A}$) and associative (($\vec{A} + \vec{B}) + \vec{C} = \vec{A} + (\vec{B} + \vec{C})$).

Subtraction is simply the opposite of addition so $\vec{B}$ can be subtracted from $\vec{A}$ by simply adding $-\vec{B}$ to $\vec{A}$. Graphical subtraction of $\vec{B}$ from $\vec{A}$ is illustrated in Figure 1.3. Notice that $\vec{A} - \vec{B} = \vec{A} + (-\vec{B})$ results in a vector directed from the head of $\vec{B}$ to the head of $\vec{A}$ (the lighter dashed arrow in Figure 1.3). Component subtraction involves

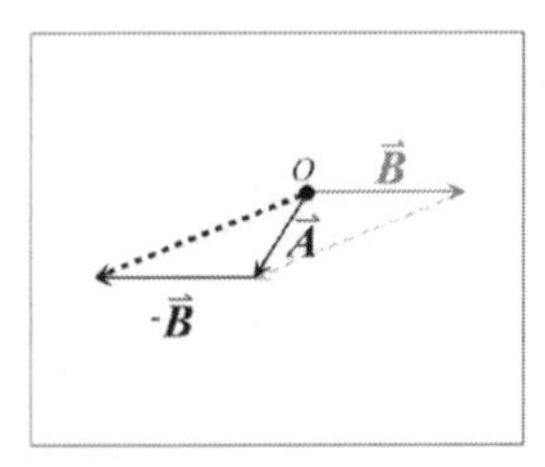

Figure 1.3 Graphical subtraction of vector $\vec{B}$ from vector $\vec{A}$

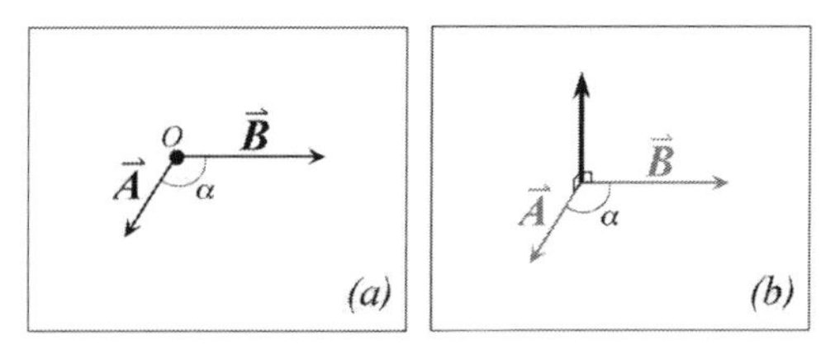

Figure 1.4 (a) Vectors $\vec{A}$ and $\vec{B}$ with an angle α between them. (b) Illustration of the relationship between vectors $\vec{A}$ and $\vec{B}$ (gray arrows) and their cross-product, $\vec{A} \times \vec{B}$ (bold arrow). Note that $\vec{A} \times \vec{B}$ is perpendicular to both $\vec{A}$ and $\vec{B}$

subtracting like components and is given by

$$\vec{A} - \vec{B} = (A_x - B_x)\hat{i} + (A_y - B_y)\hat{j} + (A_z - B_z)\hat{k}. \tag{1.3b}$$

Vector quantities may also be multiplied in a variety of ways. The simplest vector multiplication involves the product of a vector, $\vec{A}$, and a scalar, F. The resulting expression for $F\vec{A}$ is given by

$$F\vec{A} = FA_x\hat{i} + FA_y\hat{j} + FA_z\hat{k}, \tag{1.4}$$

a vector with direction identical to the original vector, $\vec{A}$, but with a magnitude F times larger than the original magnitude.

It is also possible to multiply two *vectors* together. In fact, there are two different vector multiplication operations. One such method renders a scalar as the product of the vector multiplication and is thus known as the **scalar** (or **dot**) product. The dot product of the vectors $\vec{A}$ and $\vec{B}$ shown in Figure 1.4(a) is given by

$$\vec{A} \cdot \vec{B} = |A|\,|B| \cos\alpha \tag{1.5}$$

where α is the angle between $\vec{A}$ and $\vec{B}$. Clearly this product is a scalar. Using this formula, we can determine a less mystical form of the dot product of $\vec{A}$ and $\vec{B}$. Given that $\vec{A} = A_x\hat{i} + A_y\hat{j} + A_z\hat{k}$ and $\vec{B} = B_x\hat{i} + B_y\hat{j} + B_z\hat{k}$, the dot product is given by

$$\vec{A} \cdot \vec{B} = (A_x\hat{i} + A_y\hat{j} + A_z\hat{k}) \cdot (B_x\hat{i} + B_y\hat{j} + B_z\hat{k}) \tag{1.6}$$

which expands to the following nine terms:

$$\begin{aligned}\vec{A} \cdot \vec{B} = {} & A_xB_x(\hat{i} \cdot \hat{i}) + A_xB_y(\hat{i} \cdot \hat{j}) + A_xB_z(\hat{i} \cdot \hat{k}) \\ & + A_yB_x(\hat{j} \cdot \hat{i}) + A_yB_y(\hat{j} \cdot \hat{j}) + A_yB_z(\hat{j} \cdot \hat{k}) \\ & + A_zB_x(\hat{k} \cdot \hat{i}) + A_zB_y(\hat{k} \cdot \hat{j}) + A_zB_z(\hat{k} \cdot \hat{k}).\end{aligned}$$

Now, according to (1.5), $\hat{i} \cdot \hat{i} = \hat{j} \cdot \hat{j} = \hat{k} \cdot \hat{k} = 1$ since the angle between like unit vectors is 0°. However, the dot products of all other combinations of the unit vectors are zero since the unit vectors are mutually orthogonal. Thus, only three terms survive out of the nine-term expansion of $\vec{A} \cdot \vec{B}$ to yield

$$\vec{A} \cdot \vec{B} = A_xB_x + A_yB_y + A_zB_z. \tag{1.7}$$

Given this result, it is easy to show that the dot product is commutative ($\vec{A} \cdot \vec{B} = \vec{B} \cdot \vec{A}$) and distributive ($\vec{A} \cdot (\vec{B} + \vec{C}) = \vec{A} \cdot \vec{B} + \vec{A} \cdot \vec{C}$).

Two vectors can also be multiplied together to produce another vector. This vector multiplication operation is known as the **vector** (or **cross-**)product and is signified

$$\vec{A} \times \vec{B}.$$

The magnitude of the resultant vector is given by

$$|A|\,|B| \sin \alpha \tag{1.8}$$

where α is the angle between the vectors. Note that since the resultant of the cross-product is a vector, there is also a direction to be discerned. The resultant vector is in a plane that is perpendicular to the plane that contains $\vec{A}$ and $\vec{B}$ (Figure 1.4b). The direction in that plane can be determined by using the **right hand rule**. Upon curling the fingers of one's right hand in the direction from $\vec{A}$ to $\vec{B}$, the thumb points in the direction of the resultant vector, $\vec{A} \times \vec{B}$, as shown in Figure 1.4(b). Because the resultant direction depends upon the order of multiplication, the cross-product has different properties than the dot product. It is not commutative ($\vec{A} \times \vec{B} \neq \vec{B} \times \vec{A}$; instead $\vec{A} \times \vec{B} = -\vec{B} \times \vec{A}$) and it is not associative ($\vec{A} \times (\vec{B} \times \vec{C}) \neq (\vec{A} \times \vec{B}) \times \vec{C}$) but it is distributive ($\vec{A} \times (\vec{B} + \vec{C}) = \vec{A} \times \vec{B} + \vec{A} \times \vec{C}$).

Given the vectors $\vec{A}$ and $\vec{B}$ in their component forms, the cross-product can be calculated by first setting up a 3×3 determinant using the unit vectors as the first row, the components of $\vec{A}$ as the second row, and the components of $\vec{B}$ as the third row:

$$\vec{A} \times \vec{B} = \begin{vmatrix} \hat{\imath} & \hat{\jmath} & \hat{k} \\ A_x & A_y & A_z \\ B_x & B_y & B_z \end{vmatrix}. \tag{1.9a}$$

Evaluating this determinant involves evaluating three 2×2 determinants, each one corresponding to a unit vector $\hat{\imath}$, $\hat{\jmath}$, *or* $\hat{k}$. For the $\hat{\imath}$ component of the resultant vector, only the components of $\vec{A}$ and $\vec{B}$ in the $\hat{\jmath}$ and $\hat{k}$ columns are considered. Multiplying the components along the diagonal (upper left to lower right) first, and then subtracting from that result the product of the terms along the anti-diagonal (lower left to upper right) yields the $\hat{\imath}$ component of the vector $\vec{A} \times \vec{B}$, which equals $(A_y B_z - A_z B_y)\hat{\imath}$. The same operation done for the $\hat{k}$ component yields $(A_x B_y - A_y B_x)\hat{k}$. For the $\hat{\jmath}$ component, the first and third columns are used to form the 2×2 determinant and since the columns are non-consecutive, the result must be multiplied by –1 to yield $-(A_x B_z - A_z B_x)\hat{\jmath}$. Adding these three components together yields

$$\vec{A} \times \vec{B} = (A_y B_z - A_z B_y)\hat{\imath} + (A_z B_x - A_x B_z)\hat{\jmath} + (A_x B_y - A_y B_x)\hat{k}. \tag{1.9b}$$

Vectors, just like scalar functions, can be differentiated as long as the rules of vector addition and multiplication are obeyed. One simple example is Newton's second law (which we will see again soon) that states that an object's momentum will not change unless a force is applied to the object. In mathematical terms,

$$\vec{F} = \frac{d}{dt}(m\vec{V}) \tag{1.10}$$

where m is the object's mass and $\vec{V}$ is its velocity. Using the chain rule of differentiation on the right hand side of (1.10) renders

$$\vec{F} = m\frac{d\vec{V}}{dt} + \vec{V}\frac{dm}{dt} \quad \text{or} \quad \vec{F} = m\vec{A} + \vec{V}\frac{dm}{dt} \tag{1.11}$$

where $\vec{A}$ is the object's acceleration. Exploitation of the second term of this expansion is what made Einstein famous!

Let us consider a more general example. Consider a velocity vector defined as $\vec{V} = u\hat{i} + v\hat{j} + w\hat{k}$. In such a case, the acceleration will be given by

$$\frac{d\vec{V}}{dt} = \frac{du}{dt}\hat{i} + u\frac{d\hat{i}}{dt} + \frac{dv}{dt}\hat{j} + v\frac{d\hat{j}}{dt} + \frac{dw}{dt}\hat{k} + w\frac{d\hat{k}}{dt}. \tag{1.12}$$

The terms involving derivatives of the unit vectors may seem like mathematical baggage but they will be extremely important in our subsequent studies. Physically, such terms will be non-zero only when the coordinate axes used to reference motion are not fixed in space. Our reference frame on a rotating Earth is clearly not fixed and so we will eventually have to make some accommodation for the acceleration of our rotating reference frame. Thus, all six terms in the above expansion will be relevant in our examination of the mid-latitude atmosphere.

The last stop on the review of vector calculus is perhaps the most important one and will examine a tool that is extremely useful in fluid dynamics. We will often need to describe both the magnitude and direction of the derivative of a scalar field. In order to do so we employ a mathematical operator known as the **del operator**, defined as

$$\nabla = \frac{\partial}{\partial x}\hat{i} + \frac{\partial}{\partial y}\hat{j} + \frac{\partial}{\partial z}\hat{k}. \tag{1.13}$$

If we apply this partial differential del operator to a scalar function or field, the result is a vector that is known as the **gradient** of that scalar. Consider the 2-D plan view of an isolated hill in an otherwise flat landscape. If the elevation at each point in the landscape is represented on a 2-D projection, a set of elevation contours results as shown in Figure 1.5. Such contours are lines of equal height above sea level, Z. Given such information, we can determine the gradient of elevation, ∇Z, as

$$\nabla Z = \frac{\partial Z}{\partial x}\hat{i} + \frac{\partial Z}{\partial y}\hat{j}.$$

Note that the gradient vector, ∇Z, points up the hill *from low values of elevation to high values.* At the top of the hill, the derivatives of Z in both the x and y

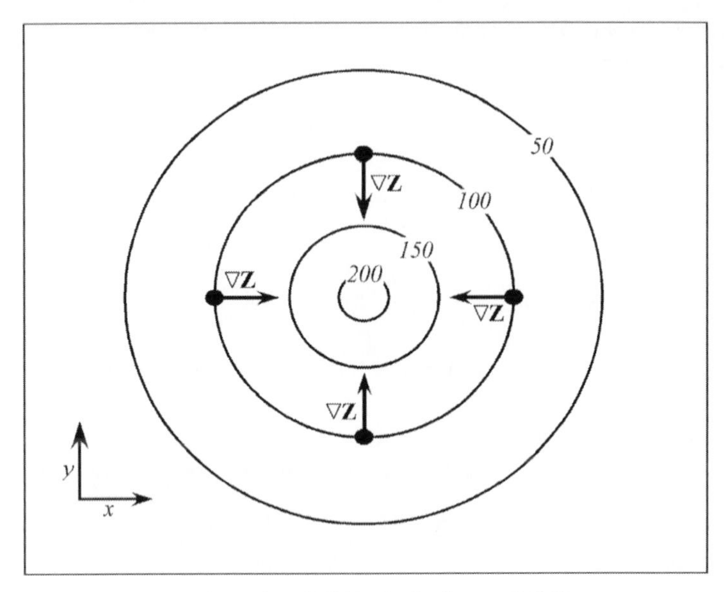

Figure 1.5 The 2-D plan view of an isolated hill in a flat landscape. Solid lines are contours of elevation (Z) at 50m intervals. Note that the gradient of Z points from low to high values of the scalar Z

directions are zero so there is no gradient vector there. Thus the gradient, ∇Z, not only measures magnitude of the elevation difference but assigns that magnitude a direction as well. Any scalar quantity, Φ, is transformed into a vector quantity, $\nabla\Phi$, by the del operator. In subsequent chapters in this book we will concern ourselves with the gradients of a number of scalar variables, among them temperature and pressure.

The del operator may also be applied to vector quantities. The dot product of ∇ with the vector $\vec{A}$ is written as

$$\nabla\cdot\vec{A}=\left(\frac{\partial}{\partial x}\hat{i}+\frac{\partial}{\partial y}\hat{j}+\frac{\partial}{\partial z}\hat{k}\right)\cdot(A_x\hat{i}+A_y\hat{j}+A_z\hat{k})$$

$$\nabla\cdot\vec{A}=\left(\frac{\partial A_x}{\partial x}+\frac{\partial A_y}{\partial y}+\frac{\partial A_z}{\partial z}\right) \tag{1.14}$$

which is a scalar quantity known as the **divergence of** $\vec{A}$. Positive divergence physically describes the tendency for a vector field to be directed away from a point whereas negative divergence (also known as **convergence**) describes the tendency for a vector field to be directed toward a point. Regions of convergence and divergence in the atmospheric fluid are extremely important in determining its behavior.

The cross-product of ∇ with the vector $\vec{A}$ is given by

$$\nabla\times\vec{A}=\left(\frac{\partial}{\partial x}\hat{i}+\frac{\partial}{\partial y}\hat{j}+\frac{\partial}{\partial z}\hat{k}\right)\times(A_x\hat{i}+A_y\hat{j}+A_z\hat{k}). \tag{1.15a}$$

The resulting vector can be calculated using the determinant form we have seen previously,

$$\nabla \times \vec{A} = \begin{vmatrix} \hat{i} & \hat{j} & \hat{k} \\ \dfrac{\partial}{\partial x} & \dfrac{\partial}{\partial y} & \dfrac{\partial}{\partial z} \\ A_x & A_y & A_z \end{vmatrix} \tag{1.15b}$$

where the second row of the 3×3 determinant is filled by the components of ∇ and the third row is filled by the components of $\vec{A}$. This vector is known as the **curl of** $\vec{A}$. The curl of the velocity vector, $\vec{V}$, will be used to define a quantity called **vorticity** which is a measure of the rotation of a fluid.

Quite often in a study of the dynamics of the atmosphere, we will encounter second-order partial differential equations. Some of these equations will contain a mathematical operator (which will operate on scalar quantities) known as the **Laplacian** operator. The Laplacian is the **divergence of the gradient** and so takes the form

$$Laplacian = \nabla \cdot (\nabla F) = \nabla^2 F = \left(\frac{\partial^2 F}{\partial x^2} + \frac{\partial^2 F}{\partial y^2} + \frac{\partial^2 F}{\partial z^2} \right). \tag{1.16}$$

It is also possible to combine the vector $\vec{A}$ with the del operator to form a new operator that takes the form

$$\vec{A} \cdot \nabla = A_x \frac{\partial}{\partial x} + A_y \frac{\partial}{\partial y} + A_z \frac{\partial}{\partial z}$$

and is known as the scalar invariant operator. This operator, which can be used with both vector and scalar quantities, is important because it is used to describe a process known as **advection**, a ubiquitous topic in the study of fluids.

1.2.2 The Taylor series expansion

It is sometimes convenient to estimate the value of a continuous function, $f(x)$, about the point $x = 0$ with a power series of the form

$$f(x) = \sum_{n=0}^{\infty} a_n x^n = a_0 + a_1 x + a_2 x^2 + \cdots + a_n x^n. \tag{1.17}$$

The fact that this can actually be done might appear to be an assumption so we must identify conditions for which this assumption is true. These conditions are that (1) the polynomial expression (1.17) passes through the point $(0, f(0))$ and (2) its first n derivatives match the first n derivatives of $f(x)$ at $x = 0$. Implicit in this second condition is the fact that $f(x)$ is differentiable at $x = 0$. In order for these conditions to be met, the coefficients $a_0, a_1, \ldots, a_n$ must be chosen properly. Substituting $x = 0$ into (1.17) we find that $f(0) = a_0$. Taking the first derivative of

(1.17) with respect to x and substituting $x = 0$ into the resulting expression we get $f'(0) = a_1$. Taking the second derivative of (1.17) with respect to x and substituting $x = 0$ into the result leaves $f''(0) = 2a_2$, or $f''(0)/2 = a_2$. If we continue to take higher order derivatives of (1.17) and evaluate each of them at $x = 0$ we find that, in order that the n derivatives of (1.17) match the n derivatives of $f(x)$, the coefficients, a_n, of the polynomial expression (1.17) must take the general form

$$a_n = \frac{f^n(0)}{n!}.$$

Thus, the value of the function $f(x)$ at $x = 0$ can be expressed as

$$f(x) = f(0) + f'(0)x + \frac{f''(0)}{2!}x^2 + \frac{f'''(0)}{3!}x^3 + \cdots + \frac{f^n(0)}{n!}x^n. \tag{1.18}$$

Now, if we want to determine the value of $f(x)$ near the point $x = x_0$, the above expression can be generalized into what is known as the Taylor series expansion of $f(x)$ about $x = x_0$, given by

$$f(x) = f(x_0) + f'(x_0)(x - x_0) + \frac{f''(x_0)}{2!}(x - x_0)^2 + \cdots + \frac{f^n(x_0)}{n!}(x - x_0)^n. \tag{1.19}$$

Since the dependent variables that describe the behavior of the atmosphere are all continuous variables, use of the Taylor series to approximate the values of those variables will prove to be a nifty little trick that we will exploit in our subsequent analyses. Most often we consider Taylor series expansions in which the quantity $(x - x_0)$ is very small in order that all terms of order 2 and higher in (1.19), the so-called **higher order terms**, can be effectively neglected. In such cases, we will approximate the given functions as

$$f(x) \approx f(x_0) + f'(x_0)(x - x_0).$$

1.2.3 Centered difference approximations to derivatives

Though the atmosphere is a continuous fluid and its observed state at any time *could theoretically* be represented by a continuous function, the reality is that actual observations of the atmosphere are only available at discrete points in space and time. Given that much of the subsequent development in this book will arise from consideration of the spatial and temporal variation of observable quantities, we must consider a method of approximating derivative quantities from discrete data. One such method is known as **centered differencing**[3] and it follows directly from the prior discussion of the Taylor series expansion.

---

[3] Centered differencing is a subset of a broader category of such approximations known as **finite differencing**.

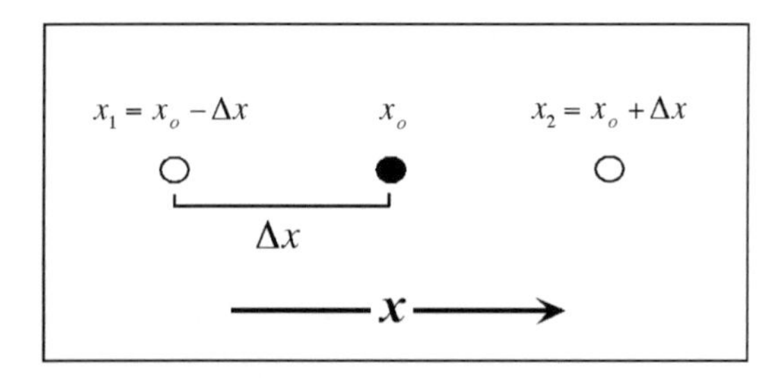

Figure 1.6 Points x_1 and x_2 defined with respect to a central point x_0

Consider the two points x_1 and x_2 in the near vicinity of a central point, x_0, as illustrated in Figure 1.6. We can apply (1.19) at both points to yield

$$f(x_1) = f(x_0 - \Delta x) = f(x_0) + f'(x_0)(-\Delta x) + \frac{f''(x_0)}{2!}(-\Delta x)^2 + \cdots + \frac{f^n(x_0)}{n!}(-\Delta x)^n \tag{1.20a}$$

and

$$f(x_2) = f(x_0 + \Delta x) = f(x_0) + f'(x_0)(\Delta x) + \frac{f''(x_0)}{2!}(\Delta x)^2 + \cdots + \frac{f^n(x_0)}{n!}(\Delta x)^n. \tag{1.20b}$$

Subtracting (1.20a) from (1.20b) produces

$$f(x_0 + \Delta x) - f(x_0 - \Delta x) = 2f'(x_0)(\Delta x) + 2f'''(x_0)\frac{(\Delta x)^3}{6} + \cdots. \tag{1.21}$$

Isolating the expression for $f'(x_0)$ on one side then leaves

$$f'(x_0) = \frac{f(x_0 + \Delta x) - f(x_0 - \Delta x)}{2\Delta x} - f'''(x_0)\frac{(\Delta x)^2}{6} - \cdots$$

which, upon neglecting terms of second order and higher in Δx, can be approximated as

$$f'(x_0) \approx \frac{f(x_0 + \Delta x) - f(x_0 - \Delta x)}{2\Delta x}. \tag{1.22}$$

The foregoing expression represents the centered difference approximation to $f'(x)$ at x_0 accurate to second order (i.e. the neglected terms are at least quadratic in Δx).

Adding (1.20a) to (1.20b) gives a similarly approximated expression for the second derivative as

$$f''(x_0) \approx \frac{f(x_0 + \Delta x) - 2f(x_0) + f(x_0 - \Delta x)}{\Delta x^2}. \tag{1.23}$$

Such expressions will prove quite useful in evaluating a number of relationships we will encounter later.

1.2.4 Temporal changes of a continuous variable

The fluid atmosphere is an ever evolving medium and so the fundamental variables discussed in Section 1.1 are ceaselessly subject to temporal changes. But what does it really mean to say 'The temperature has changed in the last hour'? In the broadest sense this statement could have two meanings. It could mean that the temperature of an individual air parcel, moving past the thermometer on my back porch, is changing as it migrates through space. In this case, we would be considering the change in temperature experienced *while moving with a parcel of air.* However, the statement could also mean that the temperature of the air parcels currently in contact with my thermometer is lower than that of air parcels that used to reside there but have since been replaced by the importation of these colder ones. In this case we would be considering the changes in temperature as measured *at a fixed geographic point.* These two notions of temporal change are clearly not the same, but one might wonder if and how they are physically and mathematically related. We will consider a not so uncommon example to illustrate this relationship.

Imagine a winter day in Madison, Wisconsin characterized by biting northwesterly winds which are importing cold arctic air southward out of central Canada. From the fixed geographical point of my back porch, the temperature (or potential temperature) drops with the passage of time. If, however, I could ride along with the flow of the air, I would likely find that the temperature does not change over the passage of time. In other words, a parcel with $T = 270°\text{K}$ passing my porch at 8 a.m. still has $T = 270°\text{K}$ at 2 p.m. even though it has traveled nearly to Chicago, Illinois by that time. Therefore, *the steady drop in temperature I observe at my porch is a result of the continuous importation of colder air parcels from Canada.* Phenomenologically, therefore, we can write an expression for this relationship we've developed:

$$\begin{matrix} \textit{Change with Time} \\ \textit{Following an Air} \\ \textit{Parcel} \end{matrix} = \begin{matrix} \textit{Change with Time} \\ \textit{at a Fixed} \\ \textit{Location} \end{matrix} - \begin{matrix} \textit{Rate of Importation} \\ \textit{of Temperature by} \\ \textit{Movement of Air.} \end{matrix} \tag{1.24}$$

This relationship can be made mathematically rigorous. Doing so will assist us later in the development of the equations of motion that govern the mid-latitude atmosphere. The change following the air parcel is called the **Lagrangian** rate of change while the change at a fixed point is called the **Eulerian** rate of change. We can quantify the relationship between these two different views of temporal change by considering an arbitrary scalar (or vector) quantity that we will call Q. If Q is a function of space and time, then

$$Q = Q(x, y, z, t)$$

and, from the differential calculus, the *total differential of* Q is

$$dQ = \left(\frac{\partial Q}{\partial x}\right)_{y,z,t} dx + \left(\frac{\partial Q}{\partial y}\right)_{x,z,t} dy + \left(\frac{\partial Q}{\partial z}\right)_{x,y,t} dz + \left(\frac{\partial Q}{\partial t}\right)_{x,y,z} dt \tag{1.25}$$

where the subscripts refer to the independent variables that are held constant whilst taking the indicated partial derivatives. Upon dividing both sides of (1.25) by dt, the total differential of t which represents a time increment, the resulting expression is

$$\frac{dQ}{dt} = \left(\frac{\partial Q}{\partial t}\right)\frac{dt}{dt} + \left(\frac{\partial Q}{\partial x}\right)\frac{dx}{dt} + \left(\frac{\partial Q}{\partial y}\right)\frac{dy}{dt} + \left(\frac{\partial Q}{\partial z}\right)\frac{dz}{dt} \tag{1.26}$$

where the subscripts on the partial derivatives have been dropped for convenience. The rates of change of x, y, or z with respect to time are simply the component velocities in the x, y, or z directions. We will refer to these velocities as u, v, and w and define them as $u = dx/dt$, $v = dy/dt$, and $w = dz/dt$, respectively. Substituting these expressions into (1.26) yields

$$\frac{dQ}{dt} = \left(\frac{\partial Q}{\partial t}\right) + u\left(\frac{\partial Q}{\partial x}\right) + v\left(\frac{\partial Q}{\partial y}\right) + w\left(\frac{\partial Q}{\partial z}\right) \tag{1.27}$$

which can be rewritten in vector notation as

$$\frac{dQ}{dt} = \left(\frac{\partial Q}{\partial t}\right) + \vec{V} \cdot \nabla Q \tag{1.28}$$

where $\vec{V} = u\hat{i} + v\hat{j} + w\hat{k}$ is the 3-D vector wind. The three terms in (1.27) involving the component winds and derivatives of Q physically represent the horizontal and vertical transport of Q by the flow. Thus, we see that dQ/dt corresponds to the Lagrangian rate of change noted in (1.24). The Eulerian rate of change is represented by $\partial Q/\partial t$. The rate of importation by the flow (recall it was subtracted from the Eulerian change on the RHS of (1.24)) is represented by $-\vec{V} \cdot \nabla Q$ (*minus* the dot product of the velocity vector and the gradient of Q). In subsequent discussions in this book, $-\vec{V} \cdot \nabla Q$ will be referred to as **advection of** Q. Next we show that the mathematical expression $-\vec{V} \cdot \nabla Q$ actually describes the rate of importation of Q by the flow.

Consider the isotherms (lines of constant temperature) and wind vector shown in Figure 1.7. The gradient of temperature (∇T) is a vector that always points from lowest temperatures to highest temperatures as indicated. The wind vector, clearly drawn in Figure 1.7 so as to transport warmer air toward point A, is directed opposite to ∇T. Recall that the dot product is given by $\vec{V} \cdot \nabla T = |\vec{V}||\nabla T| \cos\alpha$ where α is the angle between the vectors $\vec{V}$ and ∇T. Given that the angle between $\vec{V}$ and ∇T is 180° in Figure 1.7, the dot product $\vec{V} \cdot \nabla T$ returns a negative value. Therefore, the sign of $\vec{V} \cdot \nabla T$ does not accurately reflect the reality of the physical situation depicted in Figure 1.7 – that is, that importation of *warmer air* is occurring at point A.

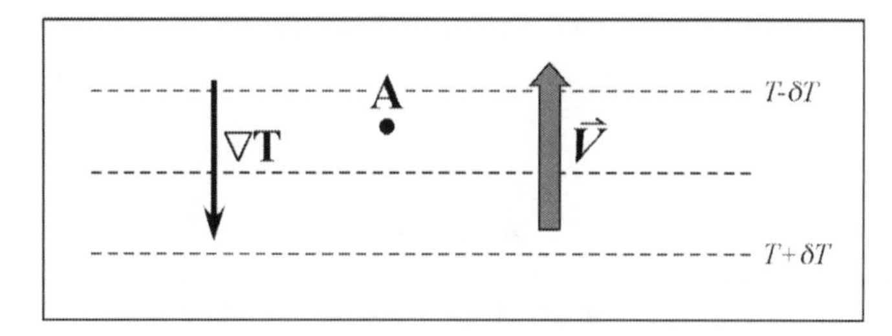

Figure 1.7 Isotherms (dashed lines) and wind vector $\vec{V}$ (filled arrow) surrounding point A. The thin black arrow is the horizontal temperature gradient vector

Thus, we define temperature advection, a measure of the rate (and sign) of importation of temperature to point A, as $-\vec{V} \cdot \nabla T$. The physical situation depicted in Figure 1.7, therefore, is said to be characterized by positive temperature (or warm air) advection.

To round out this discussion, we now return to the example that motivated the mathematical development: measuring the temperature change on my back porch. Rearranging (1.28) and substituting T (temperature) for Q we get

$$\left(\frac{\partial T}{\partial t}\right) = \frac{dT}{dt} - \vec{V} \cdot \nabla T$$

which shows that the Eulerian (fixed location) change is equal to the sum of the Lagrangian (parcel following) change and advection. In the prior example we imagined a temperature drop at my back porch. We also surmised that the temperature of individual air parcels did not undergo any change as the day wore on. Thus, the advective change at the porch must be negative – there must be negative temperature advection, or cold air advection (i.e. $-\vec{V} \cdot \nabla T < 0$), occurring in Madison on this day. Clearly, the situation of northwesterly winds importing cold air southward out of Canada fits the bill.

1.3 Estimating with Scale Analysis

In many fluid dynamical problems, it is convenient and insightful to estimate which physical terms are likely to contribute most to a particular process under study. For instance, in assessing the threat to coastal property in Hawaii in the face of a major tsunami, it is not likely that the ambient wind speed will figure into the problem in any significant way. In the development of the equations of motion in subsequent chapters, a variety of physical processes will be confronted, each of which has some bearing on the behavior of the fluid atmosphere. At many junctures, however, we will attempt to simplify those equations by estimating the magnitude of the mathematical terms that comprise them. A formal process known as scale analysis is employed in such an exercise. Here we illustrate, with a very simple example, the power of scale analysis as an analytical tool.

Imagine you are charged with filling an Olympic-sized swimming pool with water. Your boss wants to know how long it will take to get the job done and asks you for an estimate of the completion time. In order to make a reasonable approximation, you need to know a number of physical characteristics of the problem. You certainly need to know the volume of the pool and the flow rate you can expect from the hose you will use to fill the pool. You might want to know if there are cracks in the pool walls through which seepage might occur. Though it is surely physically relevant, you probably guess that you needn't concern yourself with the evaporation rate of water from the surface of the filling pool.

All four of the above-mentioned physical characteristics can be measured with varying degrees of accuracy. The volume is likely to be a fairly accurate measurement as is the flow rate from the hose. Seepage rate and evaporation rates, however, are likely to be quite difficult to measure accurately. Imagine we do, in fact, make some measurements of each of these characteristics, assigning an estimated (but characteristic) rate to each of the last three. The flow rate is found to be approximately $100\ \mathrm{m}^3\ \mathrm{h}^{-1}$, the evaporation rate $0.001\ \mathrm{m}^3\ \mathrm{h}^{-1}$, the seepage rate $0.000\,01\ \mathrm{m}^3\ \mathrm{h}^{-1}$. It is clear upon comparison of the three that the flow rate is the most important process (it is five to seven orders of magnitude larger than the others). Therefore, we could say that, subject to some small amount of error, the time needed to fill the pool is equal to

$$t_{fill} \approx \frac{Volume\ \ of\ the\ Pool}{Flow\ Rate}.$$

We will achieve a similar simplification of the equations of motion by similarly estimating the scale of various terms that appear in those equations.

1.4 Basic Kinematics of Fluids

As can be readily discerned from inspection of any satellite animation of clouds or water vapor, the wind field varies in the x and y directions. Therefore, there are x and y derivatives of the horizontal wind components, u and v. In fact, there are only four such derivatives: $\partial u/\partial x$ and $\partial u/\partial y$ along with $\partial v/\partial x$ and $\partial v/\partial y$. Let us consider all possible sums of these four derivatives with the stipulation that each sum must include a derivative of u with respect to one direction and a derivative of v with respect to the other. Under this condition there are only four independent, linear combinations of x and y derivatives of the horizontal wind, namely $\partial u/\partial x \pm \partial v/\partial y$ and $\partial v/\partial x \pm \partial u/\partial y$. We will now consider what these derivative combinations describe about the fluid flow and we will do it by considering Taylor series expansions of the functions $u(x,y)$ and $v(x,y)$. Since u and v are continuous functions of x and y

space, the expansion of each about some arbitrary point in space (say $(x,y)=(0,0)$) becomes

$$u(x,y)=u_0+\left(\frac{\partial u}{\partial x}\right)_0 x+\left(\frac{\partial u}{\partial y}\right)_0 y+\left(\frac{\partial^2 u}{\partial x^2}\right)_0 \frac{x^2}{2}+\left(\frac{\partial^2 u}{\partial y^2}\right)_0 \frac{y^2}{2}+\textit{Higher Order Terms} \tag{1.29a}$$

$$v(x,y)=v_0+\left(\frac{\partial v}{\partial x}\right)_0 x+\left(\frac{\partial v}{\partial y}\right)_0 y+\left(\frac{\partial^2 v}{\partial x^2}\right)_0 \frac{x^2}{2}+\left(\frac{\partial^2 v}{\partial y^2}\right)_0 \frac{y^2}{2}+\textit{Higher Order Terms.} \tag{1.29b}$$

If we neglect the terms of order 2 and greater (the so-called higher order terms), which is eminently defensible because they are generally very small, we have

$$u-u_0=\left(\frac{\partial u}{\partial x}\right)_0 x+\left(\frac{\partial u}{\partial y}\right)_0 y \tag{1.30a}$$

$$v-v_0=\left(\frac{\partial v}{\partial x}\right)_0 x+\left(\frac{\partial v}{\partial y}\right)_0 y \tag{1.30b}$$

where we have written $u(x,y)$ and $v(x,y)$ more conveniently as u and v, respectively.

Returning to our four independent linear combinations of x and y derivatives of the wind field, we next assign names to each combination. We will let $\partial u/\partial x+\partial v/\partial y=D$ where D is the **divergence.** We will let $\partial u/\partial x-\partial v/\partial y=F_1$ where F_1 is the **stretching deformation.** We will let $\partial v/\partial x+\partial u/\partial y=F_2$ where F_2 is the **shearing deformation.** Finally, we will let $\partial v/\partial x-\partial u/\partial y=\zeta$ where ζ is the **vorticity.** Given these definitions, we can rewrite (1.30a) and (1.30b) in terms of these quantities as

$$u-u_0=\frac{1}{2}(D+F_1)x-\frac{1}{2}(\zeta-F_2)y=\frac{1}{2}(Dx+F_1x-\zeta y+F_2y) \tag{1.31a}$$

$$v-v_0=\frac{1}{2}(\zeta+F_2)x+\frac{1}{2}(D-F_1)y=\frac{1}{2}(\zeta x+F_2x+Dy-F_1y). \tag{1.31b}$$

By assuming that u_0 and v_0 (the u and v velocities at our arbitrary origin point) are both zero we can quite readily use the expressions (1.31a) and (1.31b) to investigate what each of the four derivative fields looks like physically. We will consider each quantity in isolation even though, in nature, they all can occur simultaneously in a given observed flow.

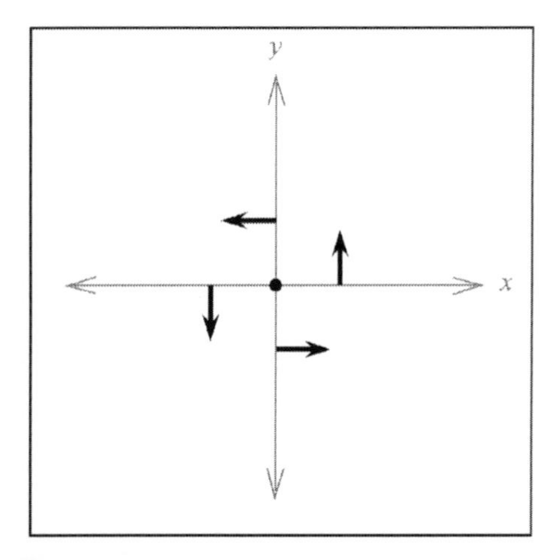

Figure 1.8 A field of pure, positive vorticity ($\zeta = 1$)

1.4.1 Pure vorticity

In order to examine what a flow with pure positive vorticity looks like, we use (1.31a) and (1.31b) and set D, F_1, and F_2 equal to zero while letting $\zeta = 1$. In such a case, (1.31a) and (1.31b) become $u = -\frac{1}{2}y$ and $v = \frac{1}{2}x$. Employing a Cartesian grid we plot u and v for the case of pure vorticity at a number of points in Figure 1.8. We find that a field of pure positive vorticity (recall that we set $\zeta = 1$) describes a *circular, counterclockwise* flow about the origin.

1.4.2 Pure divergence

An example of pure positive divergence occurs in a flow when ζ, F_1, and F_2 are all equal to zero while $D = 1$. In such a case, (1.31a) and (1.31b) become $u = \frac{1}{2}x$ and $v = \frac{1}{2}y$, respectively. Figure 1.9 illustrates the resulting flow field: a fluid moving *in all directions away from the origin*, at speeds proportional to the distance from the origin. Such a picture is consistent with the colloquial sense of the word 'divergence'. Notice that if we had assumed a value of $D = -1$ instead, we would get fluid moving toward the origin – consistent with the colloquial sense of the word 'convergence'. In fact, we will refer to negative divergence as convergence quite often in our subsequent studies.

1.4.3 Pure stretching deformation

Pure stretching deformation is obtained by setting D, ζ, and F_2 equal to zero while $F_1 = 1$. In this case, (1.31a) and (1.31b) become $u = \frac{1}{2}x$ and $v = -\frac{1}{2}y$, respectively.

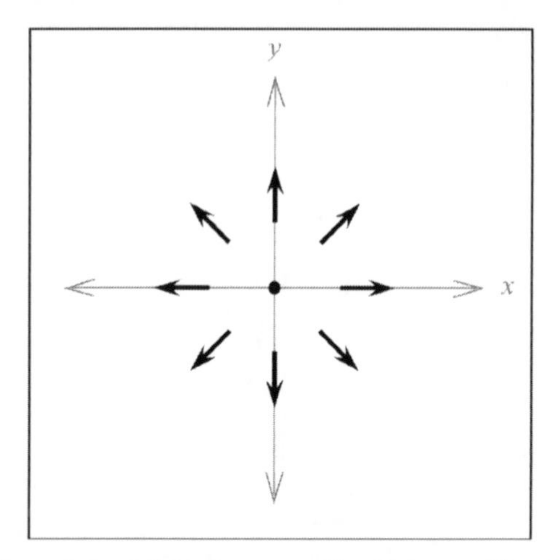

Figure 1.9 A field of pure, positive divergence ($D = 1$)

Figure 1.10 demonstrates that the resulting flow field is stretched along the x-axis and compressed along the y-axis. In fact, these two axes have special names: the flow is stretched along the **axis of dilatation** while it is compressed along the **axis of contraction**. It is important to distinguish between deformation and convergence as they are commonly confused. If we consider the area of a fluid element bounded by curve C embedded within a field of pure convergence (Figure 1.11a) we see immediately that the area will become progressively smaller under the influence of the convergent flow. If the same fluid parcel were placed in a field of pure stretching deformation, however, the shape of the originally square fluid element would be

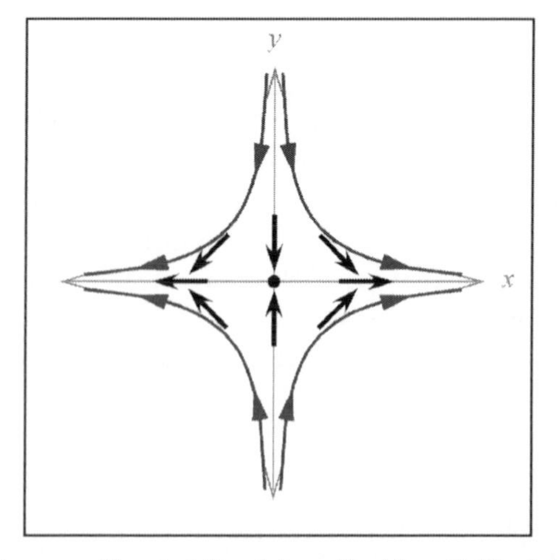

Figure 1.10 A field of pure, positive stretching deformation ($F_1 = 1$). The dark solid lines are streamlines of the deformation field. The x-axis serves as the axis of dilatation and the y-axis is the axis of contraction

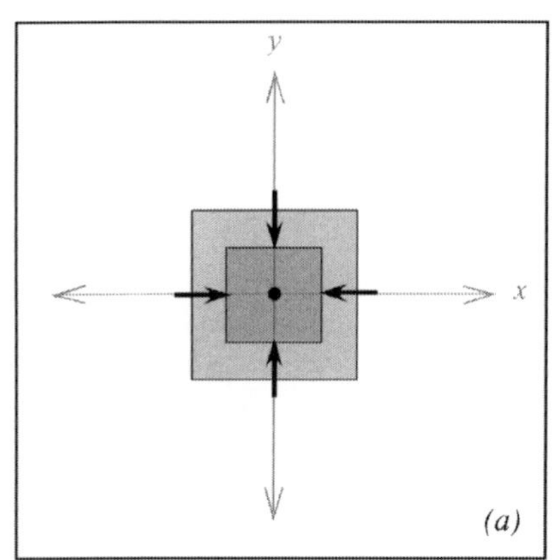

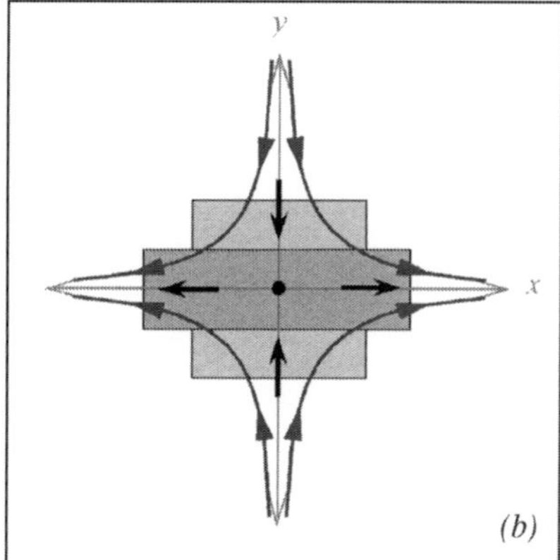

Figure 1.11 (a) A fluid element in a field of pure convergence. The lighter square represents the initially square element. Note that the area of the fluid element is decreased in a field of convergence. (b) A fluid element in a field of pure stretching deformation. The original square is deformed into a rectangle whose area is the same as that of the square

deformed into a rectangle while preserving its area (Figure 1.11b). The proof that the area is unchanged is left to the reader as an exercise. This essential physical distinction between convergence and deformation (specifically, confluence) is made manifest to the driver of an automobile using an entrance ramp to a highway. The flow of traffic is **confluent** (i.e. resembles the flow in the vicinity of the x-axis in Figure 1.10) between the entrance ramp and the highway but it is certainly not convergent. If it were, the number of accidents would be staggering!

1.4.4 Pure shearing deformation

Pure shearing deformation is obtained by letting $F_2 = 1$ while setting D, ζ, and F_1 equal to zero. By doing so, (1.31a) and (1.31b) become $u = \frac{1}{2}y$ and $v = \frac{1}{2}x$. The resulting flow field (Figure 1.12) looks like the stretching deformation rotated counterclockwise by 45°. So, how do we tell the difference between the stretching and shearing deformations and is the difference even important physically? It turns out that most often we are concerned with the **total deformation** without regard to the separate expressions for F_1 and F_2. The total deformation is given by

$$F = \left(F_1^2 + F_1^2\right)^{1/2} \tag{1.32}$$

where F represents the resultant magnitude of what appears to be a deformation vector with components, F_1 and F_2. It is clear that with a rotation of 45° of the coordinate axes, we can transform $F_1 = 1$ into $F_1' = 0$ and $F_2 = 0$ into $F_2' = 1$.

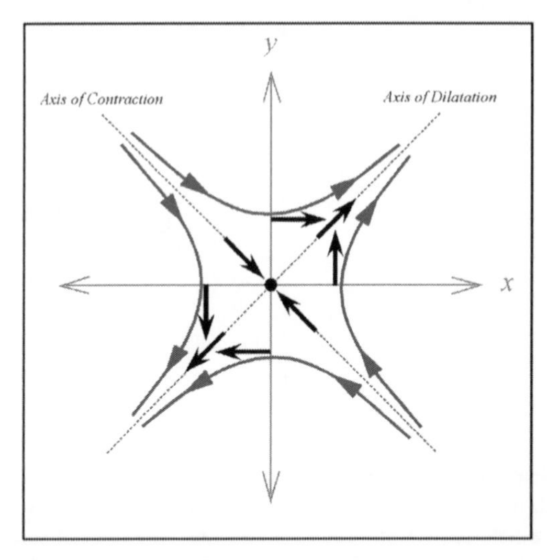

Figure 1.12 A field of pure, positive shearing deformation ($F_2 = 1$). The dark solid lines are streamlines of the deformation field. The axes of dilatation and contraction are indicated by the dashed lines

Thus, deformation is **rotationally variant**. In fact, if one rotates the coordinate axes by the angle

$$\theta = \frac{1}{2}\tan^{-1}\left(\frac{F_2}{F_1}\right) \tag{1.33}$$

then the resultant deformation has its axis of dilatation at an angle θ counterclockwise from the original x-axis. It is clear that any rotation of the x- and y-axes will have no effect whatever on the vorticity or divergence. As a result, these two properties of the flow are known as **rotationally invariant** or **Galilean invariant**. This characteristic vests the vorticity and divergence with considerable power in explaining the behavior of fluids, as we will see.

1.5 Mensuration

Before we embark upon our investigation of the forces that govern the behavior of the fluid atmosphere, we must explicitly lay out the units with which we will measure the quantities of interest. Throughout the remainder of the text we will employ the **Système Internationale** (SI) units shown in Table 1.1.

Table 1.1 Standard SI units

Property	Name	Symbol
Length	Meter	m
Mass	Kilogram	kg
Time	Second	s
Temperature	Kelvin	K

Table 1.2 Important SI derived units

Property	Name	Symbol
Frequency	Hertz	Hz (s^{-1})
Force	Newton	N (kg m s^{-2})
Pressure	Pascal	Pa (N m^{-2})
Energy	Joule	J (N m)
Power	Watt	W (J s^{-1})

Additionally, a number of derived quantities will be referenced throughout our study and they are shown in Table 1.2.

Despite the fact that we will refer to temperature in °C (or occasionally in °F when using an older diagram to illustrate a point), it is important to remember to use SI units in all calculations you may have to make.

Selected References

A complete reference list is provided in the Bibliography at the end of the book.

Spiegel, M. R., *Vector Analysis and an Introduction to Tensor Analysis*, is an outstanding, concise text on vector calculus with nearly 500 solved problems.

Thomas and Finney, *Calculus and Analytic Geometry*, provides additional detail on the Taylor series and fundamental calculus.

Hess, *Introduction to Theoretical Meteorology*, discusses the basic kinematics of fluids.

Saucier, *Principles of Meteorological Analysis*, is another fine reference on kinematics.

Problems

1.1. Let $\vec{A} = \nabla\phi = 8x\hat{i} + 3y^2\hat{j}$. If you know that $\phi(1, 1) = 8$ and $\phi(0, 1) = 4$, derive a functional expression for $\phi(x, y)$.

1.2. Prove the vector identities in (a) – (c) letting $\vec{V} = u\hat{i} + v\hat{j} + w\hat{k}$ and $\nabla = \frac{\partial}{\partial x}\hat{i} + \frac{\partial}{\partial y}\hat{j} + \frac{\partial}{\partial z}\hat{k}$:

(a) $\nabla \cdot (\nabla \times \vec{V}) = 0$

(b) $(\vec{V} \cdot \nabla)\vec{V} = (1/2)\nabla(\vec{V} \cdot \vec{V}) - \vec{V} \times (\nabla \times \vec{V})$

(c) $\nabla \cdot (f\vec{V}) = f(\nabla \cdot \vec{V}) + \vec{V} \cdot \nabla f$

(d) Prove that $\hat{k} \times (\hat{k} \times \vec{A}) = -\vec{A}$ where $\vec{A} = A_1\hat{i} + A_2\hat{j}$.

(e) Use the 'right hand rule' to verify (d) graphically.

1.3. The symbol $\vec{A}_{\vec{B}}$ stands for the projection of vector $\vec{A}$ onto vector $\vec{B}$. In other words, $\vec{A}_{\vec{B}}$ represents the component of $\vec{A}$ that is parallel to $\vec{B}$. Derive an expression for $\vec{A}_{\vec{B}}$ in terms of the vectors $\vec{A}$ and $\vec{B}$.

1.4. Show that a field of pure deformation (i.e. the combination of both components, F_1 and F_2) has no divergence and no vorticity.

1.5. Consider Figure 1.1A which shows isotherms (dashed lines) in fields of pure vorticity, pure convergence (negative divergence), and deformation. The vector ∇T has both magnitude and direction.

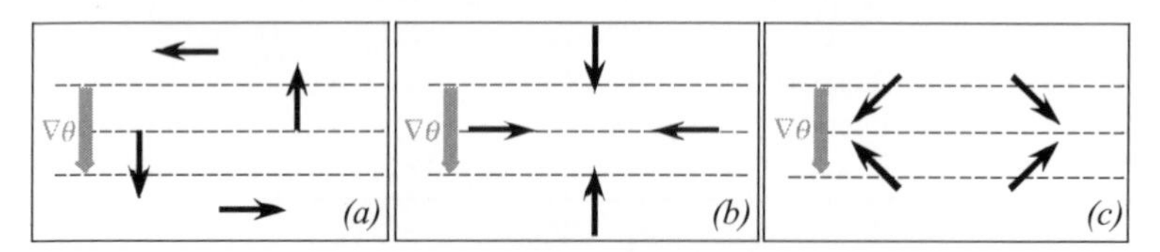

Figure 1.1A

(a) Do you think the vorticity can change both the direction and magnitude of ∇T? Does the orientation of the isotherms affect the answer to the first question? Explain.

(b) Do you think the convergence can change both the direction and magnitude of ∇T? Does the orientation of the isotherms affect the answer to the first question? Explain.

(c) Do you think the deformation can change both the direction and magnitude of ∇T? Does the orientation of the isotherms affect the answer to the first question? Explain.

1.6. Consider a fluid element with area, $A = \delta x \, \delta y$.

(a) Derive an expression for the time rate of change of this area, dA/dt. (Hint:

$$\frac{d}{dt}(\delta F) = \delta\left(\frac{dF}{dt}\right)$$

where F is any variable.)

(b) What kinematic field is represented by

$$\frac{1}{A}\frac{dA}{dt}?$$

Defend your choice.

(c) Describe (with a word) the type of flow that will result in a decrease in A. Defend your choice with a diagram and accompanying explanation.

1.7. Find the angle between the surfaces $2x^2 - y^2 + z^2 = 9$ and $3z = x^2 - 4y^2 + 5$ at the point $(2, 1, -2)$.

1.8. If $\nabla\phi = 2xyz^2\hat{i} + x^2z^2\hat{j} + 2x^2yz\hat{k}$, find $\phi(x, y, z)$ if $\phi(1, -2, 2) = 4$.

1.9. Prove that $\nabla^2(\alpha\beta) = \alpha\nabla^2\beta + 2\nabla\alpha \cdot \nabla\beta + \beta\nabla^2\alpha$ where αgnd β are scalar functions.

1.10. An automobile equipped with a thermometer is heading southward at 100 km h^{-1}, bound for a location 300 km away. During transit, the temperature drops to -5°C at the origin. If the temperature at departure was measured to be 0°C and the temperature tendency measured along the journey is $+5$°C h^{-1}, what temperature should the travelers expect at their destination?

1.11. Demonstrate that $\vec{A} \cdot (\vec{B} \times \vec{C}) = -\vec{B} \cdot (\vec{A} \times \vec{C})$.

1.12. A car is driving straight southward, past a service station, at 100 km h^{-1}. The surface pressure decreases toward the southeast at 1 Pa km^{-1}. What is the pressure tendency at the service station if the pressure measured by the car is decreasing at a rate of 50 Pa/3 h?

1.13. Imagine a stably stratified, steady-state flow in which temperature (T) is conserved. What must be the relationship between horizontal advection of T and vertical motion? Give a physical explanation of this relationship.

Solutions

1.1. $\phi(x, y) = 4x^2 + y^3 + 3$

1.7. $\alpha = 46.06^\circ$

1.8. $\phi(x, y, z) = x^2yz^2 + 12$

1.10. The temperature at the destination will be 15°C.

1.12. The pressure falls at a rate of 87.38 Pa h^{-1}.

1.13. $w = \frac{-\vec{V} \cdot \nabla T}{(\partial T / \partial z)}$

2

Fundamental and Apparent Forces

Objectives

The fluid atmosphere is a physical object and its motion is therefore governed by the laws of physics. From among these laws, Newton's second law states that the rate of change of momentum of an object (i.e. its acceleration) equals the sum of all the forces acting on that object:

$$\frac{d(Momentum)}{dt} = \sum \textit{Forces Acting on the Object.}$$

This powerful statement is valid only for motions measured in a non-accelerating coordinate system – one that is fixed in space. Such a coordinate system is known as an **inertial frame of reference**. The most convenient x, y, and z coordinates by which we measure motions on Earth refer to a grid based upon latitude and longitude (for the x and y coordinate directions) and elevation above sea level (for the z coordinate direction). Since the Earth rotates on its axis and revolves around the Sun, this Earth-based x, y, and z coordinate system undergoes constant acceleration. This fact is easily proven using a globe. After finding your location on the globe, consider the fact that what you view at that location as the immutable direction east is, in fact, constantly changing direction (to an observer fixed in space) as the Earth rotates on its axis. Thus our Earth-based coordinates are non-inertial (i.e. accelerating). This being the case, Newton's second law can only be applied to the motion of objects on Earth if we correct for the acceleration of our coordinate system.

The collection of forces required to adequately represent Newton's second law on the rotating Earth can therefore be split into two broad categories. The first of these includes forces that would affect objects even in the absence of rotation, the so-called **fundamental forces**. The most important of these fundamental forces are (1) the pressure gradient force, (2) the gravitational force, and (3) the frictional force, all of which we will investigate in this chapter. The other group of forces that we must

Mid-Latitude Atmospheric Dynamics Jonathan E. Martin

consider in a full treatment of Newton's second law arises from the need to correct for the acceleration of our terrestrial coordinate system. We will refer to such forces as **apparent forces**. Two important apparent forces to be investigated in this chapter are (1) the centrifugal force and (2) the Coriolis force. We begin this examination by considering the fundamental forces.

2.1 The Fundamental Forces

Understanding the fundamental forces is essential to gaining insight into the behavior of the fluid atmosphere. Most people have a solid intuitive feel for the gravitational and friction forces since both are so widely recognized as manifest in our daily experience. As it turns out, the effects of the often less familiar pressure gradient force are equally ubiquitous and readily detectable. We begin our examination of the fundamental forces by considering the nature of this pressure gradient force.

2.1.1 The pressure gradient force

In order to examine the pressure gradient force (PGF) we will consider the pressure exerted by the atmosphere on sides A and B of the infinitesimal fluid element illustrated in Figure 2.1. The pressure exerted on sides A and B arises from the fact

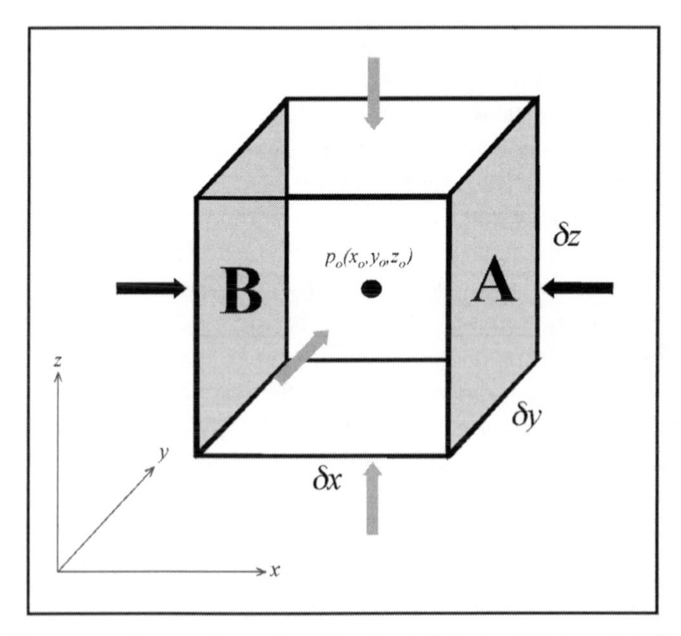

Figure 2.1 The pressure forces acting on the sides of an infinitesimal fluid element. The sides A and B are referenced in the text and the forces acting on those sides are indicated by the black arrows. The forces on other sides are indicated by the gray arrows

that random molecular motions compel molecules to strike the sides. Each time a molecule strikes the side of the fluid element, a certain amount of momentum is transferred to that side. The total momentum transfer is the sum of all the individual momentum transfers. The total momentum transferred each second defines the force exerted by the atmosphere on the side of the fluid element. Dividing this total force by the area of the side of the fluid element defines the pressure that is exerted on that side. The volume of the fluid element is given by $V = \delta x\, \delta y\, \delta z$ and its mass is given by $M = \rho \delta x\, \delta y\, \delta z$ where ρ is the density of the fluid. Let us define the pressure at the center of the fluid element to be $p(x_0, y_0, z_0) = p_0$. Assuming the pressure is continuous, we can use a Taylor series expansion to determine the pressure on sides A and B:

$$p_A = p_0 + \frac{\partial p}{\partial x}\left(\frac{\delta x}{2}\right) + \textit{Higher Order Terms} \tag{2.1a}$$

$$p_B = p_0 - \frac{\partial p}{\partial x}\left(\frac{\delta x}{2}\right) + \textit{Higher Order Terms.} \tag{2.1b}$$

Now, the x-direction pressure force acting on side A has magnitude $p_A \times (\mathit{Area\ of\ A})$ and is directed toward the center of the infinitesimal fluid element. Thus, this force can be expressed as

$$F_{A_x} = -\left(p_0 + \frac{\partial p}{\partial x}\frac{\delta x}{2}\right)\delta y\, \delta z. \tag{2.2a}$$

By similar reasoning, the x-direction pressure force acting on side B is given by

$$F_{B_x} = \left(p_0 - \frac{\partial p}{\partial x}\frac{\delta x}{2}\right)\delta y\, \delta z \tag{2.2b}$$

so that the net x-direction pressure force acting on the fluid element is

$$F_x = F_{A_x} + F_{B_x} = -\frac{\partial p}{\partial x}\delta x\, \delta y\, \delta z. \tag{2.3}$$

Thus, the net force *per unit mass* acting in the x direction on the fluid element is

$$\frac{F_x}{M} = -\frac{1}{\rho}\frac{\partial p}{\partial x}. \tag{2.4}$$

Similar expressions can be derived in exactly the same way for the y- and z-direction components of the pressure gradient force per unit mass. Therefore, the total pressure gradient force per unit mass can be expressed as

$$\frac{\vec{F}}{M} = -\frac{1}{\rho}\nabla p. \tag{2.5}$$

2.1.2 The gravitational force

Newton's law of universal gravitation says that any two elements of mass in the universe attract each other with a force proportional to their masses and inversely

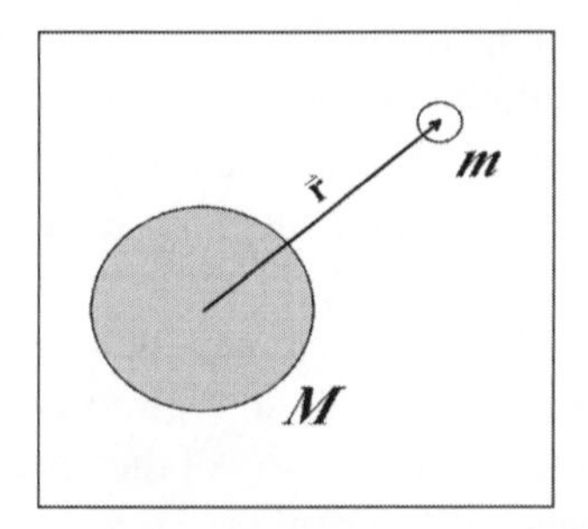

Figure 2.2 Two masses, *M* and *m*, used to illustrate Newton's law of universal gravitation. The vector $\vec{r}$ is the position vector directed from the center of mass of *M* to the center of mass of *m*

proportional to the distance between their centers of mass. This is represented symbolically, with the aid of the illustration in Figure 2.2, as

$$\vec{F}_g = -\frac{GMm}{r^2}\left(\frac{\vec{r}}{r}\right) \tag{2.6}$$

where $G = 6.673 \times 10^{-11}\ N\,m^2\,kg^{-2}$ is the universal gravitational constant, and M pulls m toward its center. For a fluid parcel of the atmosphere, M is the mass of the Earth and m is the mass of the fluid parcel. Thus, we can express the gravitational force per unit mass as

$$\frac{\vec{F}_g}{m} = -\frac{GM}{r^2}\left(\frac{\vec{r}}{r}\right). \tag{2.7}$$

Many applications in atmospheric dynamics use height above sea level (Z) as the vertical coordinate. This suggests that a parcel of air at a high elevation in the atmosphere might experience a smaller gravitational force than one located at sea level (i.e. nearer the center of gravity of the Earth). Though this conjecture is strictly true, the difference is very small from the surface to any level in the troposphere (lowest 10–12 km of the atmosphere) and we use a constant value of the gravitational force, g_0^*, where

$$g_0^* = -\frac{GM}{a^2}\left(\frac{\vec{r}}{r}\right) \tag{2.8}$$

with a being the radius of the Earth, as a consequence. It is left to the reader to demonstrate that this is an entirely reasonable simplification.

2.1.3 The frictional force

Most of us have some conceptual understanding of friction and its effect on the behavior of solids. A textbook, for instance, that is pushed across a table feels the effect of the friction between itself and the tabletop and begins to decelerate immediately. In fact, the only reason the textbook does not continue to slide along the table for ever is that a force, the friction force, is applied opposite to its motion. The frictional force in this simple example is quantified in terms of a **coefficient of friction** which is a measure of the resistance to motion that results from pushing the book over the table. This simplistic view of friction has to be modified when one considers the frictional force acting on a fluid parcel. Fluids, being collections of discrete atoms or molecules, are subject to internal friction among these particles which cause the fluid to resist the tendency to flow. We will try to gain some insight into the nature of this resistance and how to express the physics in mathematical terms.

Another analogy here may help set the stage for our more formal exploration of friction in fluids. Nearly all of us have, at one time or another, experienced traffic on a multi-lane highway. Generally cars in such traffic may pass other cars in a passing lane (on the left in North America) while passing on the right (in the cruising lanes) is discouraged. Occasionally, a driver who has just used the passing lane will decide to move to the adjacent cruising lane, in which the average speed is lower. When this happens, the passer's car imports high momentum into the cruising lane, often upsetting the smooth flow of traffic. A similar disruption occurs when a driver enters the passing lane at an insufficient speed. In the worst case (i.e. when a number of passers decide to change lanes simultaneously), the rapid flux of momentum from the passing lane to the cruising lane can cause a slowdown of the entire flow of traffic. If one considers the individual cars in this example as molecules in a fluid flow, one can see that momentum transfer between layers of a fluid (accomplished by molecules or clumps of molecules) may lie at the conceptual heart of fluid friction.

Consider, for instance, the situation depicted in Figure 2.3 in which a plate, moving at speed u_0, is placed atop a column of fluid with depth, l. The top layer of fluid moves at the velocity of the plate while the fluid at the bottom of the column has zero motion. Thus, a shearing stress exists in the fluid and a force must be exerted on the plate in order that it be kept moving at speed u_0 along the top surface of the fluid. The requisite force is proportional to u_0 since a greater force will be required for a greater speed. Additionally, since molecules of fluid that reside at the bottom of the column

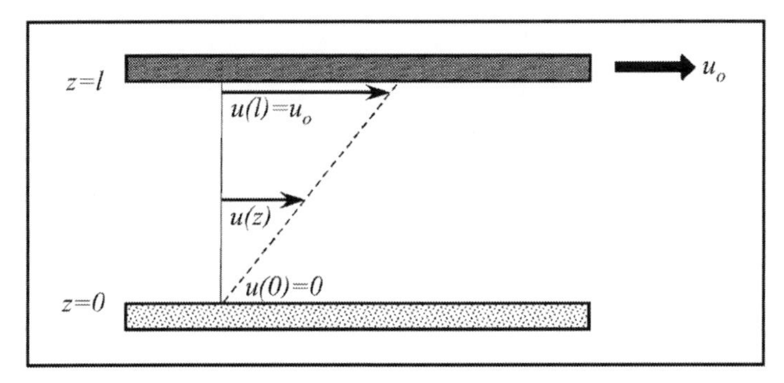

Figure 2.3 Flow beneath a moving plate illustrating 1-D, steady-state, viscous shear flow. The top plate, at height $z = l$, is moving across the top of the fluid with speed u_0 while the bottom plate is fixed. The vertical shear of the flow speed is indicated with arrows between the plates

can influence the movement of the plate through momentum transport in the fluid column, the requisite force is also inversely proportional to the depth of the fluid. The force is also proportional to the area of the plate since a larger plate makes contact with more fluid than a smaller one. The actual force required to keep the plate moving can therefore be written as $F = \mu A u_0/l$, where μ is the dynamic viscosity coefficient measured empirically and expressed in kg m^{-1} s^{-1}. If we represent the vertical shear within the fluid as $\delta u/\delta z = u_0/l$, then the force can be expressed as

$$F = \mu A \frac{\delta u}{\delta z}. \tag{2.9a}$$

Here F represents the x-direction force required to overcome the viscous effect of the vertical shear of the x-direction velocity component. Hence, as $\delta z \to 0$, the shearing stress, or viscous force per unit area, is given by

$$\tau_{zx} = \mu \frac{\partial u}{\partial z}, \tag{2.9b}$$

where the subscript '*zx*' indicates that this is the component of the shearing stress (in the x direction) that arises from the vertical shear (z) of the x-direction (x) velocity component. From the molecular viewpoint, a molecule moving to smaller z (i.e. toward the bottom of the fluid column) transports high momentum that it acquired from the motion of the plate to the surrounding fluid. Thus, there is a net downward transport of x-direction momentum and this momentum transport per unit time per unit area is the shearing stress, τ_{zx}.

The prior example considered the *steady* movement of a plate across the top of a fluid column. In nature, viscous forces result from *non-steady* shear flows. In recognition of this fact, let us consider the volume element depicted in Figure 2.4 which represents the case of non-steady, 2-D shear flow in a fluid of constant density. Analogous to our treatment of the pressure gradient force, we expand the

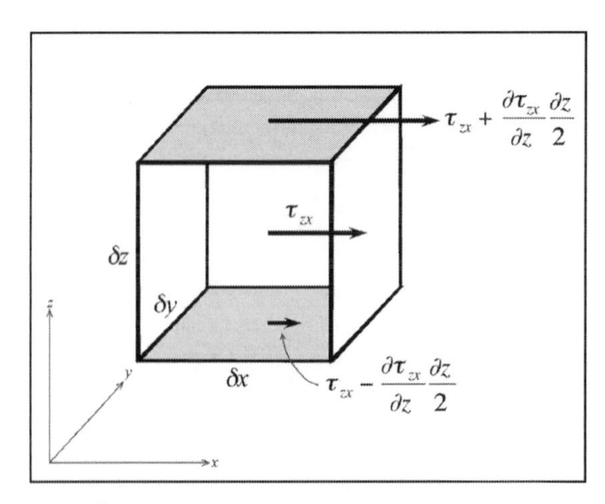

Figure 2.4 Illustration of the *x* component of the vertical shearing stress on a fluid element

shearing stress in a Taylor series in order to determine its value at the top and bottom (z-direction) facing sides of the volume element. The stress acting across the upper boundary on the fluid *below* it can be approximated as

$$\tau_{zx} + \frac{\partial \tau_{zx}}{\partial z}\frac{\delta z}{2} \tag{2.10a}$$

while the stress acting across the bottom boundary on the fluid *below* it can be approximated as

$$\tau_{zx} - \frac{\partial \tau_{zx}}{\partial z}\frac{\delta z}{2}. \tag{2.10b}$$

According to Newton's third law, this stress must be equal and opposite to the stress acting across the bottom boundary on the fluid *above* it. Since we are interested in the net stress acting on the volume element in Figure 2.4, we want to sum the forces that act on fluid *within* the volume element. Thus, we find that the net viscous force on the volume element acting in the x direction is given by

$$\left(\tau_{zx} + \frac{\partial \tau_{zx}}{\partial z}\frac{\delta z}{2}\right)\delta x \delta y - \left(\tau_{zx} - \frac{\partial \tau_{zx}}{\partial z}\frac{\delta z}{2}\right)\delta x \delta y = \frac{\partial \tau_{zx}}{\partial z}\delta x \delta y \delta z. \tag{2.11a}$$

Dividing this expression by the mass of volume element, $\rho \delta x \delta y \delta z$, we have the viscous force per unit mass arising from the vertical shear of the x-direction motion:

$$\frac{1}{\rho}\frac{\partial \tau_{zx}}{\partial z} = \frac{1}{\rho}\frac{\partial}{\partial z}\left(\mu \frac{\partial u}{\partial z}\right). \tag{2.11b}$$

If μ is constant, (2.11b) can be simplified to

$$\frac{1}{\rho}\frac{\partial}{\partial z}\left(\mu \frac{\partial u}{\partial z}\right) = \upsilon \frac{\partial^2 u}{\partial z^2} \tag{2.12}$$

where $\upsilon = \mu/\rho$ is known as the **kinematic viscosity coefficient** and has an empirically determined value of 1.46×10^{-5} m^2 s^{-1}.

Analogous derivations can be performed to determine the viscous stresses acting in the other directions. The resulting frictional force components per unit mass in the x, y, and z directions are

$$\begin{aligned} F_{rx} &= \upsilon \left(\frac{\partial^2 u}{\partial x^2} + \frac{\partial^2 u}{\partial y^2} + \frac{\partial^2 u}{\partial z^2}\right) \\ F_{ry} &= \upsilon \left(\frac{\partial^2 v}{\partial x^2} + \frac{\partial^2 v}{\partial y^2} + \frac{\partial^2 v}{\partial z^2}\right) \\ F_{rz} &= \upsilon \left(\frac{\partial^2 w}{\partial x^2} + \frac{\partial^2 w}{\partial y^2} + \frac{\partial^2 w}{\partial z^2}\right). \end{aligned} \tag{2.13}$$

For the lowest 100 km of the atmosphere, υ is so small that molecular viscosity is entirely negligible except within a few millimeter of the Earth's surface where the vertical shear is very large (on the order of 10^3 s^{-1}!). Above about 10 mm we need an entirely separate treatment of fluid friction in which it is useful to conceptualize eddies as discrete 'blobs' of fluid which move around like molecules and transfer momentum toward or away from the surface of the Earth in a manner analogous to molecules in molecular viscosity. A mixing length, defined as the average length through which an eddy can travel before mixing out its momentum, can be defined by analogy to the mean free path for molecular diffusion. With this adjustment, the dissipative effects of small-scale turbulence can be represented by defining an eddy viscosity coefficient so that

$$\frac{1}{\rho}\frac{\partial \tau_{zx}}{\partial z} \approx K\frac{\partial^2 u}{\partial z^2} \tag{2.14}$$

where K is the eddy viscosity coefficient.

2.2 Apparent Forces

In expressing his first law, Sir Isaac Newton states: 'Every body persists in its state of rest or of uniform motion in a straight line unless it is compelled to change that state by forces impressed on it.' In other words, a mass in uniform motion *relative to a coordinate system fixed in space* will remain in uniform motion in the absence of any forces. Any motion relative to a coordinate system fixed in space is known as **inertial motion** and the reference frame in which that motion is measured is known as an **inertial reference frame**. Most of us live at a single location long enough to become accustomed to thinking of north, south, east, and west as fixed directions. In reality, however, the direction I call 'north' at Madison, Wisconsin is not the same, as viewed from the perspective of a space traveler orbiting Earth, as the 'north' known to a resident of Jakarta, Indonesia. If one considers the intersection of latitude and longitude lines on a globe as the intersections of a Cartesian x and y grid describing the Earth, then it is clear that since the Earth rotates, this coordinate system is accelerating and thus provides us Earthlings with a *non-inertial* reference frame. It might appear that given our non-inertial reference frame we are not able to apply Newton's laws of motion to motion relative to the Earth. Of course, this is not true, but we do have to make some correction for the non-inertial nature of the reference frame by which we measure all such motion. We will make the necessary corrections by introducing the centrifugal and Coriolis forces, the so-called 'apparent forces'. But first, it is instructive to consider physically why the coordinate system matters at all. We can do this by considering application of Newton's laws to experiments conducted inside a closed elevator car.

In the first case, let us imagine that the car is stationary *or* moving with a constant velocity, $\vec{V}$. Under such conditions imagine that a weight is dropped within the moving car. Upon making the appropriate measurements and calculations, you would determine that the weight had fallen toward the floor of the car with a measurable, constant acceleration of 9.81 m s^{-2}. This acceleration would be observed relative to the walls and floor of the elevator car in a Cartesian coordinate system defined by the dimensions of the elevator car. In such a case, an observer in the elevator car would note complete agreement between the results of the experiment and Newton's laws of motion since the *constant velocity* elevator car provides an inertial reference frame for this experiment.

In the second case, we remotely observe the elevator car falling freely through the elevator shaft. If a similar weight is dropped within the car the weight appears to remain suspended in mid-air, at a constant elevation above the floor of the car. Measured relative to the coordinate frame of the car, the weight has zero acceleration even though to us remote observers it is clearly accelerating toward the ground at a rate of 9.81 m s^{-2}. Viewed from inside the car, Newton's laws seem to fail here, but this is because the coordinate system itself is accelerating and is therefore non-inertial.

The latitude/longitude coordinate system on a rotating Earth is also accelerating and so we have to take that acceleration into account in order to apply Newton's laws accurately to objects moving relative to that Earth-based coordinate system.

2.2.1 The centrifugal force

Each of us is located a certain distance from the axis of rotation of the Earth. Depending upon the exact distance, we are rotating around that axis at a very high, but constant speed (at Madison, Wisconsin that speed is 330 m s^{-1}!). Each of us is, therefore, not unlike the ball on the end of the string depicted in Figure 2.5. The speed of the ball is constant, equal to the rotation rate, ω, times the radius of rotation, r ($r = |\vec{r}|$). The direction of the ball changes continuously, however, and so, as viewed from the perspective of the ball, there is a uniform acceleration directed toward the axis of rotation equal to

$$\frac{d\vec{V}}{dt} = -\omega^2\vec{r}. \tag{2.15}$$

This acceleration is called the **centripetal acceleration** and is caused by the force of the string pulling on the ball. Suppose you are on the ball and rotating with it. From

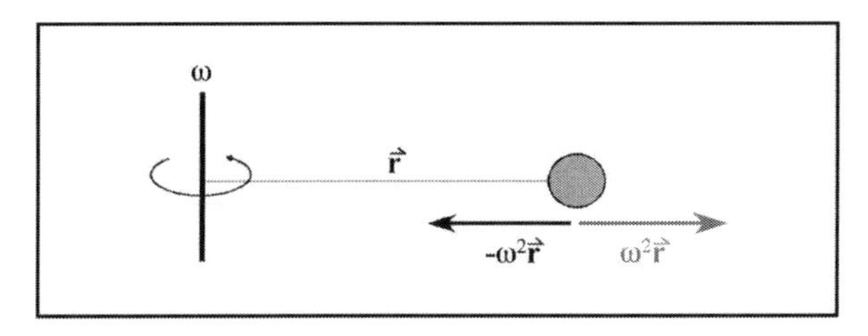

Figure 2.5 The rotating ball on a string experiences an inward-directed centripetal acceleration, indicated by the dark arrow. To the observer on the ball, a compensating centrifugal force, indicated by the gray arrow, must be included to describe accurately motions on the ball itself

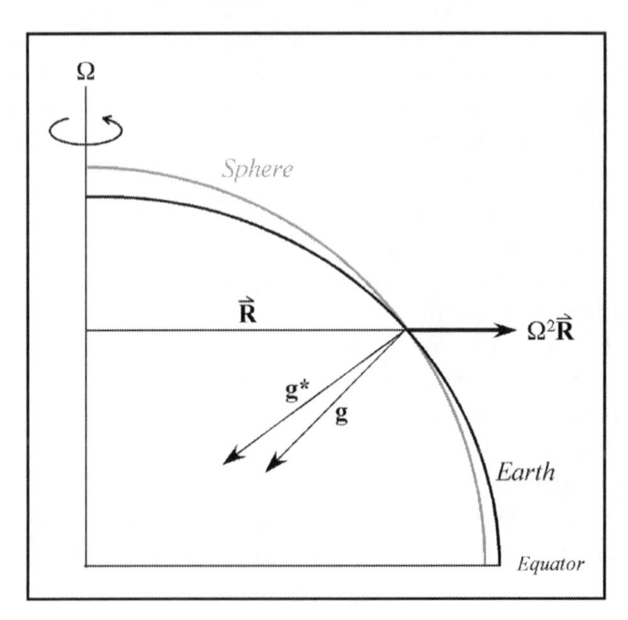

Figure 2.6 Relationship between the centrifugal force, gravitation (g^*), and effective gravity (g). The effect of the centrifugal force is to deform the Earth's shape into an oblate spheroid on which the local vertical direction is perpendicular to effective gravity as shown

your perspective the ball is stationary but, in reality, a centripetal acceleration is still being exerted upon it. In order for a person on the ball to apply Newton's laws under this condition, an apparent force that exactly balances the true centripetal force must be included in the physics; this apparent force is known as the **centrifugal force.**

In order to balance the centripetal acceleration, the centrifugal acceleration is directed *outward* along the radius of rotation and is given by

$$CEN = \omega^2 \vec{r}. \tag{2.16}$$

As depicted in Figure 2.6, on a rotating Earth, the centrifugal force affects the vertical force balance. When the centrifugal force and gravitational forces (g^*) are added, the result is called **effective gravity** (g) and is given by

$$g = g^* + \Omega^2 \vec{R} \tag{2.17}$$

where Ω is the rotation rate of the Earth and $\vec{R}$ is the position vector from the axis of rotation to the object in question. Note that effective gravity, thus defined, is directed perpendicular to the local tangent of the surface of the Earth – not necessarily toward the center of the Earth. In fact, since $\Omega^2 \vec{R}$ is directed away from the axis of rotation, g is *not* directed toward the center of the Earth except at the poles and the equator! Were the Earth a perfect sphere, this fact would result in the existence of a horizontal, equatorward-directed component of gravity. The relatively malleable crust of the Earth has long since responded to this circumstance and adopted its oblate spheroidal shape with an equatorial radius some 21 km larger that its polar radius. Given such a slightly distorted shape, the local vertical direction everywhere on Earth is defined parallel to g. The centrifugal force component of effective gravity is an example of the effect of rotation on objects *at rest* with respect to the Earth-based rotating frame of reference. In order to apply Newton's laws accurately to the motion of objects relative to that rotating frame an additional apparent force, the Coriolis force, must be considered.

2.2.2 The Coriolis force

Consider a dynamics field experiment in which one student takes a position on a merry-go-round and another student takes a position some distance above the ground in an adjacent tree. The merry-go-round is set spinning and a ball is pushed from the center of the merry-go-round toward the spinning student. From the vantage point of the tree, the motion of the ball appears as a straight line, as it should since a uniform force was administered to it. But from the perspective of the rotating frame, the ball appears to accelerate in a curved path, away from the observer in a direction opposite to the direction of rotation. Upon consulting each other's notes, the students conclude that an apparent force, arising from the rotation of the merry-go-round, deflects the ball from its path. This apparent force is the Coriolis force. How can the Coriolis force be quantified on the rotating Earth?

Suppose a hockey puck is given an impulse in the eastward direction on a frozen, frictionless Earth. Under these circumstances, the puck is rotating faster than the solid Earth beneath it so that, for its latitude, the centrifugal force acting on the puck will be increased to

$$CEN = \left(\Omega + \frac{u}{R}\right)^2 \vec{R} = \Omega^2 \vec{R} + 2\Omega u \frac{\vec{R}}{R} + \frac{u^2 \vec{R}}{R^2} \tag{2.18}$$

where u/R represents the incremental change in rotation rate resulting from the eastward impulse. The first term on the RHS of (2.18) is the already familiar centrifugal force, included in effective gravity. The second and third terms, however, are deflecting forces acting outward along $\vec{R}$ (perpendicular to the axis of rotation). For normal synoptic-scale motions on Earth, $u \ll \Omega R$ (remember, $\Omega R = 330\,\mathrm{m\,s^{-1}}$ at Madison), allowing neglect of the third term to hardly compromise the result. The remaining term, $2\Omega u \vec{R}/R$ (the excess centrifugal force), represents the Coriolis force resulting from relative motion *parallel to a latitude circle.* This Coriolis force has two components as suggested by Figure 2.7. The vertical and horizontal components are given by

$$\frac{dw}{dt} = 2\Omega u \cos\phi \text{ and } \frac{dv}{dt} = -2\Omega u \sin\phi, \tag{2.19}$$

respectively, where ϕ is the latitude. Using a shorthand in which f, the so-called Coriolis parameter, is given by $f = 2\Omega \sin\phi$, we can rewrite the horizontal component of the Coriolis force resulting from relative zonal motion as $dv/dt = -fu$. We see

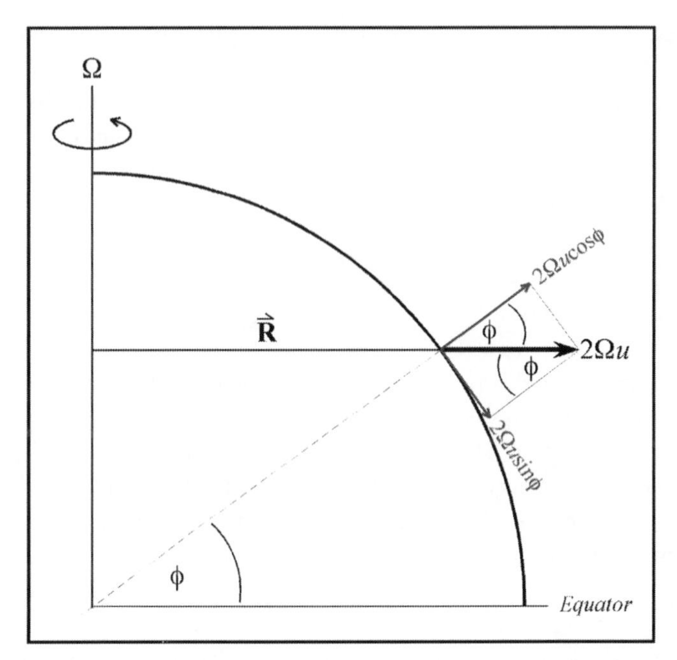

Figure 2.7 For east to west relative motions on Earth, the Coriolis force arises as excess centrifugal force

that given an eastward (westward) directed impulse, the Coriolis force will deflect the object to the south (north), or to the right of its original path, in the northern hemisphere (where ϕ is positive by convention).

What happens if we consider the hockey puck moving equatorward relative to the Earth? In the absence of applied forces, it must conserve angular momentum ($\Omega \vec{R}^2$). Upon being pushed equatorward in the northern hemisphere, the radius of rotation of the puck begins to increase. Consequently, an anti-rotational relative motion develops in order to conserve angular momentum. We can quantify this simple physics by considering a balance between the initial angular momentum of the puck and its angular momentum after displacement equatorward toward larger $\vec{R}$. (Note that displacement toward larger $\vec{R}$ also occurs if the puck is compelled to move in the relative *vertical* direction.) If we let δu signify the induced westward motion at the new radius of rotation, $\vec{R} + \delta R$, then conservation of angular momentum is given by

$$\Omega \vec{R}^2 = \left(\Omega + \frac{\delta u}{R + \delta R} \right) (\vec{R} + \delta R)^2. \tag{2.20a}$$

Expansion of (2.20a) yields

$$\Omega \vec{R}^2 = \left(\Omega + \frac{\delta u}{R + \delta R} \right) (\vec{R}^2 + 2\vec{R}\delta R + \delta R^2). \tag{2.20b}$$

Since δR (and δu) are so small, we will neglect the products of such differential terms so that (2.20b) becomes

$$\Omega \vec{R}^2 = \left(\Omega + \frac{\delta u}{R + \delta R}\right)(\vec{R}^2 + 2\vec{R}\,\delta R) \quad (2.20c)$$

or

$$\Omega \vec{R}^2 = \Omega \vec{R}^2 + 2\Omega \vec{R}\,\delta R + \frac{\vec{R}^2 \delta u}{R + \delta R} \quad (2.20d)$$

which reduces to

$$2\Omega \vec{R}\delta R = -\frac{\vec{R}^2 \delta u}{R + \delta R} \quad \text{or} \quad 2\Omega \vec{R}\,\delta R = -\vec{R}\delta u. \quad (2.20e)$$

In the end, we find that

$$\delta u = -2\Omega\,\delta R. \quad (2.21)$$

The incremental zonal velocity δu can be induced by both meridional (i.e. north/south) motion or by vertical motion as illustrated in Figure 2.8. The incremental radius of rotation, δR, has components in the vertical and meridional directions. By the similar triangles in Figure 2.8, we see that $\sin\phi = \delta R/-\delta y$ and $\cos\phi = \delta R/\delta z$. Thus, for meridional motions,

$$\delta u = -2\Omega(-\delta y \sin\phi) = 2\Omega \sin\phi(\delta y). \quad (2.22a)$$

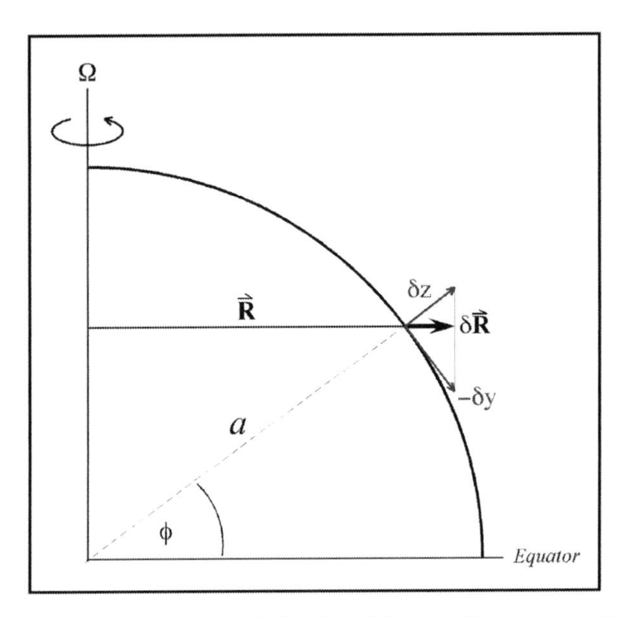

Figure 2.8 Illustration of the effect of vertical and meridional motions on the radius of rotation, $\vec{R}$. Upward and equatorward displacements produce an incremental increase in $\vec{R}$, indicated by $\delta\vec{R}$

As can be seen in Figure 2.8, however, $\delta y = a\delta\phi$, so (2.22a) can be rewritten as

$$\delta u = 2\Omega \sin\phi a\delta\phi. \tag{2.22b}$$

If we divide both sides of (2.22b) by the incremental δt and take the limit as $\delta t \to 0$, we get

$$\frac{du}{dt} = 2\Omega \sin\phi \left(a\frac{d\phi}{dt}\right). \tag{2.23a}$$

Since $ad\phi/dt = v$ and $f = 2\Omega \sin\phi$, (2.23a) can be rewritten as

$$\frac{du}{dt} = fv. \tag{2.23b}$$

It is clear from (2.23b) that equatorward motion in the northern hemisphere ($v < 0$) will induce a westward-directed zonal motion in accord with our physical intuition in the face of angular momentum conservation. Such a circumstance implies that the Coriolis force, in this instance, again compels an object to the right of its intended path.

Considering Figure 2.8, and (2.21), we see that for vertical motions

$$\delta u = -2\Omega \cos\phi\delta z. \tag{2.24a}$$

Once again, dividing both sides by δt and taking the limit as $\delta t \to 0$ results in

$$\frac{du}{dt} = -2\Omega \cos\phi \left(\frac{dz}{dt}\right) \quad \text{or} \quad \frac{du}{dt} = -2\Omega \cos\phi w. \tag{2.24b}$$

Thus, the full expression for the Coriolis force arising from meridional motions is given by

$$\frac{du}{dt} = fv - 2\Omega \cos\phi w \tag{2.25}$$

while the full 3-D Coriolis force is given by

$$\begin{aligned} \frac{du}{dt} &= fv - 2\Omega \cos\phi w \\ \frac{dv}{dt} &= -fu \\ \frac{dw}{dt} &= 2\Omega \cos\phi u. \end{aligned} \tag{2.26}$$

The Coriolis parameter, $f = 2\Omega \sin\phi$, is worth some special consideration before we leave this subject. The Coriolis parameter's dependence on latitude squares with our intuitive sense that the effect of rotation does indeed vary with latitude. We notice that the Coriolis parameter is identically zero at the equator and is a maximum at the poles. Since the Coriolis force is an apparent force arising from the acceleration of our Earth-based coordinate system, assigning a value for Ω, the rotation rate, is rather more involved than you might think.

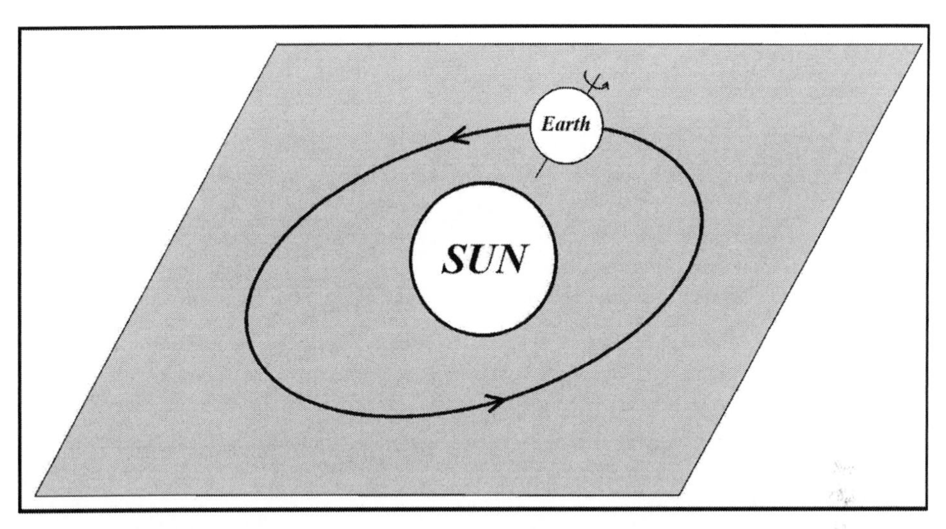

Figure 2.9 Illustration of the rotation of Earth on its axis with respect to its revolution around the Sun. The thick black line represents the Earth's revolution while the curved thin arrow represents the rotation. Gray shading is the plane of the ecliptic

The **solar day** represents the amount of time between successive local noontimes (i.e. moments at which the Sun in highest in the sky at a given location) and is 24 h long. As shown in Figure 2.9, the Earth revolves around the Sun in a counterclockwise fashion as viewed from above the plane of the ecliptic. Even if the Earth were not rotating on its axis, the revolution would provide one rotation each year – from east to west! In addition, during the year the Earth rotates (from west to east) through 365.25 solar days. Thus, as viewed from the perspective of the distant, fixed stars, the Earth must actually rotate 366.25 times (from west to east) on its axis in one year's time. Each rotation with respect to the fixed stars is therefore completed in

$$\frac{(365.25\ solar\ days) \times (24 \cdot 3600\ s\ solar\ day^{-1})}{366.25\ rotations} = 86\,156.09\ s\ \ rotation^{-1},$$

the length of the **sidereal day**. In order to apply Newton's laws accurately, we have to correct for the acceleration of our Earth-based coordinate system as viewed from the perspective of the fixed stars. Thus, Ω is determined using the length of the sidereal day as

$$\Omega = \frac{2\pi}{86\,156.09\ s} = 7.292 \times 10^{-5}\ s^{-1}.$$

Finally, it is important to note that since the Coriolis force always acts perpendicular to the motion vector, it can do no work on the moving particle since work is the scalar product of a force and a vector distance. Thus, the Coriolis force can only change the direction of motion but cannot initiate motion in an object at rest. We have now considered all the forces necessary to formulate the equations of motion on the rotating Earth from which we will investigate the fluid dynamics of the mid-latitude atmosphere. We will see in the next chapter that these equations are simply an expression of the conservation of momentum in the fluid atmosphere.

Selected References

Holton, *An Introduction to Dynamic Meteorology*, provides a thorough discussion and derivation of the fundamental and apparent forces.

Hess, *Introduction to Theoretical Meteorology*, offers similar material conveyed lucidly.

Problems

2.1. So long as it is shallow, water is a fluid with constant density. Use this fact to help solve the following problem.

(a) Develop a relationship for the horizontal pressure gradient force in terms of depth (h) of water in a shallow container.

A cylindrical tank of water is set on a turntable. The radius of the tank is r_0 and the depth of the water is z_0.

(b) The turntable is turned on (with rotation rate ω) and the system is allowed to equilibrate. Derive an expression for the height of the water surface, h, as a function of radius.

(c) Express $h(r)$ in terms of z_0 (Hint: consider the volume of fluid in the container.)

(d) If $r_0 = 1$ m, what rotation rate is required to raise the water level on the outer edge of the tank to $h = 2z_0$?

2.2. A baseball player at 30°N latitude throws a ball northward a horizontal distance of 75 m in 2 s. In what direction, and by how much, is the ball deflected laterally as a result of the rotation of the Earth?

2.3. Given the picture in Figure 2.1A, prove that $\alpha = \beta$.

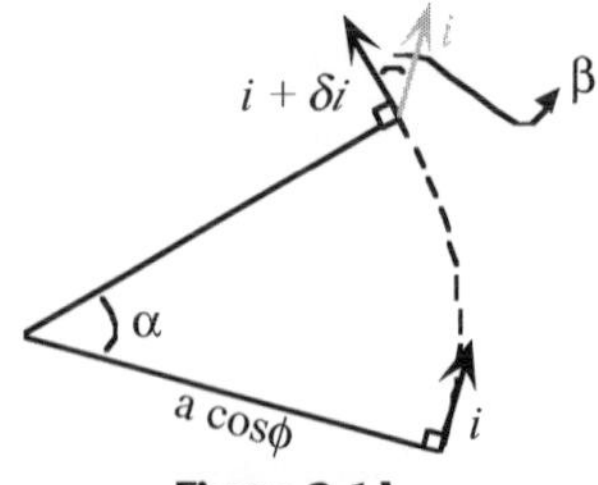

Figure 2.1A

2.4. While taking an eastbound train to work, a passenger of fixed mass finds that she weighs 542 N. On the way home she weighs herself again while the train is at full speed and finds she weighs 543 N. If she works 50 km from home, how long is her commute if she lives at 40°S? (You may assume that the average speed of the train is its full speed.)

2.5. On a typical day in the middle latitudes, the density of the air at sea level is approximately 1.25 kg m^{-3}. What would the sea-level pressure difference over a distance of 100 km have to be in order that the horizontal pressure gradient force be equal to the vertical pressure gradient force? Is this possible on Earth?

2.6. A projectile is fired vertically upward with a velocity w_0 from a point on Earth. Express the westward displacement experienced by the projectile in terms of latitude, w_0, and the rotation rate of the Earth.

2.7. What is the radius of the orbit of an equatorial, geosynchronous Earth satellite (i.e. one that remains above the same spot on Earth)?

2.8. An object at rest on the equator experiences three accelerations: one toward the center of the Earth because the Earth rotates, one toward the Sun because the Earth revolves around the Sun in a nearly circular orbit, and one toward the center of our galaxy. If the distance from the Earth to the Sun is 150×10^6 km and the period of the Sun's rotation about the galactic center is 2.5×10^8 year at a distance of 2.4×10^{17} km, compare the magnitudes of these three accelerations. Does this comparison justify assigning the value $\Omega = 7.292 \times 10^{-5}\ s^{-1}$ for the Earth's rotation rate?

2.9. Imagine that a geosynchronous meteorological satellite at 90°W must be moved to 105°W in an emergency deployment. (a) Does the satellite's orbital radius need to be increased or decreased to accomplish this task? Explain. (b) If the deployment must be completed in 3 h, calculate the exact change required to the satellite's orbital radius during deployment.

2.10. At a certain station the surface wind has a speed of 15 m s^{-1} and is directed across the isobars from high to low pressure at an angle $\alpha = 25°$. Assuming the flow is balanced, calculate the magnitude of the horizontal pressure gradient and the frictional forces per unit mass. Assume the station is located at 40°N.

2.11. Is there any place on Earth at which the atmospheric flow can be the result of a balance between friction and the pressure gradient force? Justify your answer.

2.12. Assume Jupiter is a sphere with a radius of 71 500 km. Calculate the angle between the gravitational force ($\vec{g}^*$) and the gravity force ($\vec{g}$) vectors near the top of the Jovian atmosphere as a function of latitude. Jupiter makes one rotation every 9 h 48 min 36 s.

Solutions

2.1. (a) $PGF_x = -g\frac{\partial h}{\partial x}$ (b) $h = h_0 + \frac{\omega^2 r^2}{2g}$ (c) $h(r) = z_0 + \frac{\omega^2}{2g}\left(r^2 - \frac{r_0^2}{2}\right)$ (d) $\omega = 2\sqrt{g z_0}$

2.2. $x = 0.547$ cm

2.4. 10 minutes, 18 seconds

2.5. $\partial p = 12\,262.5\,\text{hPa}$, physically impossible on Earth.

2.6. $x = \frac{4\Omega w_0^3}{3g^2} \cos\phi$

2.7. $z = 35\,804\,\text{km}$

2.8. $3.369 \times 10^{-2}\,\text{m}\,\text{s}^{-2}$, $5.946 \times 10^{-3}\,\text{m}\,\text{s}^{-2}$, $1.522 \times 10^{-8}\,\text{m}\,\text{s}^{-2}$

2.9. (a) Increase in orbital radius (b) $\Delta z = 1058.78\,\text{km}$

2.10. $F = 6.557 \times 10^{-4}\,\text{m}\,\text{s}^{-2}$ and $PGF = 1.551 \times 10^{-3}\,\text{m}\,\text{s}^{-2}$

2.11. At the equator.

2.12. $\alpha \approx (1.1316)\frac{\sin 2\phi}{g}$

3

Mass, Momentum, and Energy: The Fundamental Quantities of the Physical World

Objectives

Study of the physical world tends to be focused on the quantities known as **mass, momentum,** and **energy**. The behavior of the atmosphere is no exception to this rule. In this chapter we will investigate the manner in which these quantities and their various interactions serve to describe the building blocks of a dynamical understanding of the atmosphere at middle latitudes. We must first consider the distribution of mass in the atmosphere and the force balance that underlies this distribution. A number of insights concerning the vertical structure of the atmosphere proceed directly from this understanding.

Beginning with Newton's second law, we will construct expressions for the conservation of momentum in the three Cartesian directions. These expressions are commonly known as the equations of motion and will serve as *the* fundamental set of physical relationships for all subsequent inquiry in this book. Scale analysis of the horizontal equations of motion will reveal that a simple diagnostic relationship between the mass and momentum fields, geostrophy, characterizes the mid-latitude atmosphere on Earth. Finally, employing these equations of motion we will develop expressions for the conservation of mass and the conservation of energy. We begin by considering the distribution of mass in the atmosphere.

3.1 Mass in the Atmosphere

For our purposes, we shall define mass as *the measure of the substance of an object* and make that measurement in kilograms (kg). Though it was not clear to ancient

Mid-Latitude Atmospheric Dynamics Jonathan E. Martin

thinkers like Aristotle,[1] the atmosphere has mass. In fact the Earth's atmosphere has a mass of 5.265×10^{18} kg! The pressure exerted by this object decreases with increasing distance away from the surface as the depth of the fluid decreases. As a consequence, there is a vertical pressure gradient force given by

$$PGF_{vertical} = -\frac{1}{\rho}\frac{\partial p}{\partial z}\hat{k} \tag{3.1}$$

which compels atmospheric fluid from higher pressure (near the surface) to lower pressure (above the surface) and so is directed upward. The fact that the atmosphere does not race away into space under this forcing is a consequence of the fact that there is also the force of effective gravity acting on the fluid parcel, pulling it downwards. This force is given by

$$Gravity = -g\hat{k}. \tag{3.2}$$

The sum of the vertical pressure gradient force and gravity is zero for an atmosphere at rest. In mathematical terms

$$0 = \left(-g - \frac{1}{\rho}\frac{\partial p}{\partial z}\right)\hat{k}$$

or, after rearranging the terms and dropping the $\hat{k}$ designation for notational simplicity,

$$\frac{\partial p}{\partial z} = -\rho g. \tag{3.3}$$

This expression is known as the hydrostatic equation and represents a fundamental balance characteristic of the Earth's atmosphere: namely, that the vertical pressure gradient force is perfectly balanced by gravity. Though strictly true only for an atmosphere at rest (hence the *static* portion of the name), this **hydrostatic balance** is obeyed to great accuracy under nearly all conditions in the Earth's atmosphere.

In order to construct a vertical equation of motion we must take account of all the forces with components in the local vertical direction. The vertical pressure gradient force and gravity (combined in the hydrostatic balance) comprise the largest fraction of these forces. Surely friction, slight though it may be, will also affect motions in the vertical direction. Also, we have already shown that there is a vertical Coriolis acceleration induced by zonal motions. Thus, we can write a first approximation to the vertical equation of motion as

$$\frac{dw}{dt} = -\frac{1}{\rho}\frac{\partial p}{\partial z} - g + \vec{F}_z + (2\Omega\cos\phi)u. \tag{3.4}$$

---

[1] The theories of the ancient Greek natural philosopher Aristotle (384–322 BC) held sway in many disciplines for nearly 2000 years! He reputedly conducted an experiment to determine the weight of air. Undoubtedly using a crude scale, he 'filled' a leather bag with air, weighed it, and then compared that measurement to the weight of an 'empty' leather bag. Noting no difference between the two, he concluded that air had no weight.

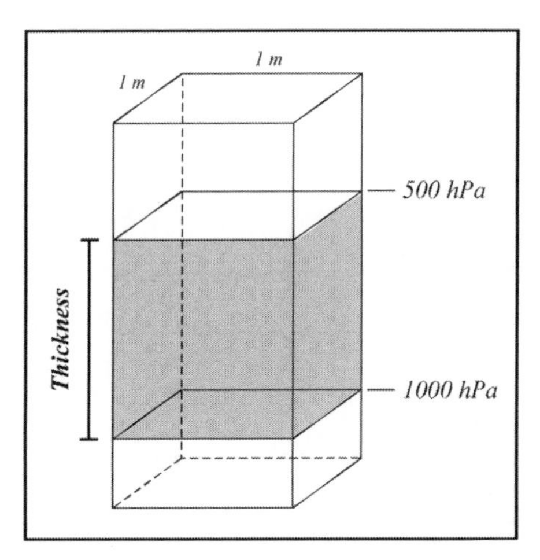

Figure 3.1 The amount of mass between any two isobaric surfaces is the same regardless of the thickness of the layer

3.1.1 The hypsometric equation

Consider the unit area column of atmosphere contained between pressure levels 1000 and 500 hPa shown in Figure 3.1. Since pressure is defined as force per unit area, we have isolated in that column an atmospheric mass sufficient to exert 500 hPa of pressure. Such a slab of the atmosphere has a unique mass whether it extends from 1000 to 500 hPa or from 812 to 312 hPa. In fact, the mass of this column can be precisely calculated as

$$Mass = (500\ hPa) \times \left(\frac{100\ N m^{-2}}{hPa}\right) \times (1\ m^2) \times \left(\frac{1}{9.81\ m\, s^{-2}}\right) = 5102.04\ kg.$$

Though the mass of a 500 hPa, unit area slab of the atmosphere is unique, its depth might be different from one day to the next. We will refer to this geometric depth as the *thickness* between two isobaric surfaces. Clearly, if the thickness varies, then so does the volume of the unit area slab. The variation of the volume of the slab dictates that the density of the air contained within the slab varies as well: less (more) dense air corresponding to a greater (smaller) thickness. By the ideal gas law, less (more) dense air will correspond to a higher (lower) column average virtual temperature, $\overline{T_v}$.[2] Thus, column average virtual temperature should have a bearing on the thickness between two isobaric levels.

Combining the hydrostatic equation with the ideal gas law provides convincing evidence to support this supposition. Recall that the ideal gas law can be written as $p = \rho R_d T_v$ where p is the pressure, ρ is the density, R_d is the gas constant for dry

[2] See Appendix A for a discussion and derivation of virtual temperature, T_v.

air,[3] and T_v is the virtual temperature. Using this expression, the hydrostatic equation can be rewritten as

$$\frac{\partial p}{\partial z} = -\frac{pg}{R_d T_v} \tag{3.5a}$$

which can be rearranged into

$$-\frac{R_d T_v}{g} \partial \ln p = \partial z. \tag{3.5b}$$

If we integrate this expression between pressure levels p_1 and p_2 ($p_1 > p_2$) at which the heights are z_1 and z_2 ($z_2 > z_1$) we get

$$-\int_{p_1}^{p_2} \frac{R_d T_v}{g} \partial \ln p = \int_{z_1}^{z_2} \partial z. \tag{3.5c}$$

Inverting the order of integration on the LHS of (3.5c) yields

$$\int_{p_2}^{p_1} \frac{R_d T_v}{g} \partial \ln p = \int_{z_1}^{z_2} \partial z$$

which can be integrated to give

$$\frac{R_d \overline{T_v}}{g} \ln\left(\frac{p_1}{p_2}\right) = z_2 - z_1 = \Delta z \tag{3.6}$$

where $\overline{T_v}$ is the pressure-weighted, column average virtual temperature, given by

$$\overline{T_v} = \frac{\int_{p_2}^{p_1} T_v \partial \ln p}{\int_{p_2}^{p_1} \partial \ln p}.$$

The foregoing expression is known as the **hypsometric equation** and it quantifies and verifies our suspicion regarding the influence of column average temperature on the thickness of an isobaric column.

We can express the hypsometric equation (and, therefore, the hydrostatic equation also) in terms of a quantity called **geopotential**, Φ. The geopotential is defined as the work required to raise a unit mass a distance dz above sea level. It quantifies the work (per unit mass) that is done against gravity in doing so. Mathematically, therefore, geopotential is given as $d\Phi = g dz$. Employing this expression, we can rewrite the

[3] R_d has a value of $287\,\mathrm{J\,kg^{-1}K^{-1}}$ and is equal to the universal gas constant ($R^* = 8.3143 \times 10^3\,\mathrm{J\,K^{-1}\,kmol^{-1}}$) divided by the molecular weight of the atmospheric mixture ($28.97\,\mathrm{kg\,kmol^{-1}}$). 'Dry' air refers to the mixture without the variable water vapor included.

hydrostatic equation as

$$\partial p = -\rho \partial \Phi \quad \text{or} \quad \frac{\partial \Phi}{\partial p} = -\alpha = -\frac{R_d T_v}{p}.$$

Correspondingly, the hypsometric equation can also be written as

$$R_d \overline{T}_v \ln\left(\frac{p_1}{p_2}\right) = \Phi_2 - \Phi_1 = \Delta\Phi.$$

We will often refer to geopotential height (Z) in subsequent discussions. The geopotential height is simply given by

$$Z = \frac{\Phi}{g_0} \tag{3.7}$$

where g_0 is the global average gravity at sea level (9.81 m s^{-2}). Thus, geometric height (z) and Z are just about equal in the troposphere.

There are several important applications of the hydrostatic and hypsometric equations that have a bearing on the analysis and understanding of mid-latitude weather systems. One of the most common analysis products used to characterize and understand the weather is a sea level pressure map. This is a map on which isobars of sea-level pressure are contoured in an attempt to identify and characterize the major circulation systems in a given location at a given time. In geographical regions characterized by high terrain, such as the Rocky Mountains of North America or the high steppe of Mongolia, the elevation is so far above sea level that use of the station pressure (i.e. the pressure actually measured with a barometer at the station) does not effectively contribute to this goal. In such regions the hypsometric equation can be used to calculate a **reduced sea-level pressure** (i.e. an estimate of what the sea-level pressure would be were the surface elevation 0 m). Consider the following example.

The station pressure at St Louis, Missouri (STL), a city close to sea level, on a certain day is measured to be 995 hPa. Meanwhile, the station pressure at Denver, Colorado (DEN), whose elevation is 1609 m above sea level, is measured at 825 hPa. There is not a *horizontal* pressure difference of 180 hPa between STL and DEN. Most of the observed pressure difference is a consequence of the *vertical* variation of pressure. By reducing the station pressure to sea level at DEN, we attempt to discover how much of the observed pressure difference actually is a horizontal pressure difference.

We begin with the hypsometric equation,

$$\frac{R_d \overline{T}_v}{g} \ln\left(\frac{p_1}{p_2}\right) = z_2 - z_1 = \Delta z$$

with $z_2 = z_{DEN}$ and $z_1 = 0$ (the geometric height at sea level). Correspondingly, $p_2 = p_{STA\,at\,DEN}$ (observed station pressure) and $p_1 = p_{SLP\,at\,DEN}$ (the desired value we will calculate as sea level pressure at DEN). Finally, $\overline{T}_v$ represents the average column temperature between sea level at DEN and the station elevation. This is clearly a fictitious quantity but we can estimate it by assuming the standard atmosphere lapse rate ($6.5\,\mathrm{K\,km^{-1}}$) throughout the fictitious column. Rearranging the hypsometric equation using the given definitions we have

$$\frac{g z_{DEN}}{R_d \overline{T}_v} = \ln\left(\frac{p_{SLP\,at\,DEN}}{p_{STA\,at\,DEN}}\right). \tag{3.8a}$$

Taking anti-logs of both sides yields

$$\left(\frac{p_{SLP\,at\,DEN}}{p_{STA\,at\,DEN}}\right) = e^{\frac{g z_{DEN}}{R_d \overline{T}_v}}$$

so that

$$p_{SLP\,at\,DEN} = p_{STA\,at\,DEN}\, e^{\frac{g z_{DEN}}{R_d \overline{T}_v}}. \tag{3.8b}$$

The above expression is known as the **altimeter equation** and is the standard expression for reducing station pressure to sea level. Supposing that the surface T_v at Denver is 20°C, we find that the reduced sea-level pressure at Denver would be 998.6 hPa. This value can be usefully compared to the sea-level pressure at St Louis on a synoptic weather chart.

The hypsometric equation can also be used to gain insights into the large-scale structure of mid-latitude weather systems. If, for instance, we consider the thickness between 1000 and 500 hPa at a given station, then (3.6) becomes

$$\Delta z = \frac{R_d \overline{T}_v}{g}\ln\left(\frac{1000}{500}\right) = \frac{R_d \overline{T}_v}{g}\ln(2) = 20.3\,\overline{T}_v. \tag{3.9}$$

Thus, a change of 60 m in the 1000–500 hPa thickness corresponds to a 2.96°C mean temperature change. This fact implies that pressure drops off more rapidly with height in a cold column of air than in a warm column. The ramifications of this fact are illustrated in Figure 3.2. in which a cold core cyclone is depicted in a vertical cross-section. Since the air column in the middle of the cyclone is colder relative to its surroundings at all levels, the thickness in that column is smaller than anywhere else. Consequently, the horizontal pressure gradient force, directed inward toward the center of the cyclone, increases in magnitude with increasing height. Thus, cold core cyclones, like those that populate the mid-latitudes on Earth, intensify with height. This characteristic of mid-latitude cyclones will prove to be a major influence on the dynamics of the cyclone life cycle.

Now that we have acquired a perspective on the distribution of mass in the atmosphere, we turn to an investigation of the basic conservation laws that govern its behavior. The atmosphere, like all physical systems, obeys the laws of conservation of energy and mass, as well as the slightly more restrictive conservation of momentum. We begin by considering the conservation of momentum.

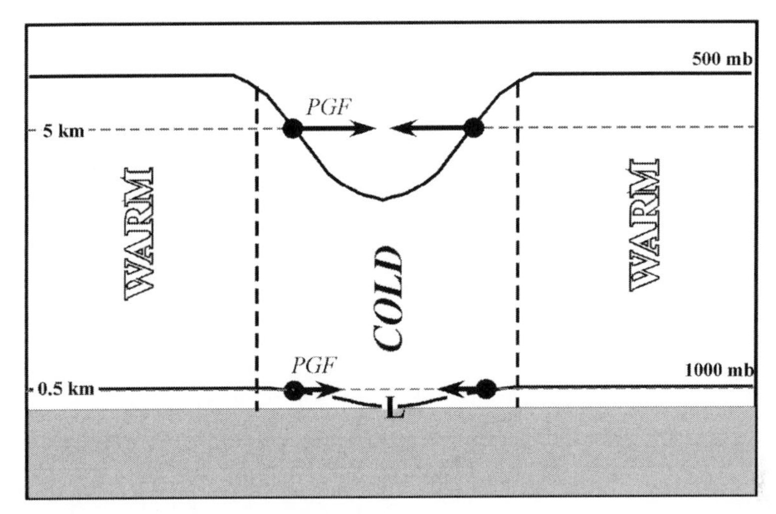

Figure 3.2 Vertical cross-section through a cold core cyclone. 'Warm' and 'Cold' refer to the column average temperatures in the three columns. Solid lines are isobars, thin dashed lines are the 0.5 km and 5 km elevation lines. The thick arrows represent the PGF, which is much larger at the top black dots. 'L' is the location of the lowest sea-level pressure

3.2 Conservation of Momentum: The Equations of Motion

Newton's second law is a statement of the conservation of momentum:

$$\frac{d}{dt}(m\vec{V}) = \sum \textit{Forces Acting on a Parcel},$$

but it is strictly true, as we have already considered, only in an inertial frame of reference. Since we will find it most convenient to use the x, y, and z coordinates fixed to Earth for our descriptions of motions, and these coordinates are accelerating, we have to relate the Lagrangian derivative of a vector in an inertial frame to the corresponding Lagrangian derivative in a rotating frame. Let $\vec{A}$ be an arbitrary vector whose Cartesian components in an inertial frame are

$$\vec{A} = A_x\hat{i} + A_y\hat{j} + A_z\hat{k}$$

and whose components in a coordinate frame rotating with an angular velocity $\vec{\Omega}$ are

$$\vec{A} = A'_x\hat{i}' + A'_y\hat{j}' + A'_z\hat{k}'.$$

Now, let $d_a\vec{A}/dt$ be the total derivative of $\vec{A}$ in the inertial (absolute) frame, expressed as

$$\frac{d_a\vec{A}}{dt} = \frac{dA_x}{dt}\hat{i} + \frac{dA_y}{dt}\hat{j} + \frac{dA_z}{dt}\hat{k}.$$

Notice that in the inertial frame the coordinate directions $\hat{i}$, $\hat{j}$, and $\hat{k}$ are unchanging. Taking the same derivative in the rotating frame, however, yields

$$\frac{d_a \vec{A}}{dt} = \frac{dA'_x}{dt}\hat{i}' + \frac{dA'_y}{dt}\hat{j}' + \frac{dA'_z}{dt}\hat{k}' + A'_x\frac{d\hat{i}'}{dt} + A'_y\frac{d\hat{j}'}{dt} + A'_z\frac{d\hat{k}'}{dt}$$

which can be rewritten as

$$\frac{d_a \vec{A}}{dt} = \frac{d\vec{A}}{dt} + A'_x\frac{d\hat{i}'}{dt} + A'_y\frac{d\hat{j}'}{dt} + A'_z\frac{d\hat{k}'}{dt} \tag{3.10}$$

given that

$$\frac{d\vec{A}}{dt} = \frac{dA'_x}{dt}\hat{i}' + \frac{dA'_y}{dt}\hat{j}' + \frac{dA'_z}{dt}\hat{k}'$$

where $d\vec{A}/dt$ represents the rate of change of $\vec{A}$ following the *relative* motion in the rotating frame.

The derivatives $d\hat{i}'/dt$, $d\hat{j}'/dt$, and $d\hat{k}'/dt$ on the RHS of (3.10) represent the rates of change of the unit vectors $\hat{i}'$, $\hat{j}'$, and $\hat{k}'$ that arise because the coordinate system is accelerating. It is important to note that each of these derivative terms describes only the change in *direction* of the unit vectors since, by definition, the vector magnitudes are always equal to one. Thus, full expressions for these derivatives are achieved upon describing the change in direction experienced by each of the unit vectors as a result of rotation of the Earth.

Figure 3.3(a) illustrates a view of the change of $\hat{i}'$ as viewed from the North Pole. The rotation vector, $\vec{\Omega}$, points upward out of the page. By similar triangles, we find that $\delta\hat{i}' = \hat{i}'\delta\theta$. Now, upon dividing both sides of this equality by the amount of time (δt) it takes to rotate through $\delta\theta$ degrees, and taking the limit of the resulting

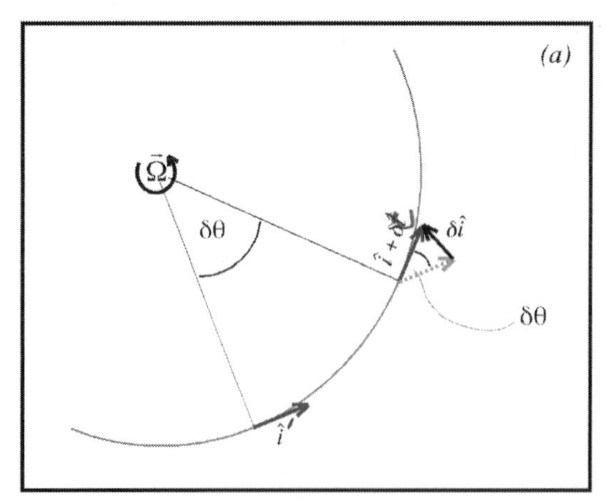

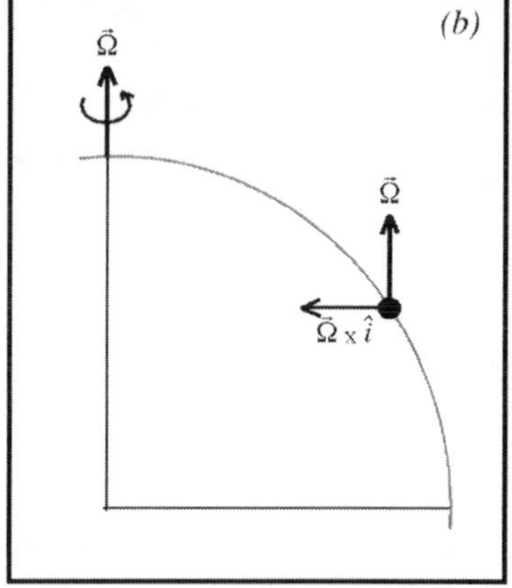

Figure 3.3 (a) View from the North Pole of the change in the $\hat{i}$ unit vector ($\delta\hat{i}$) and (b) cross-sectional view of the same vector, $\delta\hat{i}$. $\vec{\Omega}$ is the rotation vector

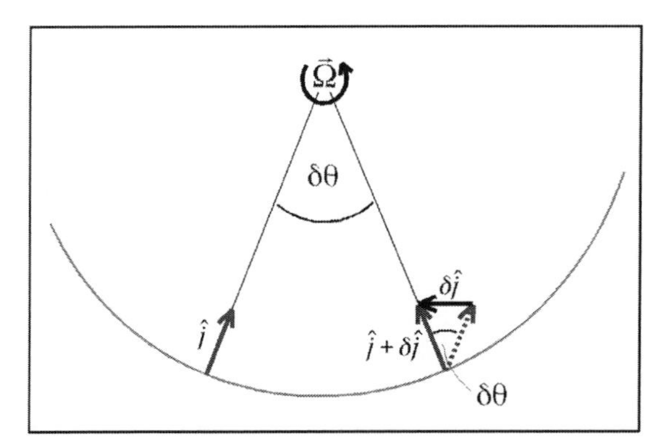

Figure 3.4 View from the North Pole of the change in the $\hat{j}$ unit vector $(\delta\hat{j})$. $\vec{\Omega}$ is the rotation vector

expression as $\delta t \to 0$, we get

$$\lim_{\delta t \to 0} \frac{\delta \hat{i}'}{\delta t} = \left|\frac{d\hat{i}'}{dt}\right| = \left|\hat{i}'\frac{d\theta}{dt}\right| = \left|\hat{i}'\vec{\Omega}\right| \tag{3.11}$$

so the magnitude of the vector $d\hat{i}'/dt$ is equal to $|\vec{\Omega}|$. It is clear from Figure 3.3(b), however, that the vector $d\hat{i}'/dt$ is directed inward toward the axis of rotation. Knowing that $d\hat{i}'/dt$ is a vector that is both perpendicular to $\hat{i}'$ and has magnitude $|\vec{\Omega}|$, we find that its full expression is given by

$$\frac{d\hat{i}'}{dt} = \vec{\Omega} \times \hat{i}'. \tag{3.12}$$

Similar relationships exist for $d\hat{j}'/dt$ and $d\hat{k}'/dt$ as can be seen in Figures 3.4 and 3.5. Consequently, we can rewrite the last three terms on the RHS of (3.10) as

$$A'_x\frac{d\hat{i}'}{dt} = A'_x(\vec{\Omega} \times \hat{i}') = \vec{\Omega} \times (A'_x\hat{i}'),$$

$$A'_y\frac{d\hat{j}'}{dt} = A'_y(\vec{\Omega} \times \hat{j}') = \vec{\Omega} \times (A'_y\hat{j}'), \text{ and}$$

$$A'_z\frac{d\hat{k}'}{dt} = A'_z(\vec{\Omega} \times \hat{k}') = \vec{\Omega} \times (A'_z\hat{k}'),$$

so that

$$A'_x\frac{d\hat{i}'}{dt} + A'_y\frac{d\hat{j}'}{dt} + A'_z\frac{d\hat{k}'}{dt} = \vec{\Omega} \times (A'_x\hat{i}' + A'_y\hat{j}' + A'_z\hat{k}') = \vec{\Omega} \times \vec{A}. \tag{3.13}$$

As a result, (3.10) can be rewritten as

$$\frac{d_a\vec{A}}{dt} = \frac{d\vec{A}}{dt} + \vec{\Omega} \times \vec{A} \tag{3.14}$$

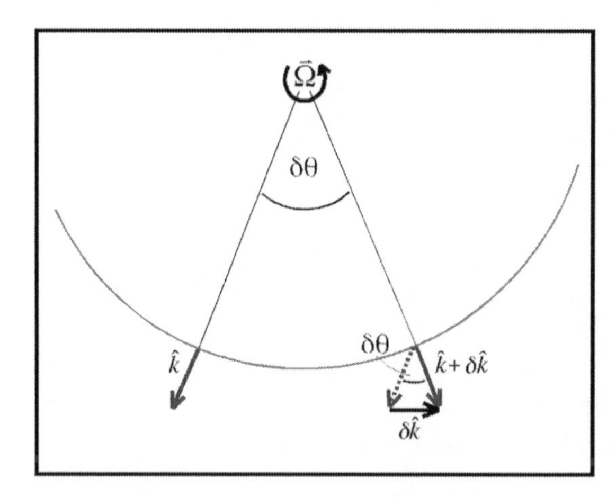

Figure 3.5 View from the North Pole of the change in the $\hat{k}$ unit vector ($\delta\hat{k}$). $\vec{\Omega}$ is the rotation vector

for any vector $\vec{A}$. This expression describes the relationship between the total derivative of a vector in inertial coordinates and its associated derivative in a coordinate system rotating with angular velocity $\vec{\Omega}$.

Employing (3.14), let us now find a relationship between the absolute velocity of an air parcel ($\vec{U}_a$) and the velocity of the same air parcel relative to Earth ($\vec{U}$). We can do this by applying (3.14) to the position vector $\vec{r}$ (where $\vec{r}$ is a vector perpendicular to the axis of rotation with magnitude equal to the distance from the surface of the Earth to the axis of rotation), for a parcel of air on Earth:

$$\frac{d_a\vec{r}}{dt} = \frac{d\vec{r}}{dt} + \vec{\Omega} \times \vec{r}. \tag{3.15a}$$

By definition, $d_a\vec{r}/dt = \vec{U}_a$ and $d\vec{r}/dt = \vec{U}$ so the desired relationship is simply

$$\vec{U}_a = \vec{U} + \vec{\Omega} \times \vec{r} \tag{3.15b}$$

which states that the absolute velocity of an object on the rotating Earth is equal to the sum of its velocity relative to the Earth ($\vec{U}$) and the velocity of the rotating Earth itself ($\vec{\Omega} \times \vec{r}$).

Now if we reapply the previous result to the vector $\vec{U}_a$ we get

$$\frac{d_a\vec{U}_a}{dt} = \frac{d\vec{U}_a}{dt} + \vec{\Omega} \times \vec{U}_a. \tag{3.16a}$$

Substituting (3.15b) for $\vec{U}_a$ above yields

$$\begin{aligned}\frac{d_a\vec{U}_a}{dt} &= \frac{d}{dt}(\vec{U} + \vec{\Omega} \times \vec{r}) + \vec{\Omega} \times (\vec{U} + \vec{\Omega} \times \vec{r}) \\ &= \frac{d\vec{U}}{dt} + \vec{\Omega} \times \frac{d\vec{r}}{dt} + \vec{\Omega} \times \vec{U} + \vec{\Omega} \times \vec{\Omega} \times \vec{r}.\end{aligned} \tag{3.16b}$$

Since $d\vec{r}/dt = \vec{U}$ and $\vec{\Omega} \times \vec{\Omega} \times \vec{r} = -\Omega^2\vec{r}$, this can be simplified to

$$\frac{d_a \vec{U}_a}{dt} = \frac{d\vec{U}}{dt} + 2\vec{\Omega} \times \vec{U} - \Omega^2\vec{r}. \tag{3.17}$$

Equation (3.17) states that the Lagrangian acceleration in an inertial system is equal to the sum of (1) the Lagrangian change of relative $\vec{U}$, plus (2) the Coriolis acceleration from relative motion in the relative frame, plus (3) centripetal acceleration resulting from the rotation of the coordinates. Recalling Newton's second law and the fact that we will consider the pressure gradient force, the frictional force, and gravitational force as the only real forces acting on the atmospheric fluid, we find that

$$\frac{d_a \vec{U}_a}{dt} = \frac{d\vec{U}}{dt} + 2\vec{\Omega} \times \vec{U} - \Omega^2\vec{r} = -\frac{1}{\rho}\nabla p + \vec{g}^* + \vec{F}$$

or, upon rearranging terms,

$$\frac{d\vec{U}}{dt} = -2\vec{\Omega} \times \vec{U} - \frac{1}{\rho}\nabla p + \vec{g} + \vec{F} \tag{3.18}$$

where the centripetal force has been combined with the gravitational force ($\vec{g}^*$) in the gravity term ($\vec{g}$). This expression states that the acceleration following the relative motion in a rotating reference frame is equal to the sum of (1) the Coriolis force, (2) the pressure gradient force, (3) effective gravity, and (4) the friction force. This is a major result but it remains in vectorial form only – a form not particularly amenable to analysis. Since the Earth is nearly a sphere, it will turn out to be quite convenient to recast this vector expression into spherical coordinates.

3.2.1 The equations of motion in spherical coordinates

Spherical coordinates treat the three dimensions in terms of longitude, latitude, and geometric height above sea level (λ, ϕ, z) using unit vectors $\hat{\imath}$, $\hat{\jmath}$, and $\hat{k}$ in the description of motions. The relative velocity vector becomes $\vec{V} = u\hat{\imath} + v\hat{\jmath} + w\hat{k}$ where the components are defined as

$$u \equiv a\cos\phi\frac{d\lambda}{dt}, \quad v \equiv a\frac{d\phi}{dt}, \quad \text{and} \quad w \equiv \frac{dz}{dt}$$

where a is the radius of the Earth.[4] Distances in the zonal and meridional directions are given by $dx = a\cos\phi\, d\lambda$ and $dy = a\, d\phi$, respectively. It is important to note that this coordinate system is not a Cartesian system because the unit vectors are not constant; they are, in fact, functions of position on Earth. A simple way of conceptualizing this fact is to consider that all longitude lines converge at the pole. Therefore, the direction 'north' is not pointed in the same absolute direction at every

[4] Formally, a should be replaced with $(r + a)$ where r is the distance above sea level and a is the radius of the Earth. However, for all tropospheric, and nearly all atmospheric, applications, $r \ll a$ so we simply use a.

longitude on Earth. This position dependence must be taken into account when the acceleration vector is expanded into its components

$$\frac{d\vec{V}}{dt} = \frac{du}{dt}\hat{i} + \frac{dv}{dt}\hat{j} + \frac{dw}{dt}\hat{k} + u\frac{d\hat{i}}{dt} + v\frac{d\hat{j}}{dt} + w\frac{d\hat{k}}{dt}. \tag{3.19}$$

We now must determine expressions for the last three terms on the RHS of (3.19).

Beginning with $d\hat{i}/dt$, we simply expand it like any other total derivative to get

$$\frac{d\hat{i}}{dt} = \frac{\partial\hat{i}}{\partial t} + u\frac{\partial\hat{i}}{\partial x} + v\frac{\partial\hat{i}}{\partial y} + w\frac{\partial\hat{i}}{\partial z}. \tag{3.20}$$

We know that $\partial\hat{i}/\partial t = 0$ as there is no local change in the coordinate direction (i.e. at any given location, east *always* points in the same direction). The $\hat{i}$ direction experiences no change as one moves north or south along a given longitude line, nor as one moves up or down in elevation so that $\partial\hat{i}/\partial y$ and $\partial\hat{i}/\partial z$ are both zero. However, as we saw already in Figure 3.3(a), the $\hat{i}$ direction does change as one moves along a latitude circle so that (3.20) can be simplified to

$$\frac{d\hat{i}}{dt} = u\frac{\delta\hat{i}}{\delta x}. \tag{3.21}$$

The problem becomes one of determining the magnitude and direction of $\partial\hat{i}/\partial x$. We can make this determination by considering a horizontal cross-section viewed from the North Pole as shown in Figure 3.6. It is evident that $\delta x = a\cos\phi\delta\lambda$ and that $|\delta\hat{i}| = |\hat{i}|\delta\lambda = \delta\lambda$ since $\hat{i}$ has unit magnitude. Therefore,

$$\left|\frac{\delta\hat{i}}{\delta x}\right| = \frac{\delta\lambda}{a\cos\phi\delta\lambda} = \frac{1}{a\cos\phi} \tag{3.22a}$$

with $\delta\hat{i}$ directed toward the axis of rotation. Thus, we must split $\delta\hat{i}$ into components in order to determine the direction (in terms of λ, ϕ, z) of $\delta\hat{i}/\delta x$. With the help

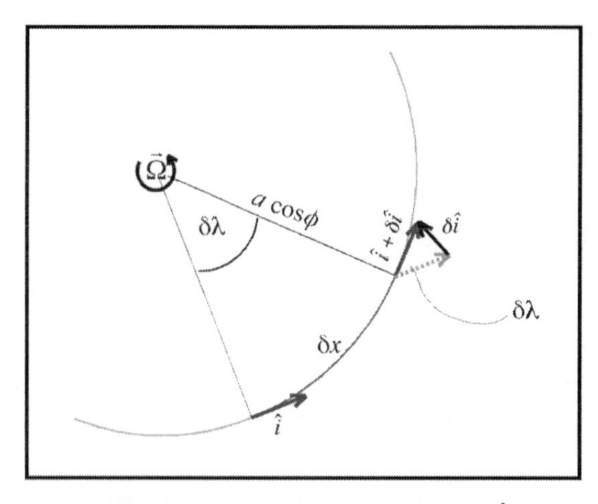

Figure 3.6 Illustration of the derivative $\frac{\delta\hat{i}}{\delta x}$

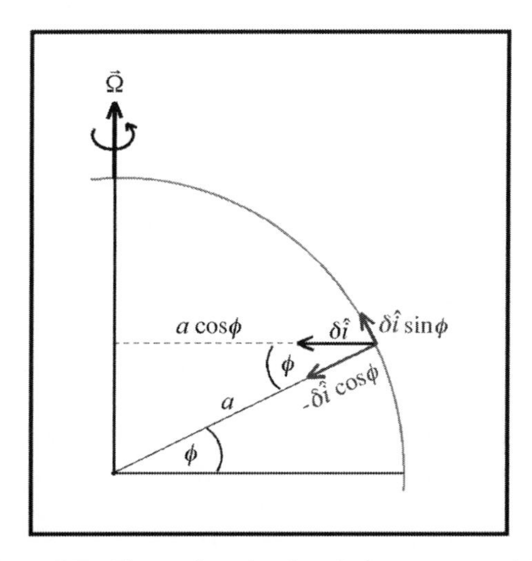

Figure 3.7 The northward and vertical components of $\delta\hat{\imath}$

of Figure 3.7, we see that $\delta\hat{\imath}$ has components in the $\hat{\jmath}$ and $-\hat{k}$ directions. The $\hat{\jmath}$ component is a function of $\sin\phi$ while the $-\hat{k}$ component is a function of $\cos\phi$. We find, therefore, that

$$\frac{\delta\hat{\imath}}{\delta x} = \frac{(\sin\phi\hat{\jmath} - \cos\phi\hat{k})}{a\cos\phi} \tag{3.22b}$$

so that, taking the limit as $\delta x \to 0$,

$$\frac{d\hat{\imath}}{dt} = \frac{u(\sin\phi\hat{\jmath} - \cos\phi\hat{k})}{a\cos\phi}. \tag{3.22c}$$

Next we consider the component form of $d\hat{\jmath}/dt$. Once again, this term must be expanded like any other Lagrangian derivative into

$$\frac{d\hat{\jmath}}{dt} = \frac{\partial\hat{\jmath}}{\partial t} + u\frac{\partial\hat{\jmath}}{\partial x} + v\frac{\partial\hat{\jmath}}{\partial y} + w\frac{\partial\hat{\jmath}}{\partial z}. \tag{3.23}$$

As was the case with $\hat{\imath}$, there is no local time derivative of $\hat{\jmath}$ nor is there any change in $\hat{\jmath}$ resulting from a change in elevation. There are, however, changes in $\hat{\jmath}$ that arise from changing position in the x or y direction. Figure 3.8(a) illustrates the geometry involved in determining $\partial\hat{\jmath}/\partial x$. The hypotenuse β of the lightly shaded triangle is given by $\beta = a/\tan\phi$ since $\sin\phi = (a\cos\phi)/\beta$. Knowing this dimension, the darker shaded triangle, shown independently in Figure 3.8(b), can be used to find $\partial\hat{\jmath}/\partial x$. It is clear from Figure 3.8(b) that $\delta x = (a/\tan\phi)\delta\alpha$ and that $\delta\hat{\jmath} = \hat{\jmath}\delta\alpha$ with $\delta\hat{\jmath}$ directed in the $-x$ direction. Thus,

$$\left|\frac{\delta\hat{\jmath}}{\delta x}\right| = \frac{\tan\phi}{a}$$

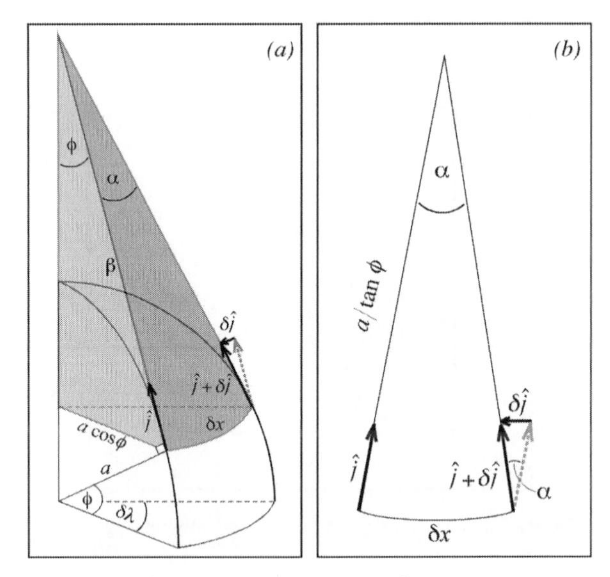

Figure 3.8 Illustration of the x variation of the unit vector $\hat{j}$. (a) A 3-D view of the plane on which $\hat{j}$ sits (darker shading). Dark-shaded triangle in (a) is illustrated in (b)

or, taking the limit as $\delta x \to 0$ and incorporating the direction,

$$\frac{\partial \hat{j}}{\partial x} = -\frac{\tan\phi}{a}\hat{\imath}. \tag{3.24}$$

Figure 3.9 illustrates the dependence of $\hat{j}$ on the y direction. We find that $\delta y = a\delta\phi$ and that $|\delta\hat{j}| = |\hat{j}\delta\phi| = \delta\phi$. Thus, $|\delta\hat{j}/\delta y| = 1/a$ with $\delta\hat{j}$ directed in the $-\hat{k}$

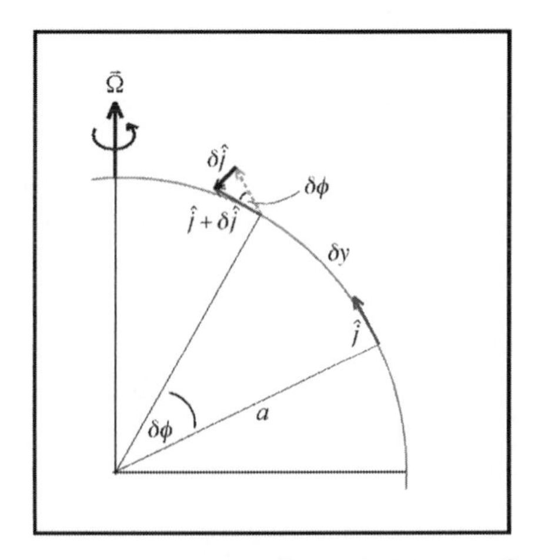

Figure 3.9 The y-direction dependence of $\hat{j}$

direction. Again, taking the limit of this expression as $\delta y \to 0$ yields

$$\frac{\partial \hat{j}}{\partial y} = -\frac{1}{a}\hat{k} \tag{3.25}$$

which, combined with (3.23) and (3.24), results in an expression for $d\hat{j}/dt$:

$$\frac{d\hat{j}}{dt} = \frac{-u \tan\phi}{a}\hat{i} - \frac{v}{a}\hat{k}. \tag{3.26}$$

Finally, we turn to $d\hat{k}/dt$ and, recognizing that $\hat{k}$ has no local time derivative nor any vertical derivative, obtain that

$$\frac{d\hat{k}}{dt} = u\frac{\partial \hat{k}}{\partial x} + v\frac{\partial \hat{k}}{\partial y}. \tag{3.27}$$

Figure 3.10 illustrates the x-direction dependence of $\hat{k}$. Since the triangle of interest represents a cross-section originating at the center of the Earth, we find that $\delta x = a\delta\lambda$ and that $|\delta\hat{k}| = |\hat{k}\delta\lambda| = \delta\lambda$ directed in the positive x direction. Consequently, $|\delta\hat{k}/\delta x| = 1/a$ which leads to the differential expression

$$\frac{\partial \hat{k}}{\partial x} = \frac{1}{a}\hat{i}. \tag{3.28}$$

Using a cross-section like that shown in Figure 3.9, but concentrating on the change in $\hat{k}$ over the distance δy, yields the expression $\partial\hat{k}/\partial y = (1/a)\hat{j}$. Thus, a complete expression for $d\hat{k}/dt$ is given by

$$\frac{d\hat{k}}{dt} = \frac{u}{a}\hat{i} + \frac{v}{a}\hat{j}. \tag{3.29}$$

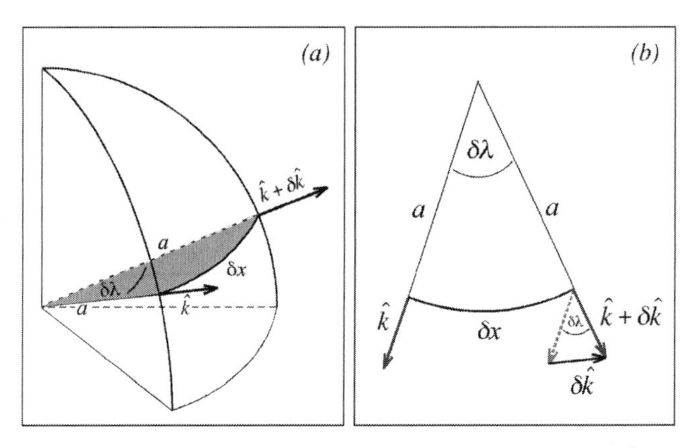

Figure 3.10 The x-direction dependence of the unit vector $\hat{k}$

Combining (3.22c), (3.26), and (3.29) we can rewrite (3.19) in its fully expanded component form as

$$\frac{d\vec{V}}{dt} = \left(\frac{du}{dt} - \frac{uv\tan\phi}{a} + \frac{uw}{a}\right)\hat{i} + \left(\frac{dv}{dt} + \frac{u^2\tan\phi}{a} + \frac{vw}{a}\right)\hat{j} + \left(\frac{dw}{dt} - \frac{u^2+v^2}{a}\right)\hat{k}. \quad (3.30)$$

This expression describes only the spherical coordinate components of the Lagrangian derivative of the relative motion. Recall that our vector expression for the equations of motion (3.18) included reference to the pressure gradient, Coriolis, gravity, and friction forces. In order to obtain a complete component expansion of the equations of motion in spherical coordinates we must expand the force terms as well.

The Coriolis force term is given by $-2\vec{\Omega} \times \vec{U}$. Figure 3.11 demonstrates that the rotation vector, $\vec{\Omega}$, is perpendicular to the x direction and so has components only in the positive $\hat{j}$ and positive $\hat{k}$ directions. Considering the trigonometry in Figure 3.11, it is clear that the $\hat{k}$ component of $\vec{\Omega}$ has magnitude $\Omega\sin\phi$ while the $\hat{j}$ component has magnitude $\Omega\cos\phi$. Thus, the component expansion of the Coriolis force term can be determined by assessing the following determinant:

$$-2\vec{\Omega} \times \vec{U} = \begin{vmatrix} \hat{i} & \hat{j} & \hat{k} \\ 0 & -2\Omega\cos\phi & -2\Omega\sin\phi \\ u & v & w \end{vmatrix} = -(2\Omega\cos\phi w - 2\Omega\sin\phi v)\hat{i} - 2\Omega\sin\phi u\hat{j} + 2\Omega\cos\phi u\hat{k}. \quad (3.31)$$

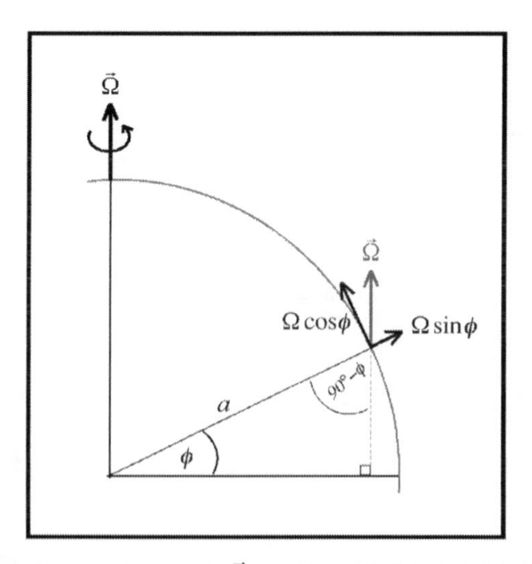

Figure 3.11 Partition of the rotation vector, $\vec{\Omega}$, into its vertical and meridional components

The component form of the pressure gradient force is given by

$$-\frac{1}{\rho}\nabla p = -\frac{1}{\rho}\frac{\partial p}{\partial x}\hat{i} - \frac{1}{\rho}\frac{\partial p}{\partial y}\hat{j} - \frac{1}{\rho}\frac{\partial p}{\partial z}\hat{k}. \tag{3.32}$$

Gravity, which acts downward in the local vertical direction, is represented by

$$\vec{g} = -g\hat{k} \tag{3.33}$$

while friction can be represented as

$$\vec{F} = F_x\hat{i} + F_y\hat{j} + F_z\hat{k}. \tag{3.34}$$

Combining (3.30), (3.31), (3.32), (3.33), and (3.34) and separating the component expression we get the three component equations of motion for flow on the rotating Earth:

$$\frac{du}{dt} - \frac{uv\tan\phi}{a} + \frac{uw}{a} = -\frac{1}{\rho}\frac{\partial p}{\partial x} + 2\Omega\sin\phi v - 2\Omega\cos\phi w + F_x \tag{3.35a}$$

$$\frac{dv}{dt} + \frac{u^2\tan\phi}{a} + \frac{vw}{a} = -\frac{1}{\rho}\frac{\partial p}{\partial y} - 2\Omega\sin\phi u + F_y \tag{3.35b}$$

$$\frac{dw}{dt} - \frac{u^2 + v^2}{a} = -\frac{1}{\rho}\frac{\partial p}{\partial z} - g + 2\Omega\cos\phi u + F_z. \tag{3.35c}$$

The various terms in (3.35) involving $1/a$ arise from the non-flatness of the Earth and are consequently known as **curvature terms**. Each of the curvature terms is quadratic in the dependent variables $(u,\ v,\ w)$ and is thus non-linear and presents difficulty in analysis. It will soon be demonstrated, however, that these curvature terms are entirely negligible in any discussion of the dynamics of mid-latitude weather systems. However, even in the absence of these particular non-linear terms, the remaining elements of (3.35) also contain non-linear elements since, for instance, in the expansion of du/dt we get

$$\frac{du}{dt} = \frac{\partial u}{\partial t} + \underline{u\frac{\partial u}{\partial x} + v\frac{\partial u}{\partial y} + w\frac{\partial u}{\partial z}}.$$

The underlined terms are also clearly quadratic in $(u,\ v,\ w)$. These terms are known as the advective acceleration terms and they are comparable to the local acceleration term (in this case, $\partial u/\partial t$). The presence of such non-linear advection processes is one reason why dynamic meteorology is so fascinating (and difficult)!

The equations of motion (3.35) are a complicated set of expressions and it is logical to inquire whether or not they can be simplified. The answer is yes and we will use the method of scale analysis, introduced in Chapter 1, to accomplish this simplification. In order to do so, we must first assign observationally based characteristic values for the set of variables involved in the equations of motion. Considering just the horizontal velocity, which appears in (3.35) as both u and v,

Table 3.1 Characteristic scales of the various terms in the horizontal equations of motion

	1	2	3	4	5	6	7
x equation	$\frac{du}{dt}$	$-2\Omega\sin\phi v$	$2\Omega\cos\phi w$	$\frac{uw}{a}$	$-\frac{uv\tan\phi}{a}$	$-\frac{1}{\rho}\frac{\partial p}{\partial x}$	F_x
y equation	$\frac{dv}{dt}$	$2\Omega\sin\phi u$		$\frac{uv}{a}$	$\frac{u^2\tan\phi}{a}$	$-\frac{1}{\rho}\frac{\partial p}{\partial y}$	F_y
Scales	$\frac{U^2}{L}$	$f_0 U$	$f_0 W$	$\frac{UW}{a}$	$\frac{U^2}{a}$	$\frac{\delta p}{\rho L}$	$\frac{\nu U}{H^2}$
Magnitude ($\mathrm{m\,s^{-2}}$)	10^{-4}	10^{-3}	10^{-6}	10^{-8}	10^{-5}	10^{-3}	10^{-12}

we know from observations that characteristically the horizontal velocity at middle latitudes is not as small as $1\ \mathrm{m\,s^{-1}}$ nor is it as large as $100\ \mathrm{m\,s^{-1}}$. Therefore, a characteristic *scale* for the horizontal velocity is something close to $10\ \mathrm{m\,s^{-1}}$. Performing a similar analysis for the other variables in (3.35) results in the following reasonable set of characteristic values for the relevant variables:

$U \sim 10\ \mathrm{m\,s^{-1}}$ characteristic horizontal velocity

$W \sim 1\ \mathrm{cm\,s^{-1}}$ characteristic vertical velocity

$L \sim 10^6\ \mathrm{m}$ characteristic length scale of synoptic-scale features

$H \sim 10^4\ \mathrm{m}$ characteristic depth (i.e. depth of the troposphere)

$\frac{\delta p}{\rho} \sim 10^3\ \mathrm{m^2\,s^{-2}}$ characteristic horizontal pressure fluctuation

$\frac{L}{U} \sim 10^5\ \mathrm{s}$ characteristic time scale.

Of the above values, the one that seems most foreign is the characteristic horizontal pressure fluctuation. If the characteristic length scale of synoptic-scale features is 10^6 m, what this variable says is that the ratio of the pressure difference between adjacent synoptic-scale features is characteristically of order 1000 Pa (10 mb).[5] The density of the air is order $1\ \mathrm{kg\,m^{-3}}$, so the characteristic ratio across the size of a typical synoptic-scale disturbance is $\sim 1000\ \mathrm{m^2\,s^{-2}}$. Given such characteristic values, we are able to estimate the scale of all terms appearing in (3.35). Since our entire analysis is designed to uncover a simplification of (3.35) that is valid for mid-latitude synoptic-scale disturbances, we will assume a latitude (ϕ_0) of 45° implying that a characteristic Coriolis parameter is given by $f_0 = 2\Omega\sin\phi_0 = 2\Omega\cos\phi_0 \cong 10^{-4}\ \mathrm{s^{-1}}$. Table 3.1 lists the approximate magnitude of each term in (3.35) based upon the characteristic scales just described. Note that the friction term is represented by (2.7) and so involves ν, the kinematic viscosity coefficient, in its formulation. Recall that this parameter has a value of $\sim 1.5 \times 10^{-5}\ \mathrm{m^2\,s^{-1}}$ at sea level.

It is clear from Table 3.1 that with scaling appropriate for mid-latitude synoptic-scale motions, only two terms in the horizontal equations of motion are of order

[5] This is consistent with synoptic experience in which the pressure difference between adjacent sea-level high- and low-pressure centers is not as small as 1 hPa nor as large as 100 hPa!

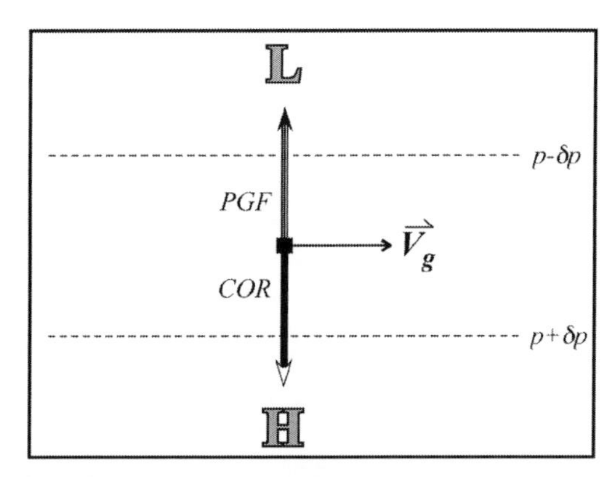

Figure 3.12 Illustration of the force balance resulting in the geostrophic wind, V_g. Arrow *PGF* represents the pressure gradient force and arrow *COR* represents the Coriolis force. The thin dashed lines are isobars and H and L represent regions of high and low pressure, respectively

10^{-3} or larger: the pressure gradient force and Coriolis force terms. This result implies that, as a first approximation to the full equations of motion (3.35), we can consider the PGF and Coriolis force terms to be in approximate balance with one another. This balance is known as the **geostrophic balance** and it represents the fundamental diagnostic balance for mid-latitude synoptic-scale flow. What kind of flow does this geostrophic balance describe? We can get some insight into this question by considering the balance of forces involved. Consider the set of sea-level isobars depicted in Figure 3.12. As we noted in Chapter 2, the PGF vector is always directed from high to low pressure, perpendicular to the isobars as depicted in Figure 3.12. In order that there be a force balance between the pressure gradient and Coriolis forces, the Coriolis force vector must be equal and opposite to the PGF vector as depicted. Since Figure 3.12 represents a hypothetical situation in the northern hemisphere, we know that the Coriolis force must be directed perpendicular to the motion of the air parcel and *to the right*. Consequently, as shown in Figure 3.12, the resulting geostrophic wind flows parallel to the isobars. Were the isobars more closely spaced in the horizontal, the magnitude of the PGF vector would be larger and a correspondingly larger Coriolis force would be required to achieve geostrophic balance. Therefore, the resulting geostrophic wind, though still oriented parallel to the isobars, would be of larger magnitude as well. Thus, to a fairly high degree of accuracy, the wind field (a vector quantity of great importance) can be uniquely specified by a 2-D representation of the scalar quantity, pressure. The mid-latitude atmosphere on Earth need not have been so accommodating to our desire for simplicity, but it is! Let us now examine the mathematical expression for the geostrophic wind.

Considering (3.35a) and (3.35b) we can write component expressions for the geostrophic balance as

$$-fv_g = -\frac{1}{\rho}\frac{\partial p}{\partial x} \quad \text{or} \quad v_g = \frac{1}{\rho f}\frac{\partial p}{\partial x} \tag{3.36a}$$

and

$$fu_g = -\frac{1}{\rho}\frac{\partial p}{\partial y} \quad \text{or} \quad u_g = -\frac{1}{\rho f}\frac{\partial p}{\partial y}. \tag{3.36b}$$

We see from (3.36) that the zonal (meridional) component of the geostrophic wind depends on the corresponding meridional (zonal) gradient of pressure in accord with our previous physical examination. In vector form, (3.36) becomes

$$\vec{V}_g = -\frac{1}{\rho f}\frac{\partial p}{\partial y}\hat{i} + \frac{1}{\rho f}\frac{\partial p}{\partial x}\hat{j} = \frac{1}{\rho f}\hat{k} \times \nabla p \tag{3.37}$$

which clearly demonstrates that the geostrophic wind ($\vec{V}_g$) must always be parallel to the isobars (i.e. perpendicular to ∇p) with a magnitude dependent on the inverse of density, the inverse of the Coriolis parameter, as well as the magnitude of the pressure gradient. Some other conclusions regarding the nature of the geostrophic flow can also be determined from (3.37). For a given magnitude of pressure gradient, the resulting geostrophic wind will be larger at lower latitude where the Coriolis parameter is smaller. However, the geostrophic balance cannot be considered at the equator (or very near it either) as at such low latitudes, the inverse of the Coriolis parameter becomes very large and the resulting $\vec{V}_g$ no longer bears a resemblance to the actual wind, $\vec{V}$. For mid-latitude flow, however, the geostrophic wind is usually within 10–15% of the observed wind. This observation does not imply that the mid-latitude atmosphere has a predilection for this simple balance, it instead testifies to the enormity of the two forces, PGF and COR, at middle latitudes.

Given that geostrophy is a balance between the PGF and Coriolis forces, we might inquire under what conditions is geostrophic balance met? Note that in (3.36) there is no reference to du/dt or dv/dt. As a consequence, the geostrophic wind is only strictly valid in regions of *zero* wind acceleration. Since the wind is a vector quantity, with magnitude and direction, if either of those properties is changed over time, the wind has been accelerated. Thus, two broad categories of flow in the atmosphere will violate the geostrophic balance: those characterized by (1) wind speed changes along the flow, and/or (2) wind direction changes along the flow. Figure 3.13 is a randomly selected northern hemisphere analysis of isobars and isotachs (lines of constant wind speed) at 9 km elevation. It is immediately clear that regions of along-flow speed variation and/or along-flow curvature are so numerous as to be the rule rather than the exception. The along-flow speed changes are most prominent in the vicinity of the local wind speed maxima known as **jet streaks**. Along-flow direction changes are most obvious in the vicinity of troughs and ridges in the pressure field. These locations, as we will show presently, are commonly associated with sensible weather in the form of circulation systems, clouds, and precipitation. The degree of departure from geostrophic balance that characterizes these regions can be assessed by considering the difference between the actual wind at a location and the calculated geostrophic wind at the same point. This difference is known as the

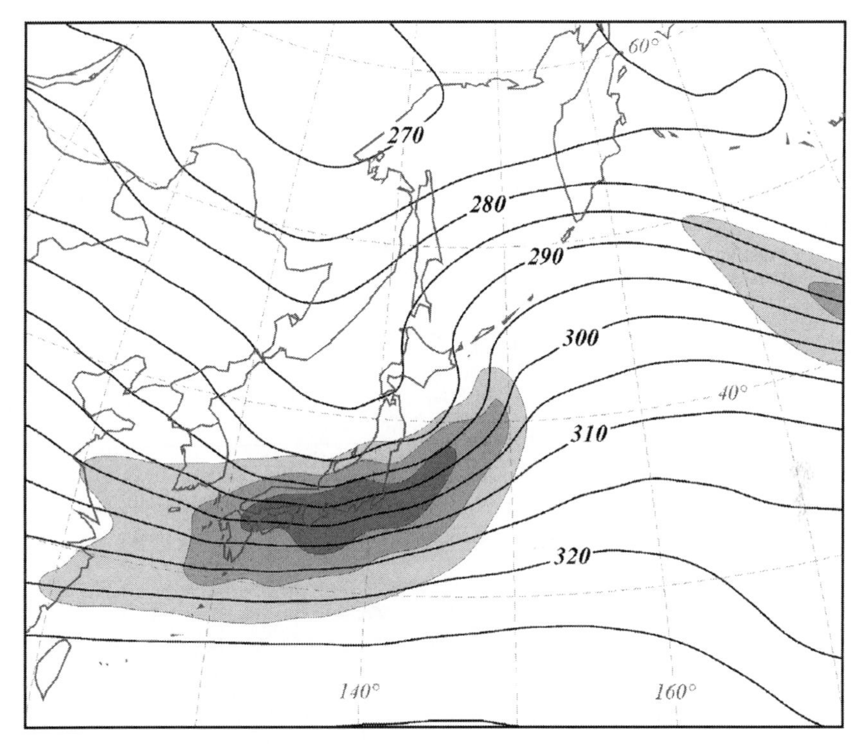

Figure 3.13 Isobars and isotachs at 9 km elevation from the National Center for Environmental Prediction's Global Forecast System initialization at 0000 UTC 23 February 2004. The isobars are labeled and contoured every 5 hPa and the isotachs are shaded every 10 m s^{-1} starting at 30 m s^{-1}

ageostrophic wind, $\vec{V}_{ag}$, and is defined mathematically as

$$\vec{V}_{ag} = \vec{V} - \vec{V}_g. \tag{3.38}$$

We can introduce some prognostic power to our simplified versions of (3.35) by retaining the next largest order terms from Table 3.1: namely, du/dt and dv/dt. The resulting expressions are

$$\frac{du}{dt} = fv - \frac{1}{\rho}\frac{\partial p}{\partial x} \tag{3.39a}$$

$$\frac{dv}{dt} = -fu - \frac{1}{\rho}\frac{\partial p}{\partial y}. \tag{3.39b}$$

If we now substitute (3.36) into (3.39) we get

$$\frac{du}{dt} = fv - fv_g = f(v - v_g) = fv_{ag} \tag{3.40a}$$

$$\frac{dv}{dt} = -fu + fu_g = -f(u - u_g) = -fu_{ag} \tag{3.40b}$$

Table 3.2 Characteristic scales for the terms in the vertical equation of motion

	1	2	3	4	5	6
	$\frac{dw}{dt}$	$-2\Omega\cos\phi u$	$\frac{-(u^2+v^2)}{a}$	$-\frac{1}{\rho}\frac{\partial p}{\partial z}$	$-g$	F_z
Characteristic scales	$\frac{UW}{L}$	$f_0 U$	$\frac{U^2}{a}$	$\frac{p}{\rho H}$	g	$\frac{\nu W}{H^2}$
Magnitudes (m s^{-2})	10^{-7}	10^{-3}	10^{-5}	10	10	10^{-15}

which can be written in vector form as

$$\frac{d\vec{V}}{dt} = -f\hat{k} \times \vec{V}_{ag}. \tag{3.41}$$

This expression clearly shows that the ageostrophic flow is associated with regions of Lagrangian acceleration of the wind. In the next section we will demonstrate why this ageostrophic wind is of such vital importance to understanding the dynamics of the mid-latitude atmosphere.

Given that geostrophic balance is such a strong constraint in the middle latitudes, there are many settings in which the ageostrophic wind is a very small portion of the actual wind. Therefore, it would be convenient if there were some easy way to characterize a flow to determine if it is likely to be nearly in geostrophic balance. Physically, a given flow will be nearly in geostrophic balance if the Lagrangian acceleration term (du/dt or dv/dt) is small compared to the Coriolis force term, as suggested by our scaling and Table 3.1. Recalling that the acceleration term is represented as U^2/L and the Coriolis force is scaled as $f_0 U$, then the ratio of these two accelerations is given by

$$\frac{\textit{Lagrangian Accel.}}{\textit{Coriolis Accel.}} = \frac{U^2/L}{f_0 U} = \frac{U}{f_0 L}. \tag{3.42}$$

Notice that this ratio is non-dimensional (i.e. it is just a number without units) and that if it is less than 0.1 for a given flow it testifies to the fact that the Coriolis acceleration is at least 10 times larger than the Lagrangian acceleration. In such a case, it is quite reasonable to approximate the flow as nearly geostrophic. The ratio defined in (3.42) is known as the **Rossby number** (R_0), after the famous atmospheric/oceanic scientist Carl Gustav Rossby.[6] We will hereafter often refer to flows that are nearly in geostrophic balance as low-R_0 flows. High-R_0 flows will, conversely, be characterized as rather far from geostrophic balance.

Thus far we have discussed the results of a scaling of the horizontal equations of motion. A similar exercise must now be performed on (3.35c), the vertical equation of motion. Table 3.2 shows the characteristic scales of the various terms in (3.35c)

[6] Carl Gustav Rossby (1898–1957) was a Swedish–American scientist who founded the first meteorology department in the United States at the Massachusetts Institute of Technology (MIT) in 1928. Rossby uncovered many of the basic principles of modern dynamical meteorology during the decades of the 1930s and 1940s.

along with their usual magnitudes for mid-latitude weather systems. Even more robustly than was the case for the horizontal equations, the vertical equation of motion is dominated by two terms: the vertical PGF and gravity. We have already seen that these two vertical forces are combined in the hydrostatic balance. Thus, a formal scaling of the equations of motion for mid-latitude synoptic-scale motions renders the following fundamental statement regarding the nature of the mid-latitude atmosphere on Earth:

To a first order, the mid-latitude atmosphere on Earth is in hydrostatic and geostrophic balance.

3.2.2 Conservation of mass

Imagine trying to fill a small basin with water from a hose. If there is a leak in the basin then one needs to know both the inflow rate from the hose as well as the outflow rate through the leak in order to accurately gauge the filling rate. If the inflow rate is suddenly increased while the outflow rate remains the same it is simple to conclude that the mass of water in the basin will increase. If we designate the mass of water in the basin as M_w, then a simple expression of the mass continuity equation becomes

$$\frac{\partial M_w}{\partial t} = \textit{Inflow Rate} - \textit{Outflow Rate}.$$

We can think of a slightly more abstract representation of this idea, illustrated in Figure 3.14, by considering an infinitesimal cube, fixed in space, through which air flows. The x-direction **mass flux** (i.e. the product of the x-direction velocity and the density of the fluid) at the center of the cube is given by ρu. Upon expanding this

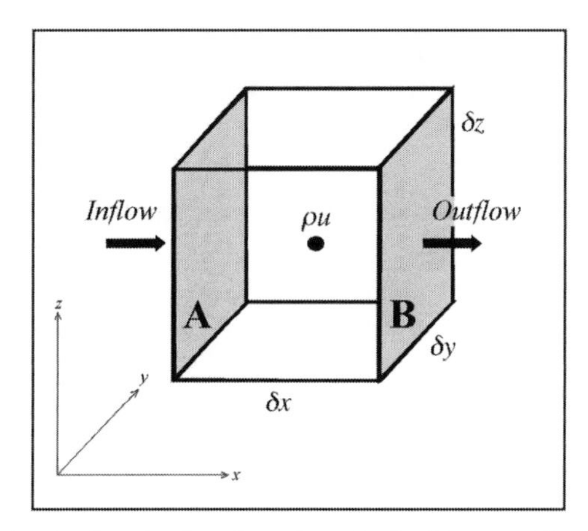

Figure 3.14 Schematic of x-direction flow through a cube fixed in space. The rate of mass flux is given by the product ρu. Accumulation of mass at the center point occurs when the inflow rate exceeds the outflow rate

function in a Taylor series about the center point we find that the rate of mass inflow through side A of the cube is given by

$$\left[\rho u - \frac{\partial}{\partial x}(\rho u)\frac{\delta x}{2}\right]\delta y \delta z \tag{3.43a}$$

while the rate of mass outflow through side B of the cube is given by

$$\left[\rho u + \frac{\partial}{\partial x}(\rho u)\frac{\delta x}{2}\right]\delta y \delta z. \tag{3.43b}$$

Now, just as in our simple example above, the rate of accumulation of mass (as a result of x-direction flow) inside the infinitesimal cube must be equal to the inflow rate minus the outflow rate. Using (3.43) this is expressed as

$$\begin{aligned}\frac{\partial M_x}{\partial t} &= \left[\rho u - \frac{\partial}{\partial x}(\rho u)\frac{\delta x}{2}\right]\delta y \delta z - \left[\rho u + \frac{\partial}{\partial x}(\rho u)\frac{\delta x}{2}\right]\delta y \delta z \\ &= -\frac{\partial}{\partial x}(\rho u)\delta x \delta y \delta z \end{aligned} \tag{3.44}$$

where M_x represents the rate of mass accumulation in the cube resulting from x-direction mass flux divergence. Similar expressions representing the rates of mass accumulation in the cube resulting from y- and z-direction mass flux divergences are given by

$$\frac{\partial M_y}{\partial t} = -\frac{\partial}{\partial y}(\rho v)\delta x \delta y \delta z \quad \text{and} \quad \frac{\partial M_z}{\partial t} = -\frac{\partial}{\partial z}(\rho w)\delta x \delta y \delta z$$

so that the net rate of mass accumulation in the cube is represented as

$$\frac{\partial M}{\partial t} = -\left[\frac{\partial}{\partial x}(\rho u) + \frac{\partial}{\partial y}(\rho v) + \frac{\partial}{\partial z}(\rho w)\right]\delta x \delta y \delta z. \tag{3.45}$$

By definition, the net mass accumulation rate per unit volume is equal to the Eulerian rate of change of the density. Thus, dividing (3.45) by the volume of the cube ($\delta x \delta y \delta z$) yields

$$\frac{\partial \rho}{\partial t} = -\left[\frac{\partial}{\partial x}(\rho u) + \frac{\partial}{\partial y}(\rho v) + \frac{\partial}{\partial z}(\rho w)\right] = -\nabla \cdot (\rho \vec{V}). \tag{3.46}$$

The expression above is known as the **mass divergence** form of the mass continuity equation. An alternative form of this expression arises by recalling that

$$\nabla \cdot (\rho \vec{V}) = \rho \nabla \cdot \vec{V} + \vec{V} \cdot \nabla \rho$$

so that (3.46) becomes

$$\frac{\partial \rho}{\partial t} + \vec{V} \cdot \nabla \rho + \rho \nabla \cdot \vec{V} = 0 \quad \text{or} \quad \frac{1}{\rho}\frac{d\rho}{dt} + \nabla \cdot \vec{V} = 0 \tag{3.47}$$

which is known as the **velocity divergence** form of the mass continuity equation.

This exact same relationship can be derived for a cube of fixed mass, δM, but varying dimensions δx, δy, and δz. Given that the mass in this example is fixed, then $d(\delta M)/dt = 0$ or

$$\frac{d(\rho\delta x\delta y\delta z)}{dt} = 0 = \frac{d\rho}{dt}\delta x\delta y\delta z + \rho\frac{d(\delta x)}{dt}\delta y\delta z + \rho\frac{d(\delta y)}{dt}\delta x\delta z + \rho\frac{d(\delta z)}{dt}\delta x\delta y \tag{3.48a}$$

by the chain rule. Now

$$\lim_{\delta x\to 0}\frac{d(\delta x)}{dt} = \partial u$$

with similar expressions applying for the last two time derivatives in (3.48a). Therefore, dividing both sides of (3.48a) by the volume of cube gives

$$\frac{d\rho}{dt} + \rho\frac{\partial u}{\partial x} + \rho\frac{\partial v}{\partial y} + \rho\frac{\partial w}{\partial z} = \frac{d\rho}{dt} + \rho\nabla\cdot\vec{V} = 0 \tag{3.48b}$$

which can be easily rearranged into (3.47).

It is instructive at this point to consider the implications of (3.47) for the fluid atmosphere. A fluid in which individual parcels experience no change of density following the motion (i.e. $d\rho/dt = 0$) is known as an **incompressible fluid**. Conversely, a compressible fluid is one in which the density can change along a parcel trajectory. As you might guess, the atmosphere is a compressible fluid, but for many atmospheric phenomena the compressibility is not a major physical consideration. In such cases, the mass continuity equation becomes a statement of zero velocity divergence. We will see later that choice of a different vertical coordinate will render the continuity equation in a much simpler, unapproximated form.

3.3 Conservation of Energy: The Energy Equation

The law of conservation of energy states that the sum of all energies in the universe is constant. This is a valuable piece of knowledge but there are many different kinds of energies manifest in the atmosphere including kinetic energy, potential energy, latent heat energy, and radiant energy to name a few. Of all these types, radiant energy from the Sun is the source of nearly all of the total energy in the atmosphere/ocean system. When solar radiation is absorbed at the Earth's surface and in the atmosphere it appears as internal energy, made manifest as a temperature change. Given the many other kinds of energy involved in the atmosphere/ocean system, one of the major problems in the atmospheric sciences is determining how this internal energy is converted into the other forms of energy.

We can get some insights into the nature of the energies in the atmosphere by taking the dot product of the acceleration vector, $d\vec{V}/dt$, with the velocity vector, $\vec{V}$. This operation is the mathematical equivalent of multiplying the component equations of motion (3.35a, b, and c) by their respective component velocities (u, v, and w). The resulting expressions are

$$\frac{1}{2}\frac{d(u^2)}{dt} - \frac{u^2 v \tan\phi}{a} + \frac{u^2 w}{a} = -\frac{u}{\rho}\frac{\partial p}{\partial x} + 2\Omega \sin\phi uv - 2\Omega \cos\phi uw + uF_x \tag{3.49a}$$

$$\frac{1}{2}\frac{d(v^2)}{dt} + \frac{u^2 v \tan\phi}{a} + \frac{v^2 w}{a} = -\frac{v}{\rho}\frac{\partial p}{\partial y} - 2\Omega \sin\phi uv + vF_y \tag{3.49b}$$

$$\frac{1}{2}\frac{d(w^2)}{dt} - \frac{w(u^2 + v^2)}{a} = -\frac{w}{\rho}\frac{\partial p}{\partial z} - gw + 2\Omega \cos\phi uw + wF_z. \tag{3.49c}$$

Summing the component expressions (3.49) together we note that all of the Coriolis and curvature terms sum to zero resulting in

$$\frac{d}{dt}\left[\frac{(u^2 + v^2 + w^2)}{2}\right] = -\frac{1}{\rho}\vec{V} \cdot \nabla p - gw + \vec{V} \cdot \vec{F}. \tag{3.50}$$

The LHS term in (3.50) represents the rate of change of the total kinetic energy (per unit mass) of the flow and so is a rate of work term. The first term on the RHS of (3.50) is pressure advection divided by density. When the velocity vector is directed across isobars from high to low (low to high) pressure, (3.50) shows that kinetic energy is produced (consumed). Note that if the flow were purely geostrophic, $\vec{V} \cdot \nabla p$ would vanish. This term is often referred to as the pressure work term and describes the rate of work done by the ageostrophic flow across isobars.

By definition, $w = dz/dt$, so that $-gw$ can be rewritten as

$$-gw = -g\frac{dz}{dt} = -\frac{d\phi}{dt}$$

where ϕ is the geopotential, a measure of the work required to raise a unit mass a distance, z, above sea level. It is instructive, therefore, to rewrite (3.50) as

$$\frac{d}{dt}\left[\frac{(u^2 + v^2 + w^2)}{2} + \phi\right] = -\frac{1}{\rho}\vec{V} \cdot \nabla p + \vec{V} \cdot \vec{F} \tag{3.51}$$

where the LHS represents the sum of the kinetic and potential energies per unit mass of an atmospheric parcel. The last term on the RHS of (3.51) represents the energy dissipated by the action of the friction force ($\vec{F}$). Note that since $\vec{V}$ and $\vec{F}$ are almost always opposite one another, the product $\vec{V} \cdot \vec{F}$ will be negative and the total kinetic and potential energies of the parcel will decrease in the presence of friction in accord with physical intuition.

Since (3.51) is derived from the equations of motion it deals only with mechanical forms of energy and is therefore referred to as the **mechanical energy equation**. In order to include thermal energy we must include the first law of thermodynamics in the form

$$\dot{Q} = c_v \frac{dT}{dt} + p\frac{d\alpha}{dt} \tag{3.52}$$

where $\dot{Q}$ represents the diabatic heating rate, c_v is the specific heat of dry air at constant volume ($717\,\mathrm{J\,kg^{-1}\,K^{-1}}$), and α is the specific volume. This expression relates the important fact that absorption of solar radiation (represented by $\dot{Q}$) can be converted to both internal energy (in the form of a temperature increase) or mechanical energy made manifest in expansion work (represented by the expansion term, $d\alpha/dt$). By rearranging (3.51) as

$$0 = \frac{d}{dt}\left[\frac{(u^2+v^2+w^2)}{2} + \phi\right] + \frac{1}{\rho}\vec{V}\cdot\nabla p - \vec{V}\cdot\vec{F}$$

we can add zero to both sides of (3.52) to yield

$$\dot{Q} = c_v\frac{dT}{dt} + p\frac{d\alpha}{dt} + \frac{d}{dt}\left[\frac{(u^2+v^2+w^2)}{2} + \phi\right] + \frac{1}{\rho}\vec{V}\cdot\nabla p - \vec{V}\cdot\vec{F}. \tag{3.53}$$

Noting that $(1/\rho)\vec{V}\cdot\nabla p$ is equal to $\alpha(dp/dt - \partial p/\partial t)$, and that

$$p\frac{d\alpha}{dt} + \alpha\frac{dp}{dt} = \frac{d}{dt}(p\alpha),$$

we can regroup terms and rewrite (3.53) as

$$\dot{Q} = \frac{d}{dt}\left[\frac{(u^2+v^2+w^2)}{2} + \phi + c_vT + p\alpha\right] - \alpha\frac{\partial p}{\partial t} - \vec{V}\cdot\vec{F} \tag{3.54}$$

which is known as the **energy equation**. This relationship implies that if the flow is frictionless ($\vec{F}=0$), adiabatic ($\dot{Q}=0$), and steady state ($\partial p/\partial t = 0$), then the quantity

$$\frac{(u^2+v^2+w^2)}{2} + \phi + c_vT + p\alpha$$

is constant. This is a special case of Bernoulli's[7] equation for an incompressible flow in which the quantity

$$\frac{(u^2+v^2+w^2)}{2} + \phi + p\alpha = \mathit{Constant}.$$

This relationship suggests that for an atmosphere at rest, any increase in elevation results, unsurprisingly, in a decrease in the hydrostatic pressure. If the atmosphere is in motion, however, a larger pressure difference will result over the same elevation interval since the difference, in this case, is a difference in the *dynamic* pressure. For

[7] Daniel Bernoulli (1700–1782) was a Swiss mathematician and fluid dynamicist. Though from an illustrious family of mathematicians, he studied medicine at his father's insistence and discovered a means to measure blood pressure that was used until the dawn of the twentieth century. When he was 25, Catherine the Great appointed him Professor of Mathematics at the Imperial Academy of St Petersburg where Leonhard Euler became one of his first students. He developed the fluid dynamical equation that bears his name at the age of 30.

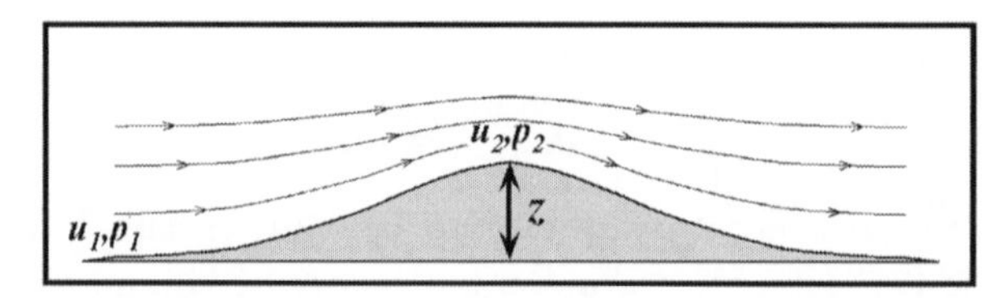

Figure 3.15 Flow over a hill illustrating the effect of dynamic pressure. Thin lines are streamlines of the flow – the closer the streamlines in the vertical, the greater the flow speed. Since $u_2 > 0$, p_2 is less than the hydrostatic pressure at height z

flow over the hill in Figure 3.15, as the air rises over the hill, the speed of the flow increases. Thus, the pressure difference between the top and the bottom of the hill ($p_2 - p_1$) must be larger than their hydrostatic pressure difference because the wind speed is higher at the top than at the bottom of the hill ($u_2 > u_1$).

Two additional relationships of meteorological consequence arise from further consideration of aspects of the energy equation. First, an illuminating alternative expression for the first law of thermodynamics arises from combining (3.52) and the ideal gas law. Differentiating the gas law with respect to time yields

$$p\frac{d\alpha}{dt} + \alpha\frac{dp}{dt} = R\frac{dT}{dt}. \tag{3.55a}$$

Substituting for $pd\alpha/dt$ (from (3.55a)) in (3.52), and recalling that $c_p = c_v + R$, yields

$$c_p\frac{dT}{dt} - \alpha\frac{dp}{dt} = \dot{Q}. \tag{3.55b}$$

If we then divide (3.55b) by T, noting that $\alpha/T = R/p$, we get

$$c_p\frac{d\ln T}{dt} - R\frac{d\ln p}{dt} = \frac{\dot{Q}}{T} \tag{3.55c}$$

where $\dot{Q}/T$ is known as the **entropy**. If the entropy is constant with time, then we have an **isentropic process** and, consequently,

$$c_p\frac{d\ln T}{dt} - R\frac{d\ln p}{dt} = 0. \tag{3.55d}$$

Integration of (3.55d) from a given p and T to a reference pressure, p_0, and a reference temperature, θ, defines what is known as the **potential temperature**. We begin by noting that

$$\int_T^\theta c_p d\ln T = \int_p^{p_0} R d\ln p$$

which yields

$$c_p(\ln\theta - \ln T) = R(\ln p_0 - \ln p).$$

Rearranging the above expression and taking anti-logs results in

$$\frac{\theta}{T} = \left(\frac{p_0}{p}\right)^{\frac{R}{c_p}} \quad \text{or} \quad \theta = T\left(\frac{p_0}{p}\right)^{\frac{R}{c_p}}, \tag{3.56}$$

known as the Poisson equation.

Physically, θ is the temperature a parcel of air would have if it were adiabatically compressed (or expanded) from its original pressure, p, to a reference pressure, p_0 (usually 1000 hPa). Every air parcel has a unique value of θ and that value is conserved for adiabatic processes (i.e. conditions in which the entropy does not change). For this reason, lines of constant θ are often referred to as **isentropes** and flow along surfaces of constant potential temperature is known as **isentropic flow**.

Finally, if we take the log differential of (3.56) with respect to height (z) we get

$$\frac{\partial \ln \theta}{\partial z} = \frac{\partial \ln T}{\partial z} + \frac{R}{c_p}\left(\frac{\partial \ln p_0}{\partial z} - \frac{\partial \ln p}{\partial z}\right). \tag{3.57a}$$

Since p_0 is a constant, its derivative is zero and (3.57a) can be rewritten as

$$\frac{1}{\theta}\frac{\partial \theta}{\partial z} = \frac{1}{T}\frac{\partial T}{\partial z} - \frac{R}{c_p p}\frac{\partial p}{dz}. \tag{3.57b}$$

Substituting for $\partial p/dz$ from the hydrostatic equation yields

$$\frac{1}{\theta}\frac{\partial \theta}{\partial z} = \frac{1}{T}\frac{\partial T}{\partial z} + \frac{R\rho g}{c_p p}. \tag{3.57c}$$

Finally, with the help of the gas law and some rearranging, (3.57c) can be written as

$$\frac{T}{\theta}\frac{\partial \theta}{\partial z} = \frac{\partial T}{\partial z} + \frac{g}{c_p} \tag{3.57d}$$

which yields an expression for the dry adiabatic lapse rate (Γ_d). If θ is constant with height (i.e. the lapse rate is dry adiabatic), then $-\partial T/\partial z = \Gamma_d = g/c_p = 9.8°\text{C km}^{-1}$. When $\partial\theta/\partial z$ is non-zero, the lapse rate ($\Gamma = -\partial T/\partial z$) is given by

$$\Gamma = \Gamma_d - \frac{T}{\theta}\frac{\partial \theta}{\partial z}. \tag{3.58}$$

Based upon (3.58), there are three conditions for stability that can be assessed. First, when $\partial\theta/\partial z > 0$, then $\Gamma < \Gamma_d$ which corresponds to a statically stable stratification. In such an environment, a lifted parcel of dry air (which must cool at the dry adiabatic rate) will always be cooler than its new environment. Second, when $\partial\theta/\partial z = 0$, then $\Gamma = \Gamma_d$ and the stratification is said to be neutral and a lifted parcel of dry air will always have the same temperature as its new surroundings. Finally, when $\partial\theta/\partial z < 0$, then $\Gamma > \Gamma_d$ which corresponds to an absolutely unstable stratification. In such a case, a lifted parcel of dry air will always be warmer than its new surroundings and will, therefore, freely convect.

In the statically stable case just described, a lifted parcel, being colder than its environment upon being lifted, will be forced back downward to its original level once the impulse that forced it to rise is exhausted. A series of oscillations about that original level will ensue. The frequency of these buoyancy oscillations is dependent on the restoring force that compels them. In this case, the restoring force (per unit volume) is the product of gravity and the density difference between the displaced parcel and its environment.

If we let δz be the vertical displacement of an air parcel about its original level, then Newton's second law tells us that

$$\frac{F_z}{Mass} = \frac{dw}{dt} = \frac{d^2(\delta z)}{dt^2}. \tag{3.59a}$$

Letting ρ (ρ') and T (T') be the density and temperature of the environment (parcel) and assuming that the pressures of the parcel and the environment are always equal, then the restoring force (per unit volume) for a displaced parcel is given by

$$\frac{F_z}{Volume} = -(\rho' - \rho)g. \tag{3.59b}$$

Thus, the restoring force per unit mass for the displaced parcel can be written as

$$\frac{F_z}{Mass} = -\frac{(\rho' - \rho)g}{\rho'}. \tag{3.59c}$$

Employing the gas law allows this expression to be rewritten as

$$\frac{F_z}{Mass} = -\left(\frac{1}{T'} - \frac{1}{T}\right) gT' = -g\left(\frac{T - T'}{T}\right). \tag{3.59d}$$

Now we can say that $(T - T')$ is equal to $(\Gamma_d - \Gamma)\delta z$ since the dry parcel cools at the dry adiabatic lapse rate and must be compared to the environment whose temperature changes at a rate described by Γ. Therefore, the restoring force per unit mass can be written as

$$\frac{F_z}{Mass} = -\frac{g}{T}(\Gamma_d - \Gamma)\delta z \tag{3.59e}$$

so that (3.59a) becomes a second-order, ordinary differential equation

$$\frac{d^2(\delta z)}{dt^2} + \frac{g}{T}(\Gamma_d - \Gamma)\delta z = 0 \tag{3.60}$$

whose solution describes a buoyancy oscillation with period $2\pi / N$ where

$$N = \left[\frac{g}{T}(\Gamma_d - \Gamma)\right]^{1/2}$$

or, substituting from (3.58),

$$N = \left[\frac{g}{\theta}\frac{\partial \theta}{\partial z}\right]^{1/2}. \tag{3.61}$$

N is known as the **Brunt–Väisälä frequency** and has units of s^{-1}. It is clear from (3.61) that for the condition of neutrality alluded to earlier (i.e. $\partial\theta/\partial z = 0$), $N = 0$ and there is no buoyancy oscillation physically consistent with a neutral displacement. For the statically stable case (i.e. $\partial\theta/\partial z > 0$), $N > 0$ and buoyancy oscillations are observed. For the absolutely unstable case (i.e. $\partial\theta/\partial z < 0$), N is imaginary and in perturbation theory such a case corresponds to a growing disturbance. Physically, this is consistent with the fact that in an absolutely unstable stratification, a lifted parcel of dry air will always be warmer than its environment and therefore, according to (3.59), experience an upward-directed buoyancy force without interruption. It is important to note that instances of absolute instability are exceedingly rare and, even when they do exist, are very short-lived as the atmosphere mixes rapidly toward neutrality in such instances.

Selected References

Hess, *Introduction to Theoretical Meteorology*, offers an alternative perspective on accelerating reference frames.
Holton, *An Introduction to Dynamic Meteorology*, provides discussion of many of the same issues.
Brown, *Fluid Mechanics of the Atmosphere*, provides illuminating discussion of the energy equation.
Acheson, *Elementary Fluid Dynamics*, discusses many of the same issues.

Problems

3.1. Assume that air flows over a broad building 10 m high. The flow is in steady state and the density is constant ($\rho = 1.3\,\mathrm{kg\,m^{-3}}$) through this depth of the atmosphere. The observed speed at ground level is $5\,\mathrm{m\,s^{-1}}$ while on the rooftop it is $9\,\mathrm{m\,s^{-1}}$.

(a) What is the pressure difference, in hPa, between the ground and roof level?
(b) How much of this pressure difference is purely hydrostatic?
(c) What is the magnitude and direction of the non-hydrostatic pressure gradient force vector generated by these circumstances?

In all of the above, you may neglect the vertical variation in temperature.

3.2. (a) Prove that the divergence of the geostrophic wind is given by

$$\nabla \cdot \vec{V}_g = -V_g(\cot\phi/a)$$

where a = radius of the earth and ϕ is latitude.
(b) Explain why (physically) this is true. (Hint: recall that the magnitude of the Coriolis force depends on wind speed.)
(c) Calculate the divergence of the geostrophic wind at 43°N at a point where $|v_g| = 20\,\mathrm{m\,s^{-1}}$.

3.3. The perturbation ocean surface height (POSH) is defined as the height of the local ocean surface above or below mean sea level (which is 0 meters). Suppose a sophisticated satellite instrument is built that can measure the local POSH to an accuracy of 1 cm. A plot of the measurements taken on a certain day is given in Figure 3.1A (solid lines are contours of POSH in cm).

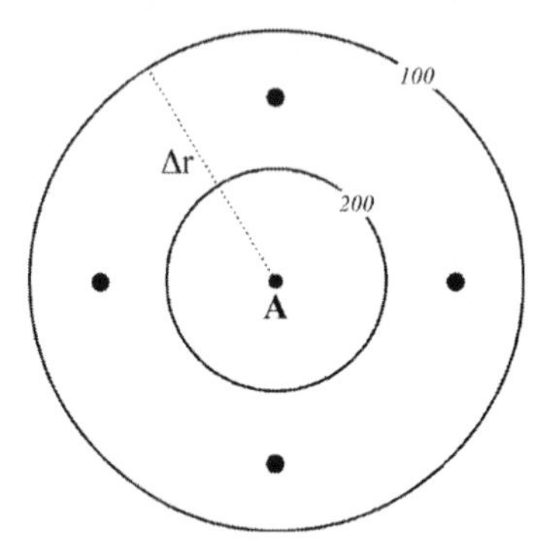

Figure 3.1A

If $\Delta r = 500\,\text{km}$ and $\rho_w = 1000\,\text{kg m}^{-3}$ (density of ocean water):

(a) What is the magnitude of the atmospheric pressure gradient just above the ocean surface?
(b) What is the departure from mean sea-level atmospheric pressure at A in Figure 3.1A?
(c) Calculate the velocity (speed and direction) of the geostrophic flow in the raised water surface at the indicated points.
(d) Is this geostrophic flow cyclonic or anticyclonic? Defend your answer.
(e) What type of surface-layer ocean circulation would develop beneath an atmospheric anticyclone? Explain your answer. (The surfacelayer is the top few meters of the ocean).

3.4. Derive a relationship for the height of a given pressure surface (P) in terms of the pressure (P_0) and temperature (T_0) at *sea level* assuming that the temperature decreases uniformly with height at a rate of Γ °C km^{-1}.

3.5. The vector form of the frictionless, horizontal equation of motion is

$$\frac{d\vec{V}}{dt} = -\frac{1}{\rho}\nabla p - f\hat{k} \times \vec{V}$$

(a) Expand this vector expression into its x and y components.
(b) Show that the *vector* ageostrophic wind can be expressed as

$$\frac{\hat{k}}{f} \times \frac{d\vec{V}}{dt} = \vec{V}_{ag}$$

Consider the 300 mb jet streak shown in Figure 3.2A (the solid lines are **isotachs**, lines of constant wind speed).

(c) Draw the ageostrophic winds at each of the shaded circles. Explain your work.
(d) Indicate the locations of upper-level divergence and convergence of the horizontal wind field. Explain.

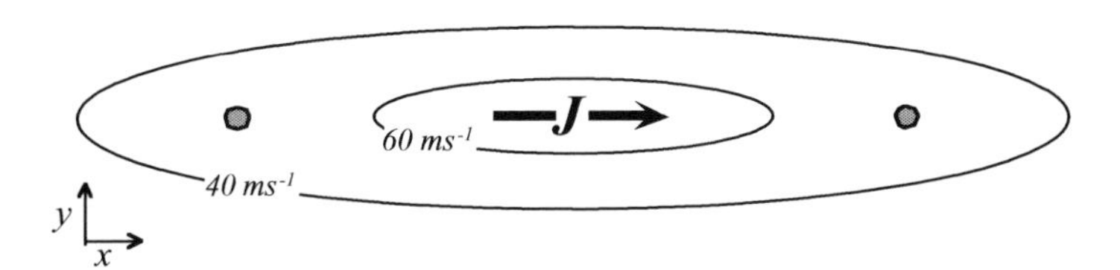

Figure 3.2A

3.6. The temperature decreases at a constant rate from 1000 to 500 hPa. At Madison, the 500 hPa temperature is −30°C and the 1000–500 hPa thickness is 5180 m. What is the 1000 hPa temperature at Madison? ($R = 287\,\text{J kg}^{-1}\,\text{K}^{-1}$, $g = 9.8\,\text{m s}^{-2}$).

3.7. During winter, Rapid City, South Dakota can witness remarkable temperature inversions as a consequence of its proximity to the Black Hills and its position in the center of the continent. It is not uncommon to find surface temperatures near −30°C as cold, arctic air funnels southward at the ground, while at the same time the temperature 100 m above the surface may be as warm 10°C as a result of strong downslope flow in the lee of the Black Hills. Calculate the frequency of a buoyancy oscillation in such an inversion layer. What might be some sensible weather consequences of such a circumstance?

3.8. A 1 m^2 column of air in the 1000–850 hPa layer is subjected to 3×10^6 J of heating. What is the change in the 850 hPa geopotential height if the 1000 hPa height tendency is zero? (The specific heat of air at constant pressure $c_p = 1004\,\text{J kg}^{-1}\,\text{K}^{-1}$).

3.9. Over low-elevation stations in the central United States, the likelihood of frozen or liquid precipitation in a winter storm is often assessed by considering the 1000–500 hPa thickness. The critical value at these locations for equal chances of snow or rain is 5400 m. Explain why this value is a reasonable one to choose. What are some possible problems with relying on this parameter to guide a precipitation-type forecast? Propose and defend an alternative thickness-based parameter for making this forecast.

3.10. By how much does the relative speed of a westward-directed parcel of air (moving at $-2\,m\,s^{-1}$) change if that parcel is moved from the equator to 30°N? Could the upper tropospheric outflow of equatorial thunderstorms be related to the presence of the so-called **subtropical** jet stream at 30°N? Explain.

3.11. Assume astronomers discover a new planet with a characteristic horizontal velocity scale (at middle latitudes), and a rotation rate equal to Earth's. How small would a planet have to be in order that the effect of curvature term on the flow would be comparable to the Coriolis force?

3.12. Show that the continuity equation for a fluid of variable density can be written as

$$\frac{1}{V}\frac{dV}{dt} = \frac{\partial u}{\partial x} + \frac{\partial v}{\partial y} + \frac{\partial w}{\partial z}$$

where V is the volume of the fluid parcel.

3.13. The potential density (D) is a useful diagnostic parameter defined as the density that an air parcel would have if it were adiabatically compressed or expanded to 1000 hPa. Derive an expression for D. Show that the product of the potential temperature (θ) and the potential density (i.e. $D\theta$) is a constant.

Solutions

3.1. (a) 131.04 Pa (b) 127.4 Pa (c) 0.28 m s^{-2} directed upward

3.2. (c) -3.367×10^{-6} s^{-1}

3.3. (a) 1.96×10^{-2} Pa m^{-1} (b) -98.1 hPa (c) 0.23 m s^{-1} (d) Anticyclonicallly (e) Cyclonically

3.6. 267.7 K

3.7. 0.121 s^{-1}

3.8. $\Delta z = 9.29$ m

3.10. 133.79 m s^{-1}

3.11. $a = 10^5$ m

4

Applications of the Equations of Motion

Objectives

In the previous chapter we derived expressions for the equations of motion as well as the continuity of mass. The present chapter will concentrate on illustrating various simple applications of these equations to observable phenomena in the mid-latitude atmosphere. Many of the applications of the equations of motion, in particular, will be greatly simplified by first recasting the expressions in one of three new Cartesian coordinate systems. The first two use pressure (p) or potential temperature (θ) as the vertical coordinate and are known as **isobaric** or **isentropic coordinates**, respectively. Upon transforming the equations to these new coordinate systems we will illustrate the utility of each with some examples. Next, by combining the geostrophic and hydrostatic balances, we will find that the vertical shear of the geostrophic wind is directly linked to the magnitude of the horizontal temperature gradient in the **thermal wind relationship**.

Finally, we shall adopt a third new Cartesian coordinate system, known as **natural coordinates**, in which we employ the direction of the flow at each point in the fluid as the basis for defining the horizontal directions. We will use this natural coordinate perspective to examine a number of additional balanced flows including inertial, cyclostrophic, and gradient balances. Applications of each of these balances will also be considered. We begin by transforming the equations of motion and the continuity equation into isobaric coordinates.

4.1 Pressure as a Vertical Coordinate

Recall that in height coordinates, the vector form of the frictionless horizontal momentum equation is

$$\frac{d\vec{V}}{dt} = -\frac{1}{\rho}\nabla p - f\hat{k} \times \vec{V}.$$

Mid-Latitude Atmospheric Dynamics Jonathan E. Martin

In order to recast this expression in isobaric coordinates we must convert the pressure gradient force term into an equivalent expression in isobaric coordinates. This is done most easily by considering the differential (dp) on a constant pressure surface:

$$(dp)_p = \left(\frac{\partial p}{\partial x}\right)_{y,z} dx_p + \left(\frac{\partial p}{\partial y}\right)_{x,z} dy_p + \left(\frac{\partial p}{\partial z}\right)_{x,y} dz_p \tag{4.1a}$$

where the subscripts indicate differentiation carried out holding the subscripted variable constant. Now, since there is no change in pressure on an isobaric (i.e. constant pressure) surface, then $(dp)_p = 0$ so that

$$0 = \left(\frac{\partial p}{\partial x}\right)_{y,z} dx_p + \left(\frac{\partial p}{\partial y}\right)_{x,z} dy_p + \left(\frac{\partial p}{\partial z}\right)_{x,y} dz_p. \tag{4.1b}$$

Next, we expand dz_p as a function of x and y to yield

$$0 = \left(\frac{\partial p}{\partial x}\right)_{y,z} dx_p + \left(\frac{\partial p}{\partial y}\right)_{x,z} dy_p + \left(\frac{\partial p}{\partial z}\right)_{x,y} \left[\left(\frac{\partial z}{\partial x}\right)_{y,p} dx_p + \left(\frac{\partial z}{\partial y}\right)_{y,p} dy_p\right]$$

which can be rearranged into

$$\begin{aligned} 0 = &\left[\left(\frac{\partial p}{\partial x}\right)_{y,z} + \left(\frac{\partial p}{\partial z}\right)_{x,y} \left(\frac{\partial z}{\partial x}\right)_{y,p}\right] dx_p \\ &+ \left[\left(\frac{\partial p}{\partial y}\right)_{x,z} + \left(\frac{\partial p}{\partial z}\right)_{x,y} \left(\frac{\partial z}{\partial y}\right)_{y,p}\right] dy_p. \end{aligned} \tag{4.1c}$$

Since this statement is true for all dx and dy, the terms in square brackets in (4.1c) must both be zero. Hence,

$$\left(\frac{\partial p}{\partial x}\right)_{y,z} = -\left(\frac{\partial p}{\partial z}\right)_{x,y} \left(\frac{\partial z}{\partial x}\right)_{y,p} \quad \text{and} \quad \left(\frac{\partial p}{\partial y}\right)_{x,z} = -\left(\frac{\partial p}{\partial z}\right)_{x,y} \left(\frac{\partial z}{\partial y}\right)_{y,p}. \tag{4.1d}$$

With the help of the hydrostatic equation, these expressions become

$$-\left(\frac{\partial p}{\partial x}\right)_{y,z} = -\rho g \left(\frac{\partial z}{\partial x}\right)_{y,p} \quad \text{and} \quad -\left(\frac{\partial p}{\partial y}\right)_{x,z} = -\rho g \left(\frac{\partial z}{\partial y}\right)_{y,p}. \tag{4.1e}$$

Dividing both sides of (4.1e) by ρ yields

$$-\frac{1}{\rho}\left(\frac{\partial p}{\partial x}\right)_z = -g\left(\frac{\partial z}{\partial x}\right)_p = -\left(\frac{\partial \phi}{\partial x}\right)_p$$

$$-\frac{1}{\rho}\left(\frac{\partial p}{\partial y}\right)_z = -g\left(\frac{\partial z}{\partial y}\right)_p = -\left(\frac{\partial \phi}{\partial x}\right)_p$$

where, for convenience, we have dropped the subscripts x and y on the LHS derivatives. The LHS expressions represent the height coordinate versions of the

x- and y-direction pressure gradient force terms. Thus, the RHS expressions represent the x- and y-direction pressure gradient force terms in isobaric coordinates. Therefore, the isobaric coordinate expression for the pressure gradient force in vector form is

$$PGF_p = -\nabla_p \phi \tag{4.2}$$

where

$$\nabla_p = \frac{\partial}{\partial x}\hat{i} + \frac{\partial}{\partial y}\hat{j}.$$

The isobaric coordinate form of the pressure gradient force involves no reference to density and is therefore much more amenable to operational use. The removal of density from the expression for the pressure gradient force is a major advantage of isobaric coordinates and provides the motivation for their use.

With the result in (4.2), the vector form of the horizontal equation of motion can be rewritten as

$$\frac{d\vec{V}}{dt} = -\nabla_p \phi - f\hat{k} \times \vec{V} \tag{4.3}$$

where, importantly,

$$\frac{d}{dt} = \frac{\partial}{\partial t} + u\frac{\partial}{\partial x}\hat{i} + v\frac{\partial}{\partial y}\hat{j} + \omega\frac{\partial}{\partial p}\hat{k}.$$

The component velocity in the last term, ω, is equal to

$$\omega = \frac{dp}{dt} \tag{4.4}$$

and is a measure of vertical velocity in units of hPa s^{-1} (or, more commonly in operations, $\mu\text{bar s}^{-1}$, where $1\ \mu\text{b} = 10^{-3}$ hPa).

By neglecting the horizontal acceleration vector, a new expression for the geostrophic balance arises from (4.3): namely,

$$f\hat{k} \times \vec{V} = -\nabla_p \phi. \tag{4.4a}$$

Taking $-\hat{k} \times$ (4.4a) and dividing by f on both sides yields an expression for the geostrophic wind in isobaric coordinates,

$$\vec{V}_g = \frac{\hat{k}}{f} \times \nabla_p \phi. \tag{4.4b}$$

Without reference to ρ, (4.4b) provides a much simpler expression for calculating the geostrophic wind from observations. The simplicity is illustrated in Figure 4.1 which shows an example of the 500 hPa geopotential height contours and actual wind vectors from the middle latitudes. The geostrophic wind is parallel to the geopotential height contours with a magnitude dependent on the magnitude of $\nabla_p \phi$. For the most

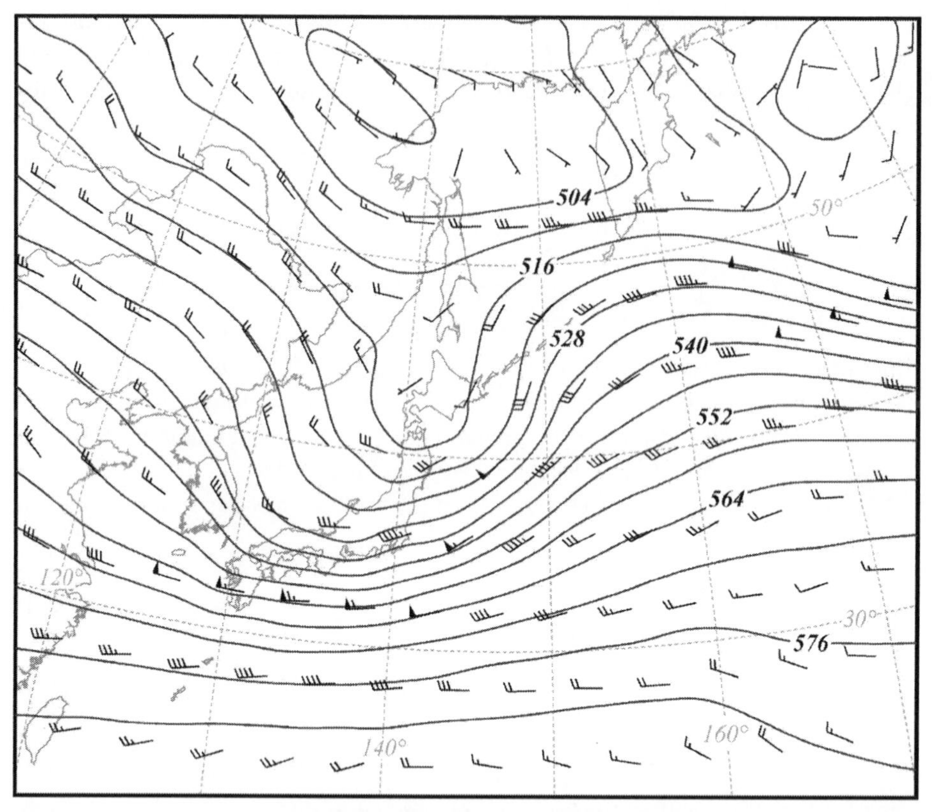

Figure 4.1 The 500 hPa geopotential height and winds at 0000 UTC 23 February 2004. Geopotential height is labeled in dam (deca meters) and contoured every 6 dam. Winds are represented as flags pointing into the direction from which the wind is coming. Speed of the winds is indicated by the barbs on the flags as: half barb, less than 5 m s^{-1}; full barb, 5 m s^{-1}; pennant, 25 m s^{-1}

part, the actual wind is close to the geostrophic wind with notable exceptions being in regions of wind speed maxima and strong curvature.

Another consequence of the simplicity of (4.4b) is that, so long as we consider f to be constant, then

$$
\begin{aligned}
\nabla \cdot \vec{V}_g &= \nabla \cdot \left(\frac{\hat{k}}{f} \times \nabla_p \phi \right) \\
&= \nabla \cdot \left(-\frac{1}{f}\frac{\partial \phi}{\partial y}\hat{i} + \frac{1}{f}\frac{\partial \phi}{\partial x}\hat{j} \right) \\
&= \frac{1}{f}\left[\frac{\partial}{\partial x}\left(-\frac{\partial \phi}{\partial y} \right) + \frac{\partial}{\partial y}\left(\frac{\partial \phi}{\partial x} \right) \right] \\
&= \frac{1}{f}\left(-\frac{\partial^2 \phi}{\partial x \, \partial y} + \frac{\partial^2 \phi}{\partial x \, \partial y} \right) = 0
\end{aligned}
\tag{4.5}
$$

so that the geostrophic wind is non-divergent. This is an extremely important property of the geostrophic wind and its importance for understanding mid-latitude weather systems will be amplified when we examine the continuity equation in isobaric coordinates.

Rather than transform the continuity equation from height to pressure coordinates, we will derive the isobaric form of the continuity equation by considering a control volume ($\delta V = \delta x \, \delta y \, \delta z$) as before. Using the hydrostatic equation ($\delta p = -\rho g \delta z$) we can rewrite δV as

$$\delta V = \frac{-\delta x \, \delta y \, \delta p}{\rho g}.$$

Let us adopt the Lagrangian perspective that the mass of the control volume (given by $\delta M = \rho \delta V = -\delta x \, \delta y \, \delta p / g$) does not change following the parcel. Then the rate of change of mass (per unit mass) is given by

$$\frac{1}{\delta M}\frac{d(\delta M)}{dt} = 0 = \frac{-g}{\delta x \, \delta y \, \delta p}\frac{d}{dt}\left(\frac{-\delta x \, \delta y \, \delta p}{g}\right). \tag{4.6a}$$

Applying the chain rule to the RHS of (4.6a) yields

$$\frac{1}{\delta x \, \delta y \, \delta p}\left[\frac{d(\delta x)}{dt}\delta y \, \delta p + \frac{d(\delta y)}{dt}\delta x \, \delta p + \frac{d(\delta p)}{dt}\delta x \, \delta y\right] = 0. \tag{4.6b}$$

Since, as we saw before,

$$\frac{d(\delta x)}{dt} = \delta u, \quad \frac{d(\delta y)}{dt} = \delta v, \quad \text{and} \quad \frac{d(\delta p)}{dt} = \delta \omega,$$

(4.6b) can be simplified to

$$\frac{\partial u}{\partial x} + \frac{\partial v}{\partial y} + \frac{\partial \omega}{\partial p} = 0 \tag{4.7}$$

as δx, δy, and δp approach zero. This is the isobaric form of the continuity equation. This form of the continuity equation is much simpler than the height coordinate version ((3.46) and (3.47)) since, similar to the isobaric expression for the pressure gradient force, density does not appear in it.

A simple rearrangement of (4.7) produces

$$\nabla \cdot \vec{V}_h = \frac{\partial u}{\partial x} + \frac{\partial v}{\partial y} = -\frac{\partial \omega}{\partial p} \tag{4.8a}$$

which relates the fact that the horizontal divergence on an isobaric surface is directly related to the vertical motion (ω), a variable of exceptional importance in creating the sensible weather. If we know the vertical (p-direction) distribution of horizontal divergence in an atmospheric column, then we can determine the vertical motion distribution in that column as well. Consider the hypothetical situation depicted in Figure 4.2 in which horizontal convergence of air occurs near the surface ($\nabla \cdot \vec{V}_h < 0$)

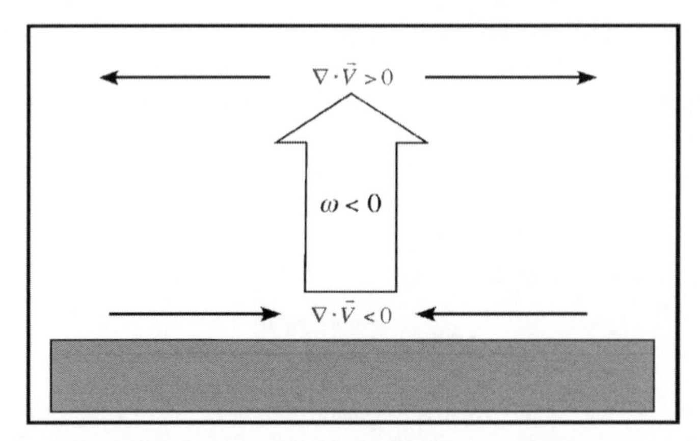

Figure 4.2 Illustration of the vertical distribution of divergence associated with upward vertical motion. The divergence values are measured on isobaric surfaces and $\omega < 0$ corresponds to ascent

and horizontal divergence of air occurs at the top of the column ($\nabla \cdot \vec{V}_h > 0$). In accord with the continuity of mass, such a circumstance must be accompanied by upward vertical motion in the intervening column of air. Integrating (4.8a) with respect to pressure yields

$$\int_{p_s}^{p_t} \left(\frac{\partial u}{\partial x} + \frac{\partial v}{\partial y} \right) \partial p = -\int_{p_s}^{p_t} \partial \omega \tag{4.8b}$$

or

$$(\nabla \cdot \vec{V}_h)_{p_t} - (\nabla \cdot \vec{V}_h)_{p_s} = -[\omega_{p_t} - \omega_{p_s}]. \tag{4.8c}$$

There can be no vertical motion precisely at ground level, so $\omega_{p_s} = 0$. Since $(\nabla \cdot \vec{V}_h)_{p_t} - (\nabla \cdot \vec{V}_h)_{p_s} > 0$, we find that $\omega_{p_t} < 0$ (i.e. there is upward vertical motion at the top of the hypothetical column) as we suspected.

Another useful physical insight arises from (4.8a) by considering the horizontal wind as the sum of its geostrophic and ageostrophic components. By substituting $\vec{V}_h = \vec{V}_g + \vec{V}_{ag}$ into (4.8a) we get

$$\nabla \cdot \vec{V}_h = \nabla \cdot (\vec{V}_g + \vec{V}_{ag}) = \nabla \cdot \vec{V}_g + \nabla \cdot \vec{V}_{ag} = -\frac{\partial \omega}{\partial p}. \tag{4.9a}$$

Recall from (4.5) that so long as f is constant, the geostrophic wind is non-divergent so that (4.9a) becomes

$$\nabla \cdot \vec{V}_{ag} = -\frac{\partial \omega}{\partial p} \tag{4.9b}$$

which states that the divergence of the ageostrophic wind determines the distribution of vertical motion in the atmosphere. Thus, it is the ageostrophic wind that is entirely responsible for the distribution of cyclones, anticyclones, clouds, and precipitation in the atmosphere. The ramifications of this statement are profound. Despite the fact that the mid-latitude atmosphere is predominantly in geostrophic balance, all of the important weather with which we are confronted develops as a direct result of the often relatively small ageostrophic portion of the wind.

Finally, we can quite easily express the thermodynamic energy equation in isobaric coordinates by writing the first law of thermodynamics in pressure coordinates as

$$c_p\left(\frac{\partial T}{\partial t}+u\frac{\partial T}{\partial x}+v\frac{\partial T}{\partial y}+\omega\frac{\partial T}{\partial p}\right)-\alpha\omega=\dot{Q}. \tag{4.10a}$$

Rearranging the LHS and dividing by c_p yields

$$\left(\frac{\partial T}{\partial t}+u\frac{\partial T}{\partial x}+v\frac{\partial T}{\partial y}\right)-\left(\frac{\alpha}{c_p}-\frac{\partial T}{\partial p}\right)\omega=\frac{\dot{Q}}{C_p} \tag{4.10b}$$

which can be rewritten as

$$\left(\frac{\partial T}{\partial t}+u\frac{\partial T}{\partial x}+v\frac{\partial T}{\partial y}\right)-\sigma_p\omega=\frac{\dot{Q}}{C_p} \tag{4.10c}$$

where

$$\sigma_p=\left(\frac{\alpha}{c_p}-\frac{\partial T}{\partial p}\right)$$

is a measure of the static stability in isobaric coordinates. If an atmospheric flow is assumed to be (1) adiabatic ($\dot{Q}=0$), (2) steady state ($\partial T/\partial t=0$), and (3) stably stratified ($\sigma_p>0$), then (4.10c) can be written in a physically illuminating manner as

$$\frac{(-\vec{V}_h\cdot\nabla T)}{-\sigma_p}=\omega. \tag{4.11}$$

This expression states that the horizontal temperature advection is related to the vertical motion such that warm (cold) air advection is associated with upward (downward) vertical motions. The cyclone depicted in Figure 4.3 illustrates that these relationships, though based upon some troubling assumptions (most notably the steady-state assumption), do tend to be observed in the real mid-latitude atmosphere. For this reason meteorologists are often very interested in the sign of the horizontal temperature advection.

4.2 Potential Temperature as a Vertical Coordinate

Though adopting pressure as a vertical coordinate simplifies a number of the basic equations by removing reference to density, air parcels are no more constrained to remain on an isobaric surface than they are to remain on a geometric height surface. In many applications it is desirable to choose potential temperature (θ) as the vertical coordinate since (1) for statically stable stratifications, θ is a monotonic

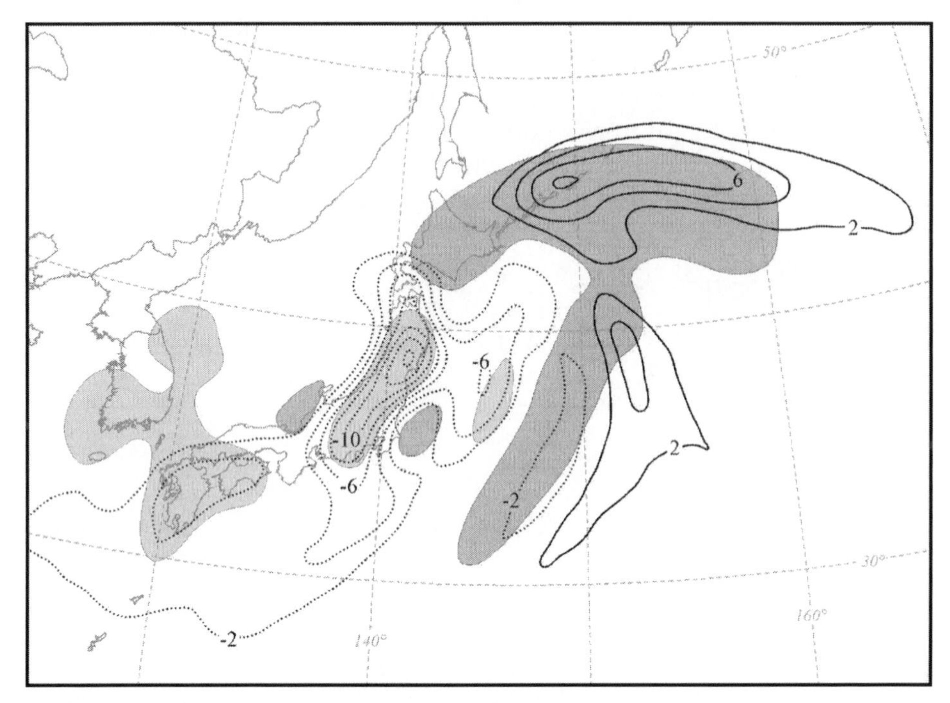

Figure 4.3 The 700 hPa temperature advection and vertical motion at 0000 UTC 23 February 2004. Solid (dotted) lines indicate positive (negative) temperature advection labeled in units of $10^{-4}\,\mathrm{K\,s^{-1}}$ and contoured every $2(-2) \times 10^{-4}\,\mathrm{K\,s^{-1}}$ starting at $2(-2) \times 10^{-4}\,\mathrm{K\,s^{-1}}$. Vertical motion (omega) is shaded dark (light) gray where $\omega < -5\mu\,\mathrm{bar\,s^{-1}}$ ($\omega > 5\mu\mathrm{bar\,s^{-1}}$)

function of height, and (2) for adiabatic processes air parcels are required to remain on the same θ surface. This second characteristic means that for adiabatic flow, the θ surface is an actual material surface along which air parcels must move. In this section we will briefly outline the basic equations in a coordinate system that uses θ as the vertical coordinate, the so-called **isentropic** coordinate system. We begin with a transformation of the pressure gradient force into isentropic coordinates.

We can convert the pressure gradient force term into isentropic coordinates by considering the differential (dp) on a surface of constant θ:

$$dp_\theta = \left(\frac{\partial p}{\partial x}\right)_{y,z,t} dx_\theta + \left(\frac{\partial p}{\partial y}\right)_{x,z,t} dy_\theta + \left(\frac{\partial p}{\partial z}\right)_{x,y,t} dz_\theta + \left(\frac{\partial p}{\partial t}\right)_{x,y,z} dt_\theta \tag{4.12a}$$

where, as in (4.1a), the subscripts refer to the differentiation carried out holding that variable constant. We will consider the x-direction pressure gradient force here and so we divide each term in (4.12a) by dx_θ to yield

$$\left(\frac{dp}{dx}\right)_\theta = \left(\frac{\partial p}{\partial x}\right)_{y,z,t} + \left(\frac{\partial p}{\partial z}\right)_{x,y,t} \left(\frac{dz}{dx}\right)_\theta \tag{4.12b}$$

since the terms $(dy/dx)_\theta$ and $(dt/dx)_\theta$ have no physical meaning. With the aid of the hydrostatic equation we can write

$$\left(\frac{dp}{dx}\right)_\theta = \left(\frac{\partial p}{\partial x}\right)_{y,z,t} - \rho g \left(\frac{dz}{dx}\right)_\theta \tag{4.12c}$$

which can be rearranged in order to isolate the x-coordinate expression for the pressure gradient force as

$$-\frac{1}{\rho}\left(\frac{\partial p}{\partial x}\right)_{y,z,t} = -\frac{1}{\rho}\left(\frac{dp}{dx}\right)_\theta - g\left(\frac{dz}{dx}\right)_\theta. \tag{4.12d}$$

In order to proceed we must consider an expression for $-(1/\rho)(dp/dx)_\theta$ which can be done by evaluating the x-direction log differential of the Poisson equation (3.56) given by

$$\frac{d\ln\theta}{dx} = \frac{d\ln T}{dx} + \frac{R}{c_p}\left(\frac{d\ln 1000}{dx} - \frac{d\ln p}{dx}\right). \tag{4.13a}$$

It is clear that $d\ln 1000/dx = 0$, so that the above expression can be written as

$$\frac{1}{\theta}\frac{d\theta}{dx} = \frac{1}{T}\frac{dT}{dx} - \frac{R}{c_p p}\frac{dp}{dx} \tag{4.13b}$$

where all derivatives are taken on an isentropic surface. In that case, $d\theta/dx$ is identically zero and we can isolate an expression for $-(1/\rho)(dp/dx)_\theta$ as

$$-\frac{1}{\rho}\left(\frac{dp}{dx}\right)_\theta = -\frac{c_p p}{\rho RT}\left(\frac{dT}{dx}\right)_\theta = -c_p\left(\frac{dT}{dx}\right)_\theta. \tag{4.13c}$$

Substituting the above expression into (4.12d) we get

$$-\frac{1}{\rho}\left(\frac{\partial p}{\partial x}\right)_{y,z,t} = -c_p\left(\frac{dT}{dx}\right)_\theta - g\left(\frac{dz}{dx}\right)_\theta = -\frac{\partial}{\partial x}(c_p T + \phi)_\theta \tag{4.14}$$

where $(c_p T + \phi)_\theta$ is known as the **Montgomery streamfunction** or **Montgomery potential**, often denoted as Ψ_M. An analogous expression can be derived for the y direction so that the horizontal pressure gradient force in isentropic coordinates is represented by

$$PGF = -\nabla_\theta \Psi_M. \tag{4.15}$$

We are now able to formulate the horizontal equation of motion in isentropic coordinates as

$$\frac{d\vec{V}_\theta}{dt} = -\nabla_\theta \Psi_M - f\hat{k} \times \vec{V}_\theta + \vec{F}_\theta \tag{4.16}$$

where the Lagrangian operator is defined as

$$\frac{d}{dt} = \frac{\partial}{\partial t} + u_\theta \frac{\partial}{\partial x} + v_\theta \frac{\partial}{\partial y} + \frac{d\theta}{dt}\frac{\partial}{\partial \theta}.$$

It is clear from the last term in the preceding expression that only diabatic heating can compel a parcel to move in the 'vertical' (i.e. θ) direction.

Next, we consider the hydrostatic balance in isentropic coordinates. As has been the case in the other coordinate systems we have thus far examined, we begin by taking the 'vertical' derivative of the Montgomery potential to get

$$\frac{\partial \Psi_M}{\partial \theta} = c_p \frac{\partial T}{\partial \theta} + g \frac{\partial z}{\partial \theta}. \tag{4.17}$$

Using the height coordinate expression for the hydrostatic equation along with the chain rule we have

$$\frac{\partial p}{\partial \theta} \frac{\partial \theta}{\partial z} = -\rho g$$

which can be expressed as

$$g \frac{\partial z}{\partial \theta} = -\frac{1}{\rho} \frac{\partial p}{\partial \theta} = -\frac{RT}{p} \frac{\partial p}{\partial \theta}. \tag{4.18}$$

Substituting (4.18) into the RHS of (4.17) and dividing by $c_p T$ yields

$$\frac{1}{c_p T} \frac{\partial \Psi_M}{\partial \theta} = \frac{1}{T} \frac{\partial T}{\partial \theta} - \frac{R}{c_p p} \frac{\partial p}{\partial \theta}. \tag{4.19}$$

Logarithmically differentiating the Poisson equation with respect to θ gives

$$\frac{1}{\theta} \frac{\partial \theta}{\partial \theta} = \frac{1}{T} \frac{\partial T}{\partial \theta} - \frac{R}{c_p p} \frac{\partial p}{\partial \theta} \tag{4.20}$$

and so the LHSs of (4.19) and (4.20) can be equated:

$$\frac{1}{c_p T} \frac{\partial \Psi_M}{\partial \theta} = \frac{1}{\theta} \frac{\partial \theta}{\partial \theta} = \frac{1}{\theta}.$$

Thus, the final expression for $\partial \Psi_M / \partial \theta$, the hydrostatic equation in isentropic coordinates, is

$$\frac{\partial \Psi_M}{\partial \theta} = \frac{c_p T}{\theta}. \tag{4.21}$$

Given that isentropic coordinates are monotonic in height under statically stable conditions, the distance between successive isentropic surfaces can be measured either as a geometric height interval or as a pressure interval. The advantage of making this measurement in terms of a pressure interval is that the measurement can then be converted into an increment of mass. Consider the cube of air depicted in Figure 4.4. The amount of mass in the volume element $\delta x\, \delta y\, \delta \theta$ is equal to

$$\delta M = -\delta p \left(\frac{\delta x\, \delta y}{g} \right) = -\frac{\delta p}{\delta \theta} \left(\frac{\delta x\, \delta y\, \delta \theta}{g} \right) \tag{4.22}$$

where the minus sign arises from the fact that p varies in the opposite sense as θ in the vertical direction (i.e. p decreases in the same direction as θ increases). By

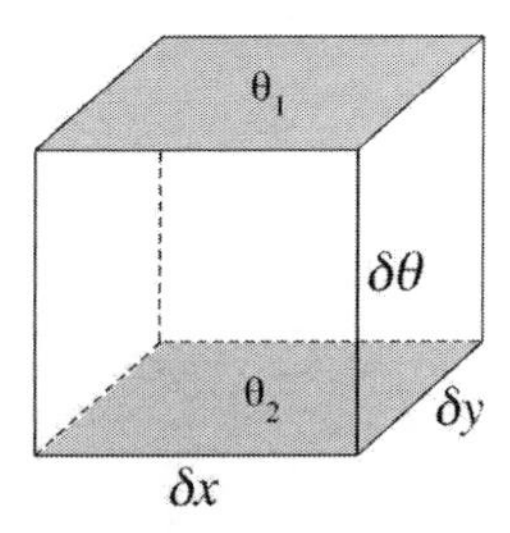

Figure 4.4 Infinitesimal cube of air in isentropic coordinates. Since θ is monotonic with height in a stably stratified atmosphere, θ_1 and θ_2 ($\theta_1 > \theta_2$) occur at different pressures. Note that $\theta_1 > \theta_2$

analogy to (4.6) we can express the continuity of mass as

$$\frac{1}{\delta M}\frac{d}{dt}(\delta M) = 0 = \left[\frac{-g}{\delta x\,\delta y\,\delta\theta}\left(\frac{\delta\theta}{\delta p}\right)\right]\frac{d}{dt}\left[-\frac{\delta p}{\delta\theta}\left(\frac{\delta x\,\delta y\,\delta\theta}{g}\right)\right]. \tag{4.23a}$$

Since g is a constant this can be rewritten as

$$\left(-\frac{\delta\theta}{\delta p}\right)\left(\frac{1}{\delta x\,\delta y\,\delta\theta}\right)\left[\frac{d}{dt}\left(-\frac{\delta p}{\delta\theta}\right)\delta x\,\delta y\,\delta\theta + \left(-\frac{\delta p}{\delta\theta}\right)\delta y\,\delta\theta\frac{d}{dt}(\delta x)\right.$$
$$\left. + \left(-\frac{\delta p}{\delta\theta}\right)\delta x\,\delta\theta\frac{d}{dt}(\delta y) + \left(-\frac{\delta p}{\delta\theta}\right)\delta x\,\delta y\frac{d}{dt}(\delta\theta)\right] = 0. \tag{4.23b}$$

Consistent with prior definitions,

$$\frac{d}{dt}(\delta x) = \delta u,\ \frac{d}{dt}(\delta y) = \delta v,\quad \text{and}\quad \frac{d}{dt}(\delta\theta) = \delta\left(\frac{d\theta}{dt}\right).$$

Employing these definitions and taking the limit as the dimensions δx, δy, and $\delta\theta$ approach zero, (4.23b) can be simplified to

$$\frac{\partial\theta}{\partial p}\frac{d}{dt}\left(\frac{\partial p}{\partial\theta}\right) + \frac{\partial u}{\partial x} + \frac{\partial v}{\partial y} + \frac{\partial}{\partial\theta}\left(\frac{d\theta}{dt}\right) = 0. \tag{4.23c}$$

Multiplying by $\partial p/\partial\theta$ and expanding the total derivative yields

$$\frac{\partial}{\partial t}\left(\frac{\partial p}{\partial\theta}\right) = -u\frac{\partial}{\partial x}\left(\frac{\partial p}{\partial\theta}\right) - v\frac{\partial}{\partial y}\left(\frac{\partial p}{\partial\theta}\right) - \left(\frac{\partial p}{\partial\theta}\right)\nabla\cdot\vec{V}_\theta - \left(\frac{\partial p}{\partial\theta}\right)\frac{\partial}{\partial\theta}\left(\frac{d\theta}{dt}\right) \tag{4.24}$$

which is the continuity equation in isentropic coordinates. The term on the LHS of (4.24), $(\partial/\partial t)(\partial p/\partial\theta)$, represents a measure of the local rate of change of mass. The first and second terms on the RHS of (4.24) represent the horizontal advection of mass on isentropic surfaces. The third term, the divergence term, describes the effect of horizontal divergence on the mass distribution. Convergence (divergence) in isentropic coordinates increases (decreases) the amount of mass contained between two θ surfaces. Finally, the fourth term on the RHS of (4.24) is the diabatic heating

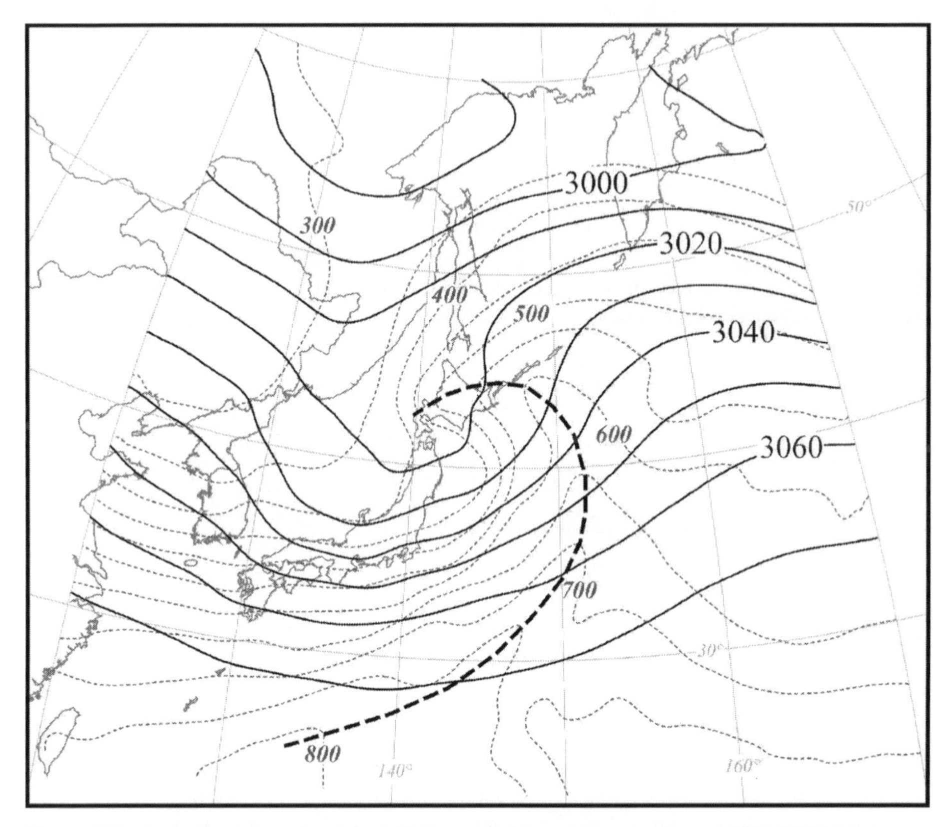

Figure 4.5 Isobaric topography of the 305 K potential temperature surface at 0000 UTC 23 February 2004. Thick solid lines are isopleths of Montgomery streamfunction labeled in $m^2 s^{-2}$ and contoured every $10 m^2 s^{-2}$. Thin dashed lines are isobars labeled in hPa and contoured every 50 hPa. The thick dashed line indicates the leading edge of negative pressure advection; to the north and east of that line positive pressure advection prevails

term which suggests that vertical gradients in diabatic heating, by changing the distance between isentropic surfaces, can contribute to local changes in the mass distribution. We will take advantage of the isentropic form of the continuity equation again in our discussion of potential vorticity.

Before leaving this brief introduction to isentropic coordinates, it is important to point out one of the most common applications of this coordinate system. We have already seen that the Montgomery streamfunction can be used to define a geostrophic 'horizontal' flow on an isentropic surface. Unlike isobaric surfaces, which are quasi-horizontal, isentropic surfaces can have considerable slope in the vertical. A simple means of portraying that slope on an isentropic surface is to plot the isobaric topography of the isentropic surface as shown in Figure 4.5. It is easy to identify regions on the example isentropic surface where the geostrophic flow is directed upward (downward) toward lower (higher) pressures. Thus, it would seem that pressure advection on an isentropic surface gives some indication of the sign of the vertical motion. We can, in fact, investigate this intriguing relationship by considering the Lagrangian derivative of pressure on an isentropic surface:

$$\frac{dp}{dt} = \omega = \frac{\partial p}{\partial t} + u\frac{\partial p}{\partial x} + v\frac{\partial p}{\partial y} + \frac{d\theta}{dt}\left(\frac{\partial p}{\partial \theta}\right) \tag{4.25}$$

where all the derivatives are taken on an isentropic surface. Under adiabatic conditions, the last term on the RHS of (4.25) can be neglected. We find, however, that an additional assumption needs to be made in order that pressure advection (the middle two terms on the RHS of (4.25)) alone can determine the sign of the vertical motion – that is, the pressure distribution has to be in steady state (i.e. $\partial p/\partial t = 0$). This condition is not often met in the atmosphere as the structure of an individual weather system is continually changing and, therefore, so is the topography of its numerous isentropic surfaces. Despite this difficulty, it is often possible to determine the sign of the vertical motion correctly by considering the pressure advection on an isentropic surface.

Finally, consider the 700 hPa isobar on the 305 K isentropic surface depicted in Figure 4.5. The Poisson equation involves temperature, pressure, and θ. Thus, there is a unique value of temperature at 700 hPa that corresponds to 305 K (namely, 275.4 K). Thus, the 700 hPa isobar on the 305 K isentropic surface corresponds exactly to the 275.4 K isotherm on the 700 hPa isobaric surface! This simple example leads us to a simple rule:

An isobar on an isentropic surface is equivalent to an isotherm on an isobaric surface.

Thus, the diagnostic of pressure advection on an isentropic surface is very similar to that of temperature advection on an isobaric surface. As a consequence, it is not unusual for synoptic meteorologists to give a rough diagnosis of the vertical motion in a mid-latitude weather system by considering the sign of the temperature advection. This diagnostic is limited in precisely the same way as the pressure advection diagnostic we just considered with respect to Figure 4.5. It is only valid for adiabatic, steady-state conditions in which the static stability is positive. This last characteristic of the adiabatic method for diagnosing vertical motions in the isobaric coordinate system is buried in the monotonic assumption that underlies the use of isentropic coordinates. We will see in later chapters that much more satisfying diagnostics of the vertical motion in mid-latitude weather systems arise from more stringent dynamical considerations.

4.3 The Thermal Wind Balance

Recall that the hypsometric equation (3.6) suggested that the thickness between two isobaric surfaces is smaller in a cold column of air than in a warm column. Consider a hypothetical example in which a cold column and a warm column are horizontally juxtaposed, as in Figure 4.6. The distance between the 1000 and 800 hPa surfaces must be larger in the warm air than the cold so that the 800 hPa surface slopes downward

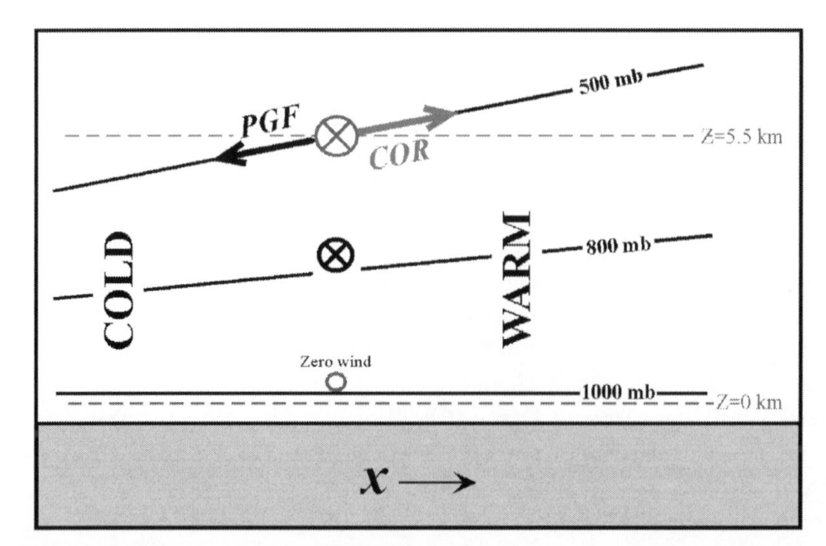

Figure 4.6 Vertical cross-section across a region of horizontal temperature contrast. Solid black lines are isobars. Dashed gray lines are elevation contours. The lack of PGF at Z = 0 km leads to 'Zero wind' there. At 5.5 km in the same vertical column there is a large geostrophic wind into the page, signified by the gray 'X' in the circle. The vertical shear of the geostrophic wind is signified by the darker 'X' in the center of the column

toward the cold air as illustrated. Similarly, the distance between the 800 and 500 hPa surfaces must be larger in the warm air than the cold and so the 500 hPa surface slopes even more dramatically downward toward the cold air. Thus, we find that the slope of the isobaric surfaces increases with increasing height in the presence of a horizontal contrast in column average temperature. Of course, the slope of an isobaric surface is equivalent to the existence of a geopotential gradient along that surface since the geopotential difference is simply $g\Delta z$. We now know that the pressure gradient force on an isobaric surface is related to the geopotential gradient on that surface. Thus, the increased slope to the isobaric surfaces in Figure 4.6 also means that the magnitude of the horizontal pressure gradient force increases with increasing height. Consequently, the geostrophic wind must be increasing with increasing height as well. Therefore, there is a physical relationship between the vertical shear of the geostrophic wind (i.e. the manner in which the geostrophic wind changes with height) and the horizontal temperature gradient. We now explore the mathematical description of this relationship by first considering the hydrostatic equation in isobaric coordinates.

Recall that the hydrostatic equation is given by $\partial p/\partial z = -\rho g$. This is easily rearranged into

$$\frac{g\partial z}{\partial p} = -\frac{1}{\rho} = -\frac{RT}{p} \tag{4.26a}$$

and, since $g\partial z = \partial\phi$, it can be expressed as

$$\frac{\partial\phi}{\partial p} = -\frac{RT}{p}, \tag{4.26b}$$

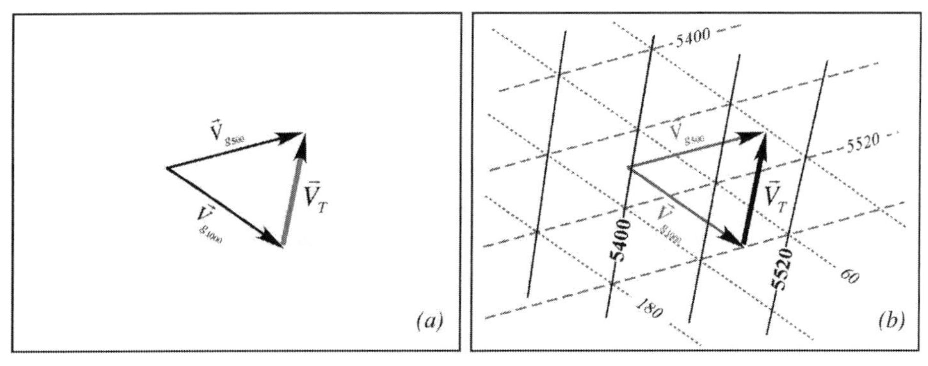

Figure 4.7 (a) Graphical depiction of the thermal wind vector obtained by subtracting the 1000 hPa geostrophic wind from the 500 hPa geostrophic wind. (b) Illustration of the relationship between the thermal wind vector and the 1000–500 hPa thickness isopleths. Dashed lines are 500 hPa geopotential heights, dotted lines are 1000 hPa geopotential heights, and the solid black lines are of 1000–500 hPa thickness. All isopleths are labeled in m and contoured every 60 m

the isobaric form of the hydrostatic equation. Now, the vertical derivative (in isobaric coordinates) of the geostrophic wind relationship ($\vec{V}_g = (\hat{k}/f) \times \nabla\phi$) is

$$\frac{\partial \vec{V}_g}{\partial p} = \frac{\hat{k}}{f} \times \nabla \frac{\partial \phi}{\partial p}. \tag{4.27a}$$

Substituting for $\partial\phi/\partial p$ from (4.26b) yields

$$\frac{\partial \vec{V}_g}{\partial p} = \frac{\hat{k}}{f} \times \nabla - \frac{RT}{p} = \left(\frac{-R}{fp}\right) \hat{k} \times \nabla T \tag{4.27b}$$

confirming the physics depicted in Figure 4.6: that the vertical shear of the geostrophic wind is directly related to the horizontal temperature gradient. Based upon this temperature gradient dependence, the vertical shear of the geostrophic wind is known as the **thermal wind**. The component form of (4.27b) yields

$$\frac{\partial u_g}{\partial p} = \frac{R}{fp}\frac{\partial T}{\partial y} \quad \text{and} \quad \frac{\partial v_g}{\partial p} = -\frac{R}{fp}\frac{\partial T}{\partial x}. \tag{4.28}$$

Returning to Figure 4.6, we find that $\partial T/\partial x > 0$ and therefore $\partial v_g/\partial p < 0$ which is consistent with an increase in v_g with height as depicted. In graphical form, the thermal wind vector is simply the vector difference between the geostrophic wind at some upper level in the atmosphere and the geostrophic wind at some lower level, as shown in Figure 4.7(a). Consequently, the thermal wind vector ($\vec{V}_T$) is actually best represented by $\vec{V}_T = -\partial\vec{V}_g/\partial p$. From (4.27a) it is clear that the thermal wind vector will be parallel to isopleths of thickness, with lower thickness to its left (right) in the northern (southern) hemisphere (Figure 4.7b). Given this physical relationship, it is possible to determine the sign of the column-averaged geostrophic temperature advection simply by knowing the direction of the thermal wind.

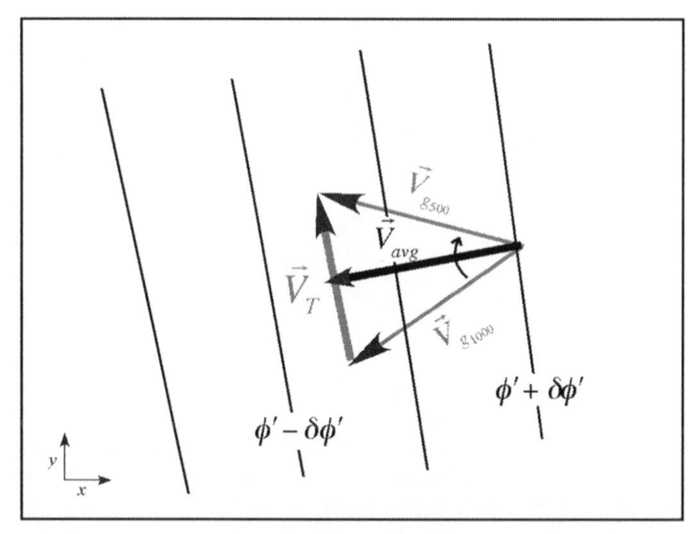

Figure 4.8 Depiction of geostrophic winds veering with height in the northern hemisphere. The thick black arrow is the column-averaged geostrophic wind. Solid lines are isopleths of 1000–500 hPa thickness. The column-averaged wind is clearly directed across thickness isopleths from large values to low values, indicative of column-averaged geostrophic warm air advection

Consider, for instance, the situation depicted in Figure 4.8 in which the geostrophic wind direction veers (turns clockwise) with height. The thermal wind vector, as represented by the thick arrow, is given by $\vec{V}_T = \vec{V}_{g_{500}} - \vec{V}_{g_{1000}}$. Assuming the situation is occurring in the northern hemisphere, the thickness isopleths must be drawn as in Figure 4.8. The column-averaged geostrophic temperature advection is given by

$$-\overline{\vec{V}}_g \cdot \nabla \overline{T} \quad \text{or} \quad -\overline{\vec{V}}_g \cdot \nabla \left(-\frac{\partial \phi}{\partial p} \right) \tag{4.29}$$

where $\overline{\vec{V}}_g$ is the column-averaged geostrophic wind ($\overline{\vec{V}}_g = (\vec{V}_{g_{500}} + \vec{V}_{g_{1000}})/2$) and the column-averaged temperature is related to the thickness by the hypsometric equation. Thus, we find that under these circumstances the average geostrophic wind is directed across the thickness isopleths from higher values to lower values, indicating column-averaged geostrophic warm air advection. Therefore, simple knowledge of the vertical distribution of the geostrophic wind at a point can be used to determine a portion of the temperature tendency in the vicinity of that point.

Another clear application of the thermal wind relationship that has a bearing on the structure and behavior of mid-latitude weather systems is consideration of the mid-latitude jet stream. The jet stream is a core of high-speed winds located at the top of the troposphere as shown in Figure 4.9(a). Given that the winds are predominantly geostrophic at middle latitudes, a large fraction of the total wind in the jet is described by the geostrophic wind. A vertical cross-section of geostrophic winds through the mid-latitude jet stream is shown in Figure 4.9(b). Note that there is considerable vertical shear of the geostrophic wind from ~700 to 350 hPa. The thermal wind relationship demands that this vertical shear be accompanied by a horizontal temperature contrast. Figure 4.9(b), which also shows the vertical cross-section of potential temperature through the jet stream core, illustrates that a significant horizontal temperature contrast is present through the entire troposphere and lower stratosphere. Such a temperature contrast is characteristic of the fronts within extratropical cyclones. Figure 4.9(a) shows where the jet stream is located in relation to an associated extratropical cyclone. Notice that the jet streak, the local portion of the broader jet stream, is in the vicinity of the surface cold front. This is, of course, not an accident but a mandate since the large vertical geostrophic shears associated with the jet streak must be associated with a large horizontal temperature contrast such as the one that characterizes the cold frontal zone. This is one reason that the position of the jet stream is so important in the discussion of mid-latitude weather systems.

From a broader perspective, the thermal wind relationship also has important dynamical consequences for the general circulation of the atmosphere. Given that the Earth is an oblate spheroid, it is not heated evenly by the Sun: the equatorial regions are warmer than the polar regions. As a consequence, there is a pole to equator temperature contrast in both hemispheres so that time-averaged thickness isopleths ring the Earth like latitude lines with a poleward-directed temperature gradient vector. The thermal wind relationship mandates that this thermal contrast be reflected by the presence of *westerly vertical shear* in both hemispheres. Thus, the fundamental fact that mid-latitude weather systems move from west to east on Earth is a direct consequence of the uneven heating of the Earth by the Sun combined with the primacy of the thermal wind balance at middle latitudes.

Finally, the thermal wind relationship forms the cornerstone of modern dynamical meteorology as well as the first-order balance for the flow in the middle latitudes on Earth. This latter point is a direct consequence of the fact that the mid-latitude atmosphere is, to first order, geostrophically and hydrostatically balanced. The combination of these balances into the thermal wind balance will provide us with a powerful diagnostic tool for understanding the structure, dynamics, and evolution of mid-latitude weather systems in subsequent chapters.

4.4 Natural Coordinates and Balanced Flows

It is probably clear by this point in our investigation of dynamics that, despite the potential for great complication, the gross behavior of the mid-latitude atmosphere can be understood in terms of relatively simple approximate force balances. Additional insight into the variety of simple force balances relevant to understanding the atmosphere can be achieved by idealizing the flow as steady state and purely horizontal (i.e. without vertical motions). Despite the unrealistic nature of these idealizations, important new insights arise through entertaining these simplifications.

In this section, we will again consider the frictionless equation of motion

$$\frac{d\vec{V}}{dt} = -\nabla_p \phi - f\hat{k} \times \vec{V} \tag{4.30}$$

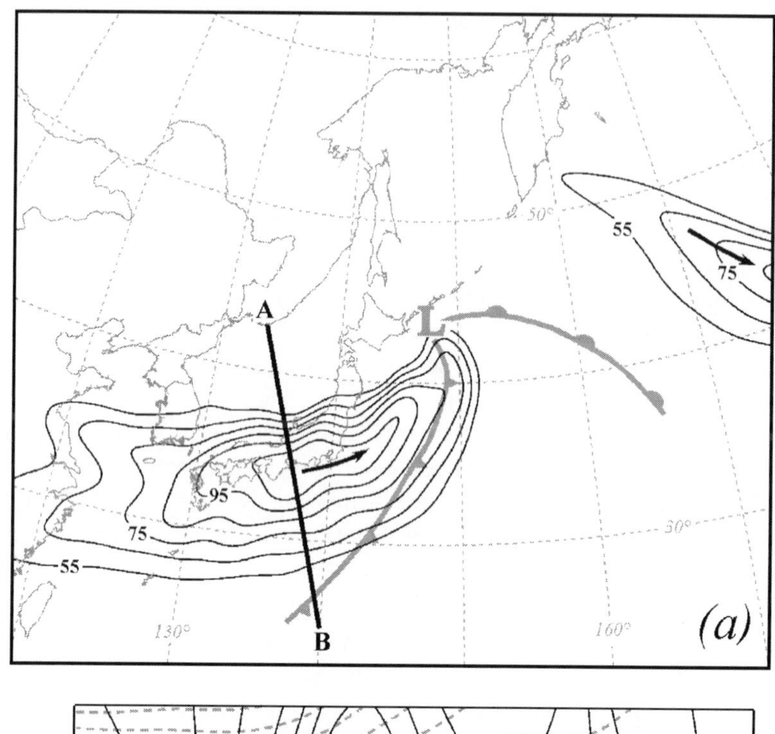

50°
55
75
A
L
95
75
55
30°
130°
B
160°
(a)

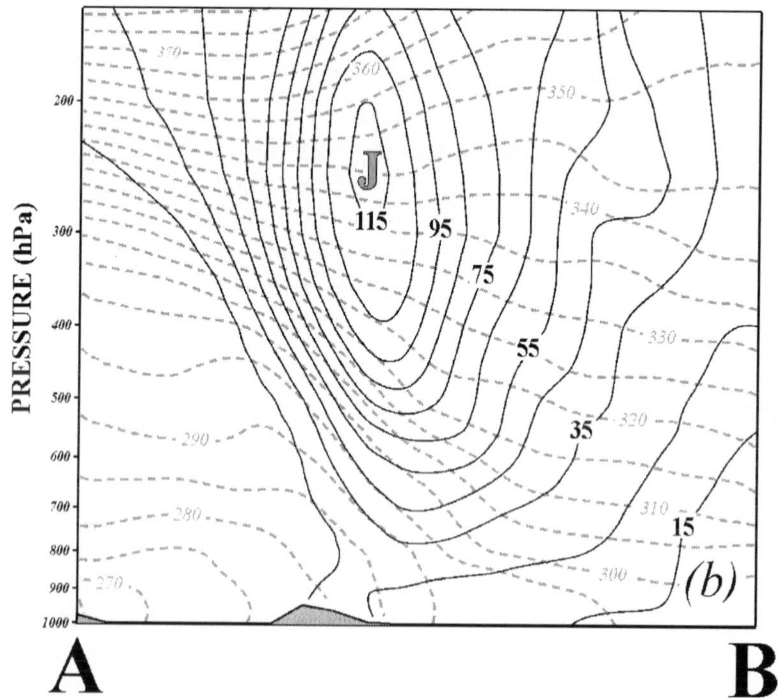

PRESSURE (hPa)
200
300
400
500
600
700
800
900
1000
370
360
350
340
330
320
310
300
290
280
270
J
115
95
75
55
35
15
(b)
A
B

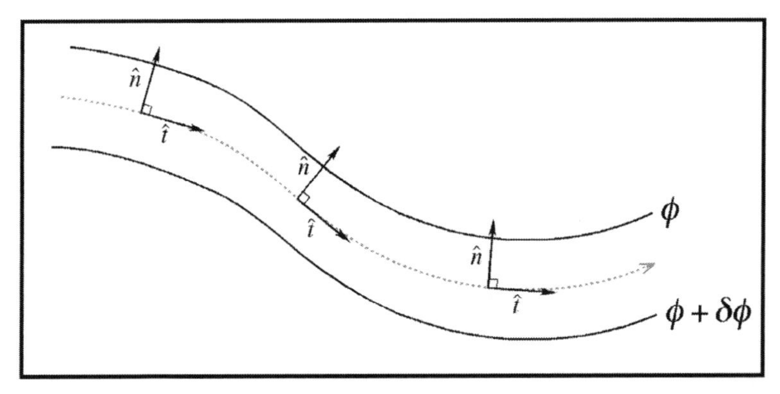

Figure 4.10 Schematic illustrating the relationship between the horizontal flow and the natural coordinate unit vectors, $\hat{t}$ and $\hat{n}$. The gray dotted line is a streamline in flow parallel to geopotential height contours

but in a Cartesian coordinate system based upon the orientation of the fluid flow. Such a system is known as **natural coordinates** and it will prove to be very useful in these investigations. One might wonder why should we bother with yet another coordinate transformation? The motivation for the adoption of natural coordinates lies in the advantage it produces in describing the acceleration term in (4.30). Acceleration is a vector quantity so it can be the result of (1) a change in flow *speed*, or (2) a change in flow *direction*, resulting from curvature in the flow. Upon expanding (4.30) in a system of natural coordinates, these aspects of acceleration can be considered separately, thus providing considerable physical insight.

We begin the transformation by defining the natural coordinate system as a Cartesian coordinate system based upon a set of orthogonal unit vectors $\hat{t}$, $\hat{n}$, and $\hat{k}$. As illustrated in Figure 4.10, $\hat{t}$ is oriented parallel to the horizontal velocity vector at each point, $\hat{n}$ is oriented normal to the horizontal flow at each point such that it is *positive to the left of the flow direction*, and $\hat{k}$ is directed upward. In this natural coordinate system the velocity vector, $\vec{V}$, is written as $\vec{V} = V\hat{t}$ where V is the *magnitude* of the velocity vector and can be expressed as $V = ds/dt$ where s is a measure of the distance in the $\hat{t}$ direction. The acceleration, $d\vec{V}/dt$, is therefore given by

$$\frac{d\vec{V}}{dt} = \frac{d}{dt}(V\hat{t}) = \hat{t}\frac{dV}{dt} + V\frac{d\hat{t}}{dt}. \tag{4.31}$$

We next need to develop an expression for the rate of change of direction, $d\hat{t}/dt$. This direction change is dependent on the presence of flow curvature as illustrated

Figure 4.9 (a) The 300 hPa isotachs of the geostrophic wind at 0000 UTC 23 February 2004. Isotachs are labeled in m s^{-1} and contoured every 10 m s^{-1} beginning at 55 m s^{-1}. Heavy arrows indicate the direction of the wind. Vertical cross-section along line A–B shown in (b). Light gray L and frontal symbols indicate position of surface cyclone at this time. (b) Vertical cross-section of geostrophic isotachs and potential temperature along line A–B in (a). Isotachs are solid black lines labeled and contoured as in (a). Gray dashed lines are isentropes labeled in K and contoured every 5 K. Note that the region of maximum vertical shear is also the region of maximum horizontal temperature gradient throughout the troposphere

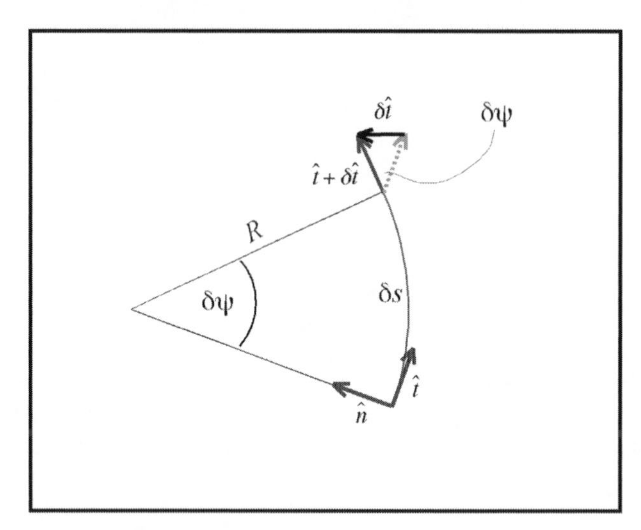

Figure 4.11 The rate of change of the natural coordinate vector $\hat{t}$ following the motion. R is the radius of curvature of the parcel trajectory

in Figure 4.11. In order to describe this curvature, we adopt the convention that the radius of curvature of parcel trajectories (i.e. $R =$ radius of curvature *following the parcel motion*) will be positive when $\hat{n}$ is directed toward the center of the curvature. Thus, $R > 0$ for counterclockwise flow and $R < 0$ for clockwise flow. For the schematic given in Figure 4.11 we see that $\delta s = R\delta\psi$ and, by similarity, $\delta\hat{t} = |\hat{t}|\,\delta\psi$. Equating the expressions for $\delta\psi$ from both expressions we get

$$\delta\psi = \frac{\delta s}{R} = \frac{\delta\hat{t}}{|\hat{t}|} = \delta\hat{t} \tag{4.32}$$

since $\hat{t}$ is a unit vector. Notice that as $\delta s \to 0$, $\delta\hat{t}$ is parallel to $\hat{n}$ so that

$$\lim_{\delta s\to 0}\frac{\delta\hat{t}}{\delta s} = \frac{d\hat{t}}{ds} = \left(\frac{1}{R}\right)\hat{n} = \frac{\hat{n}}{R}. \tag{4.33a}$$

Therefore,

$$\frac{d\hat{t}}{dt} = \frac{d\hat{t}}{ds}\frac{ds}{dt} = \left(\frac{\hat{n}}{R}\right)V = \left(\frac{V}{R}\right)\hat{n} \tag{4.33b}$$

since $V = ds/dt$ by definition. Thus, we can rewrite (4.31) as

$$\frac{d\vec{V}}{dt} = \frac{dV}{dt}\hat{t} + \frac{V^2}{R}\hat{n} \tag{4.34}$$

which demonstrates that the acceleration following the motion is the sum of (1) the rate of change of the speed of the air parcel, and (2) its centripetal acceleration arising from curvature in the flow.

Since the Coriolis force acts normal to the flow it must be in the $\hat{n}$ direction. In the northern hemisphere, the Coriolis force acts to the right of the motion, so in the $-\hat{n}$ direction. Thus, we represent the Coriolis force as $COR = -(fV)\hat{n}$. In the southern hemisphere, the Coriolis force acts to the left of the motion, the $\hat{n}$ direction. Given that latitude is positive (negative) in the northern (southern) hemisphere, by convention, the same expression for the Coriolis force

$$-f\hat{k} \times \vec{V} = -(fV)\hat{n} \tag{4.35}$$

is applicable in the southern hemisphere. The pressure gradient force has components in both the along-flow ($\hat{t}$) and across-flow ($\hat{n}$) directions so it can be rewritten as

$$-\nabla_p \phi = -\left(\frac{\partial \phi}{\partial s}\hat{t} + \frac{\partial \phi}{\partial n}\hat{n}\right). \tag{4.36}$$

Thus, the frictionless equation of motion (4.31) can be rewritten in natural coordinates as

$$\left(\frac{dV}{dt}\hat{t} + \frac{V^2}{R}\hat{n}\right) = -\left(\frac{\partial \phi}{\partial s}\hat{t} + \frac{\partial \phi}{\partial n}\hat{n}\right) - (fV)\hat{n} \tag{4.37}$$

which can be split into its along-flow component

$$\frac{dV}{dt} = -\frac{\partial \phi}{\partial s} \tag{4.38a}$$

and its across-flow component;

$$\frac{V^2}{R} + fV = -\frac{\partial \phi}{\partial n}. \tag{4.38b}$$

For motion parallel to geopotential height contours, $\partial\phi/\partial s = 0$ (i.e. there is no change in ϕ in the along-flow direction), and the *speed* of the flow is constant. In this case, the flow can be classified into a number of simple categories based upon the relative contributions of the three terms in (4.38b), the $\hat{n}$-component equation of motion.

4.4.1 Geostrophic flow

Recall that in considering the $\hat{n}$ equation of motion we are implicitly considering a flow in which the speed is constant. If we further consider a perfectly straight flow, then $|R| = \infty$. In such a case, only the Coriolis and pressure gradient forces remain from (4.38b) so that

$$fV = -\frac{\partial \phi}{\partial n}. \tag{4.39a}$$

Accordingly, the flow is in geostrophic balance and the geostrophic wind is expressed as

$$V_g = -\frac{1}{f}\frac{\partial \phi}{\partial n}. \tag{4.39b}$$

4.4.2 Inertial flow

Occasionally a fluid may be compelled to move by something other than a pressure gradient force internal to the fluid. A simple example of this is the initiation of fluid flow in a body of water. Quite commonly such motions (known as **currents**) are generated by the influence of the wind stress upon the surface of the water.[1] In such cases, the pressure gradient force is zero implying that the geopotential field is horizontally uniform on an isobaric surface. Thus only forces arising from the inertia of the fluid remain in the governing equations and the $\hat{n}$ equation of motion reduces to a balance between the centrifugal and Coriolis forces

$$\frac{V^2}{R} + fV = 0. \tag{4.40a}$$

The resulting motion is known as **inertial motion**. Solving (4.40a) for R, the radius of curvature of parcel trajectories characteristic of such inertial motion, yields

$$R = -\frac{V}{f}. \tag{4.40b}$$

Thus, so long as R is fairly small (as is nearly always the case since V is small for such motions), the trajectories of purely inertial motions follow circular, anticyclonic paths. Given that the parcel speeds are constant (i.e. $\nabla\phi = 0$), the amount of time needed to trace out a circle of radius R is

$$\begin{aligned} t &= \frac{\textit{Distance Covered by Parcel in a Circular Path}}{\textit{Speed of the Parcel}} \\ &= \left|\frac{2\pi R}{V}\right| = \left|\frac{2\pi R}{-Rf}\right| = \left|\frac{2\pi}{f}\right|. \end{aligned} \tag{4.41a}$$

Of course, this amount of time is, by definition, the period of a single oscillation of this inertial motion. Thus, we can rewrite (4.41a) as

$$P = \frac{2\pi}{2\Omega\sin\phi} = \frac{\pi}{\Omega\sin\phi}. \tag{4.41b}$$

The numerator in (4.41b) refers to π radians, equivalent to half of a complete rotation, while Ω is the rotation rate of the Earth (1 rotation per sidereal day). Thus, (4.41b) can be expressed as

$$P = \frac{\pi}{\Omega\sin\phi} = \left(\frac{1/2\ rotation}{1\ rotation/day}\right)\frac{1}{\sin\phi} = \frac{1/2 day}{\sin\phi} \tag{4.41c}$$

known as the **half-pendulum day**. An example of evidence for inertial motion in the ocean is given in Figure 4.12 which illustrates the kinetic energy spectrum in the

[1] There are, however, very few, if any, examples of motion initiated in the atmosphere in the absence of an atmospheric pressure gradient force.

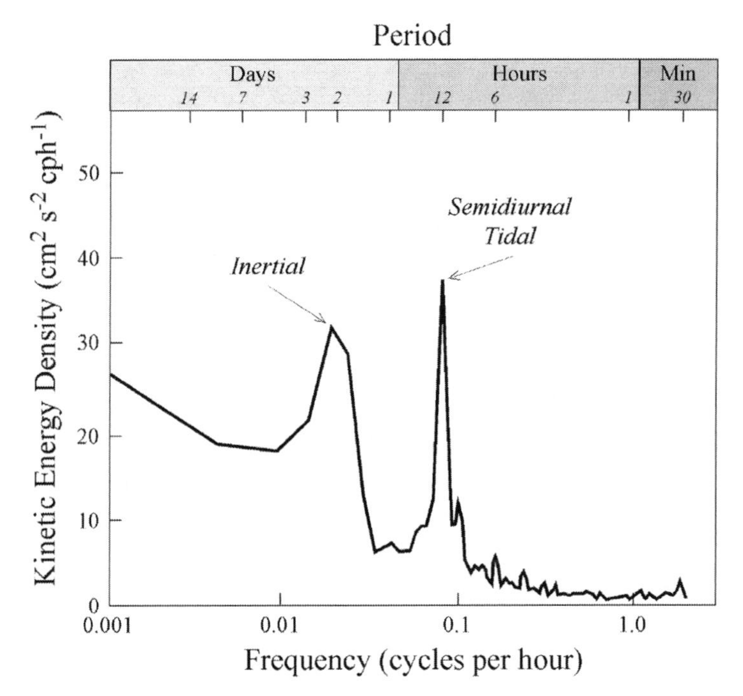

Figure 4.12 Power spectrum of kinetic energy (KE) at 30 m in the ocean near Barbados (13°N). Such a plot of KE per unit frequency interval vs. frequency illustrates the partition of total KE into oscillations of different periods. Two strong peaks are evident: the twice-daily tides and the inertial frequency. (After Warsh *et al.* 1971. Reproduced with permission of the American Meteorological Society)

ocean for a location near 13°N. It is clear that two prominent spikes in kinetic energy, the results of two prominent modes of fluid flow at that location, occur in association with (1) the twice-daily (semi-diurnal) tides, and (2) an inertial oscillation with a period of nearly 2 days. Using (4.41c) at 13°N we find that the period of the inertial oscillation there is $P = 2.2$ days.

4.4.3 Cyclostrophic flow

There are a number of circumstances under which the Coriolis force may exert very little influence on balanced motions. On Earth, fluid flows located at low latitudes and/or of small horizontal scale will not be influenced significantly by the Coriolis force. Extraterrestrially, some of the planets in the Solar System (i.e. Venus and Titan) have slow rotation rates which render the Coriolis force weak. Under any such circumstances, the $\hat{n}$ equation of motion reduces to a balance between the centrifugal force and the pressure gradient force

$$\frac{V^2}{R} = -\frac{\partial \phi}{\partial n}$$

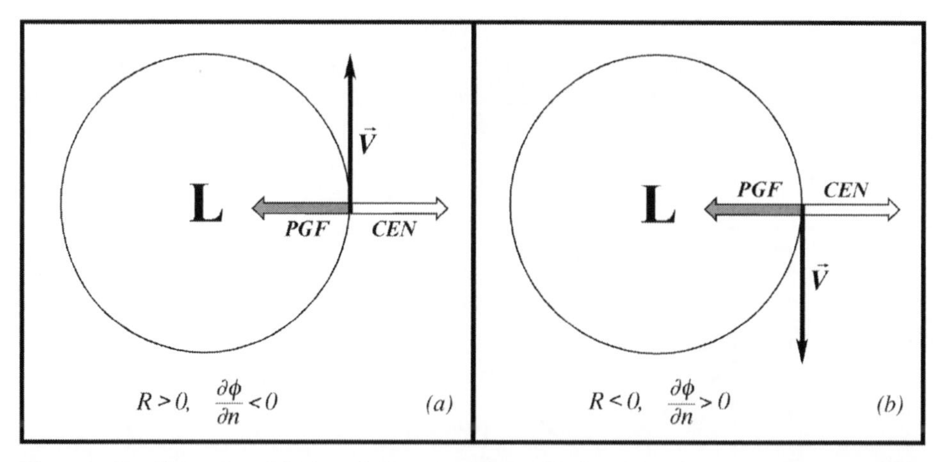

Figure 4.13 Illustration of the force balance in cyclostrophic flow. The pressure gradient force (PGF) and centrifugal force (CEN) are represented by the shaded and open arrows, respectively, while the wind vector is indicated by the bold black arrow. The signs of both the radius of curvature and the geopotential height gradients are also given for (a) cyclonic flow (b) anticyclonic flow

which can be solved for V to yield

$$V = \left(-R\frac{\partial \phi}{\partial n}\right)^{1/2}. \tag{4.42}$$

The balance between the centrifugal and pressure gradient forces is known as **cyclostrophic balance** and V is, therefore, the **cyclostrophic wind.** Given that there is no Coriolis force involved in the cyclostrophic balance, and that the centrifugal force is *always* directed outward from the center of rotation, there is no preference for cyclonic or anticyclonic flow around a region of low geopotential as illustrated in Figure 4.13.

Note that the cyclostrophic balance is a valid approximation provided that the centrifugal force (V^2/R) is much larger than the Coriolis force (fV). Making this comparison in the form of a ratio of forces yields

$$\frac{V^2/R}{fV} = \frac{V}{fR}.$$

Recall that V/fR is called the Rossby number. Thus, cyclostrophic balance is the preferred balance for large-Rossby-number flows. Tornadoes can be reasonably approximated as very small-scale, circular vortices with tangential wind speeds of the order of $100\,\mathrm{m\,s^{-1}}$ not more than 1000 m from the center of the vortex. If such a tornado occurs in the central plains of North America, where the latitude is $\sim$40°N ($f \approx 10^{-4}\,\mathrm{s^{-1}}$), then the Rossby number for such a flow is

$$R_o = \frac{V}{fR} = \frac{100\,m\,s^{-1}}{(10^{-4}s^{-1})(1000\,m)} = 1000$$

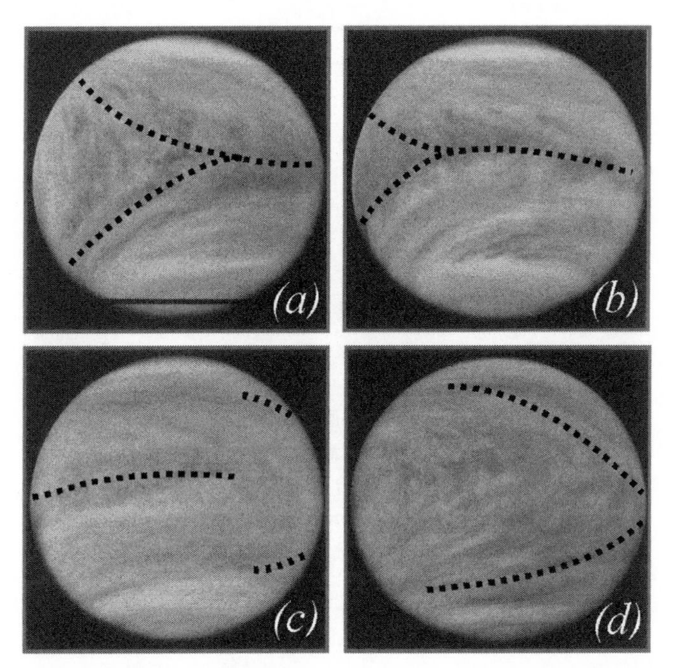

Figure 4.14 Infrared satellite images (NASA) of the Venusian cloud tops taken by Pioneer Venus over a 4 day period in April 1979: (a) 16 April 1979; (b) 17 April 1979; (c) 18 April 1979; (d) 19 April 1979. The thick dashed lines trace the Y-shaped feature in the Venusian clouds that circles the planet every 4 days, evidence of the super-rotation of the Venusian atmosphere

which certainly qualifyies as very large. Thus, the high wind speeds observed in tornadoes are, to a first order, in cyclostrophic balance. Consistent with this conclusion, anticyclonically spinning tornadic vortices are occasionally observed. The much smaller and less violent dust devils and small water spouts that share characteristics of tornadic storms exhibit even less preference for cyclonic or anticyclonic rotation, occurring frequently in both varieties.

Two striking examples of cyclostrophically balanced flows exist on other planets in the Solar System. The planet Venus has a period of rotation of 243 Earth days and it rotates east to west! Consequently, the stratospheric winds in the Venusian atmosphere are directed from east to west but these winds are in excess of 100 m s^{-1}, circumnavigating the planet in 4 days as illustrated in Figure 4.14. Since such wind speeds far exceed the angular speed of the solid planet beneath them, the Venusian atmosphere is said to be **super-rotating**, a still poorly understood circumstance. The Rossby number of such a flow is

$$\frac{V}{fR} \approx \frac{100\, m\, s^{-1}}{(10^{-7} s^{-1})(\sim 10^{7} m)} \approx 100.$$

Saturn's largest moon, Titan, takes nearly 16 Earth days to rotate once and so it also has a very small Coriolis force. The stratospheric winds remotely sensed by the

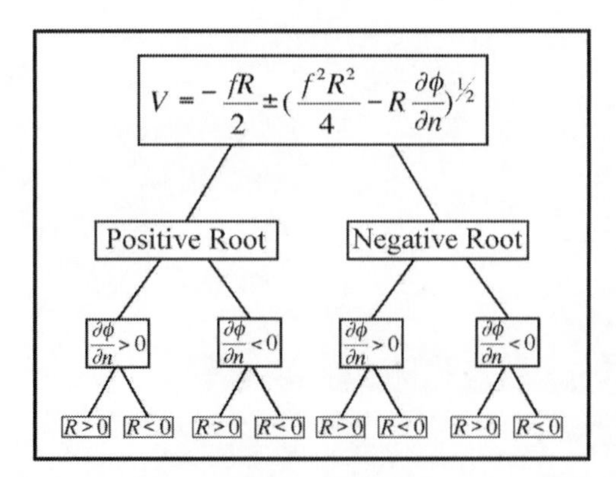

Figure 4.15 Solution tree for the gradient wind relationship

Voyager spacecraft are on the order of 100 m s^{-1}. Consequently, the Rossby number is $\sim$10 and the flow is in approximate cyclostrophic balance.

4.4.4 Gradient flow

Cursory examination of a randomly selected 500 hPa analysis of geopotential height and winds, such as in Figure 4.1, illustrates that, even in the face of variable flow curvature, horizontal, frictionless flow nearly parallel to the geopotential height lines is the rule rather than the exception at middle latitudes. Such flow is known as gradient flow and it is a balance between the pressure gradient force, the Coriolis force, and the centrifugal force arising from the flow curvature. As such, the gradient wind equation is simply the $\hat{n}$ equation of motion:

$$\frac{V^2}{R} + fV = -\frac{\partial \phi}{\partial n}.$$

This expression is quadratic in V so, using the quadratic formula, we find that

$$V = \frac{-f \pm \sqrt{f^2 - 4(1/R)(\partial \phi/\partial n)}}{(2/R)} = -\frac{fR}{2} \pm \left(\frac{f^2R^2}{4} - R\frac{\partial \phi}{\partial n}\right)^{1/2}. \qquad (4.43)$$

This complicated-looking expression has a number of mathematically possible solutions, not all of which correspond to physical reality. In order to isolate the physically relevant solutions, we must first determine how many mathematical solutions exist. There will be both a positive and a negative root to (4.43), each of which has a solution for both $\partial\phi/\partial n > 0$ and $\partial\phi/\partial n < 0$. Each of those four solutions has solutions for both $R > 0$ and $R < 0$. Thus, there are a total of eight mathematically possible solutions for (4.43) as illustrated schematically in Figure 4.15. In order to be a physically relevant solution, however, V must be a positive real number. We begin our investigation of the solution tree in Figure 4.15 by considering cases in which $\partial\phi/\partial n > 0$.

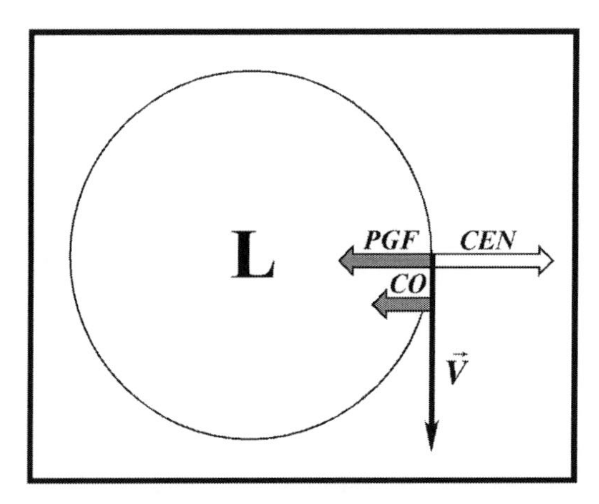

Figure 4.16 Force balance for an anomalous low in the northern hemisphere. The pressure gradient, Coriolis, and centrifugal forces are represented by PGF, CO, and CEN, respectively

(a) *Solutions for $\partial\phi/\partial n > 0$, $R > 0$*

Under the condition of $\partial\phi/\partial n > 0$ and $R > 0$ there are both positive and negative roots. Since the product $-R\partial\phi/\partial n < 0$ under the given conditions, the radical in (4.43) is rendered less than $fR/2$. Whether this radical is added to (positive root) or subtracted from (negative root) the leading $-\frac{fR}{2}$ term in (4.43), the resulting expression for V is negative and so both solutions are unphysical.

(b) *Solutions for $\partial\phi/\partial n > 0$, $R < 0$*

Under the given conditions, the product $-R\partial\phi/\partial n > 0$ so the radical in (4.43) will be greater than $|fR|/2$. Also, since $R < 0$, the leading $-fR/2$ term is positive. Thus, the positive root produces a positive V and corresponds to a physical solution. The character of the flow associated with this solution is illustrated in Figure 4.16. In the northern hemisphere $R < 0$ implies clockwise (anticyclonic) flow so $\hat{n}$ must be directed outward from the center of rotation. Given that $\partial\phi/\partial n > 0$, there must be a geopotential minimum at the center of rotation (i.e. the disturbance is a low-pressure system). Consequently, the pressure gradient force is directed inward as is the Coriolis force (which must act to the right of the wind). The centrifugal force is always directed outward from the center of rotation and must be large enough, in this case, to balance the other two forces. Since this case illustrates gradient wind balance achieved with a clockwise flow around a region of low pressure, it is known as the **anomalous low**. Though it represents a physically possible solution, it is, as you might suspect, rarely observed in nature. Recalling that under these conditions the radical term in (4.43) is greater than $|fR|/2$, the negative root produces a negative V. This corresponds to an unphysical solution.

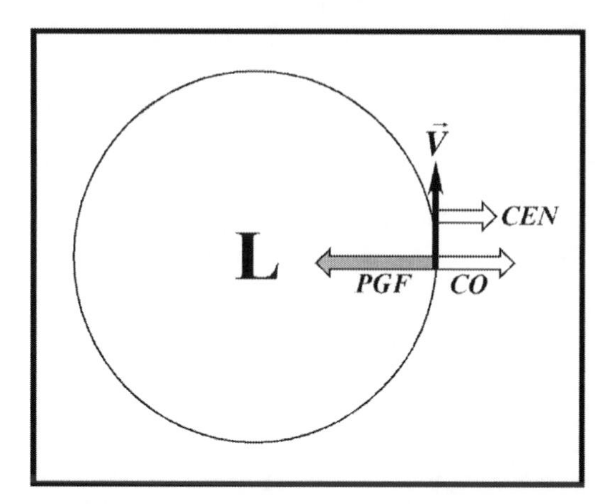

Figure 4.17 Force balance for a regular low in the northern hemisphere. The pressure gradient, Coriolis, and centrifugal forces are represented by PGF, CO, and CEN, respectively

(c) Solutions for $\partial\phi/\partial n < 0$*,* $R > 0$

Under the given conditions, the product $-R\partial\phi/\partial n > 0$ so the term in the radical of (4.43) will be larger than $fR/2$. Upon adding this to the leading $-fR/2$ term, the positive root returns a positive V corresponding to a physical solution illustrated in Figure 4.17. Once again considering flow in the northern hemisphere, $R > 0$ implies counterclockwise (cyclonic) flow around the center of rotation so that $\hat{n}$ is directed inward. Given $\partial\phi/\partial n < 0$, the flow rotates around a minimum in geopotential. Thus, the pressure gradient force is directed toward the center of rotation and is balanced by outward-directed Coriolis and centrifugal forces. Such cyclonic flow around a region of low geopotential characterizes the commonly observed **regular low**. In the negative root, the radical is subtracted from $-fR/2$ yielding a negative V which corresponds to an unphysical solution.

(d) Solutions for $\partial\phi/\partial n < 0$*,* $R < 0$

Under these conditions, the product $-R\partial\phi/\partial n < 0$. Depending on the magnitude of this product, the term in the radical of (4.43) *could be* less than zero, in which case the solution (positive or negative root) would be imaginary and consequently unphysical. If the radical term is not negative, it is certainly less than $fR/2$ but greater than zero. In that case, the solution for the positive root is $V \geq -fR/2$ which is positive since $R < 0$. Thus, the positive root corresponds to the physical solution illustrated in Figure 4.18 in which clockwise (anticyclonic in the northern hemisphere) flow around the center of rotation occurs. Given that $\partial\phi/\partial n < 0$, the center of the rotation must be a maximum in geopotential height and the pressure gradient force is directed outward along with the centrifugal force. The result is a clockwise flow around the region of high geopotential, a seemingly regular circumstance. The nature of the force

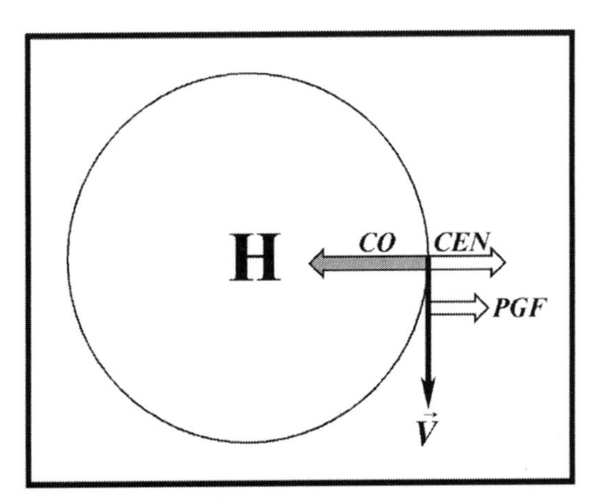

Figure 4.18 Force balance for an anomalous high in the northern hemisphere. The pressure gradient, Coriolis, and centrifugal forces are represented by PGF, CO, and CEN, respectively

balance in this case, however, suggests otherwise. Since this case is characterized by $V \geq -fR/2$, then it is also true that

$$\frac{V^2}{R} \geq -\frac{fV}{2} \quad \text{or} \quad 2\frac{V^2}{R} \geq -fV \tag{4.44}$$

which states that the Coriolis force is less than twice the centrifugal force. This implies that the centrifugal force must be larger than the pressure gradient force, a condition that earns this particular version of the gradient wind balance the title **anomalous high**.

Finally, provided that the radical term in (4.43) is real, if it is subtracted from $-fR/2$, then $V \leq -fR/2$ which is positive given that $R < 0$. This corresponds to a physical solution in the northern hemisphere in which clockwise (anticyclonic) flow around the center of rotation occurs as illustrated in Figure 4.19. The center of the rotation must be a maximum in geopotential since $\partial\phi/\partial n < 0$, so the solution again describes anticyclonic flow around a region of high geopotential. In this case, however, the outward-directed pressure gradient force is larger than the centrifugal force and the solution is therefore known as the **regular high**.

Recall that the existence of both the anomalous and regular highs is impossible unless $(f^2R^2/4 - R\partial\phi/\partial n)^{1/2} \geq 0$. This condition can be rewritten as $f^2R^2/4 \geq R\partial\phi/\partial n$. If we consider the absolute values of R and $\partial\phi/\partial n$, then

$$\left|\frac{\partial\phi}{\partial n}\right| \leq \frac{|R|\, f^2}{4} \tag{4.45}$$

suggesting that there is a constraint on the magnitude of the pressure gradient force in the vicinity of high-pressure systems such that at small radius from the center, the

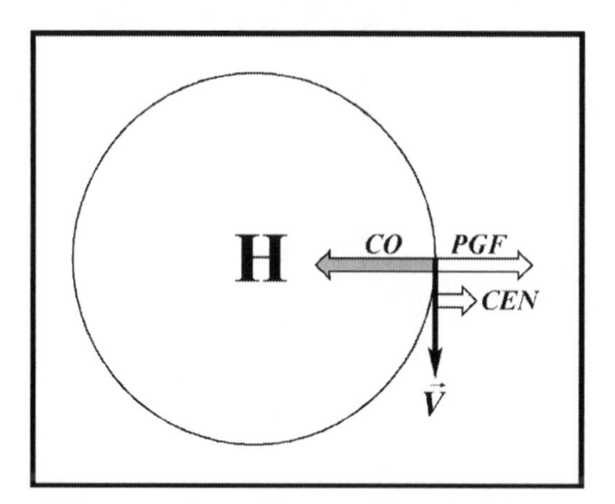

Figure 4.19 Force balance for a regular high in the northern hemisphere. The pressure gradient, Coriolis, and centrifugal forces are represented by PGF, CO, and CEN, respectively

pressure gradient must become very small, eventually vanishing at the center of the high. There is no such constraint on regions of low pressure. This stark difference is often made manifest on sea-level pressure analyses such as the one shown in Figure 4.20. Notice how amorphous the isobaric pattern becomes near the anticyclone and how different that is from the isobaric field around the cyclone. This is more than an intellectual curiosity, of course, as the weak pressure gradient near the center of the anticyclone dictates that the winds will be light to non-existent in its vicinity. Such large-scale conditions can lead to the production of devastating sensible weather events such as very low overnight temperatures in winter and fog formation at any time of year.

Daily perusal of upper tropospheric geopotential height and sea-level pressure analyses demonstrates that surface cyclones (anticyclones) are invariably located downstream (i.e. to the east) of an upper-level trough (ridge) axis. This characteristic distribution can be explained through consideration of the gradient wind balance combined with the continuity of mass. Returning to (4.38b) and substituting for $-\partial\phi/\partial n$ from the definition of the geostrophic wind,

$$V_g = -\frac{1}{f}\frac{\partial\phi}{\partial n},$$

the gradient wind balance can be rewritten as

$$\frac{V^2}{R} + fV - fV_g = 0. \tag{4.46a}$$

Thus, a ratio of the geostrophic wind to the gradient wind is given by

$$\frac{V_g}{V} = 1 + \frac{V}{fR} = 1 + R_o \tag{4.46b}$$

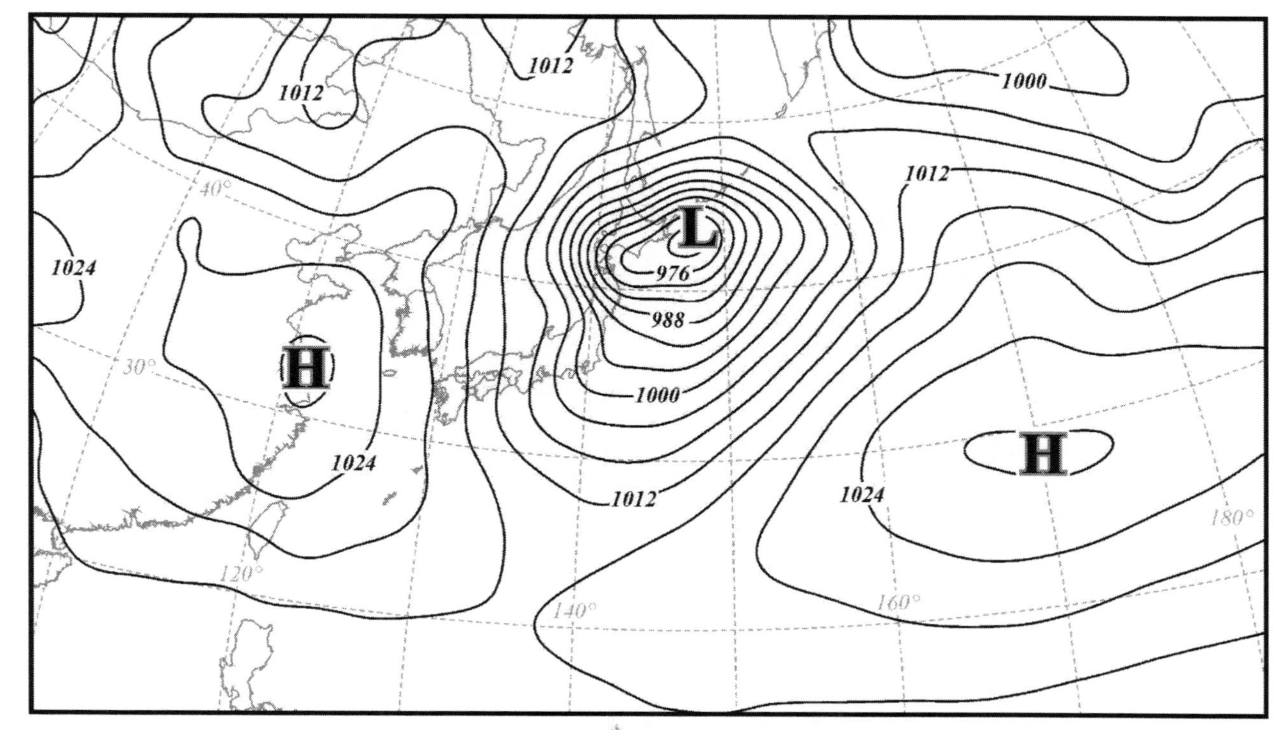

Figure 4.20 Sea-lavel pressure analysis for 0000 UTC 23 February 2004. Solid lines are isobars labeled in hPa and contoured every 4 hPa. Capital L and H represent centers of sea-level low- and high-pressure systems, respectively. Note the tight pressure gradient around the low and the much weaker pressure gradient around the highs

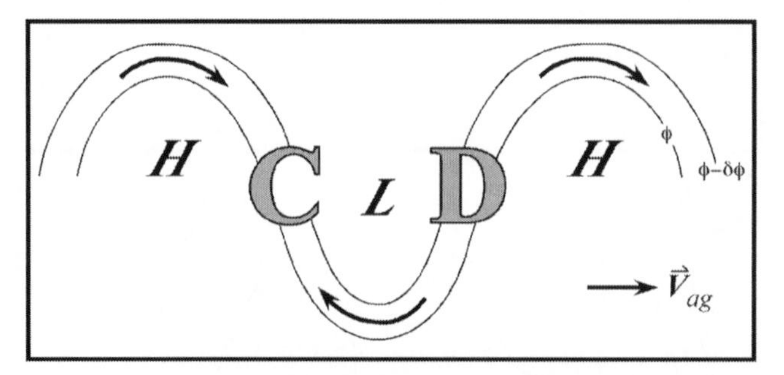

Figure 4.21 Dark arrows represent the along-flow ageostrophic winds in a 500 hPa northern hemisphere trough–ridge couplet. Regions of relatively high and low geopotential height are represented by H and L, respectively. Regions of middle tropospheric convergence and divergence of the ageostrophic winds are represented by the shaded C and D, respectively

demonstrating that for (1) cyclonic flow ($R > 0$), the geostrophic wind is larger than the gradient wind, and for (2) anticyclonic flow ($R < 0$), the geostrophic wind is smaller than the gradient wind. If one considers the gradient wind to be the real wind then the ageostrophic wind is given by

$$V_{ag} = V - V_g = -\frac{V^2}{fR}$$

and is parallel to the full wind at every point. Since the real wind is better described by the gradient wind than the geostrophic wind in regions of flow curvature, we can say that the real flow through troughs is **subgeostrophic**, whereas it is **supergeostrophic** through ridges as illustrated in the schematic 500 hPa flow shown in Figure 4.21. Notice that the ageostrophic winds at this level diverge downstream (i.e. to the right) of the trough axis and converge upstream of it. Now, the continuity equation requires that the divergence aloft on the downstream side of the trough axis be accompanied by surface convergence and upward vertical motion in the intervening column. This upward vertical motion is responsible for the production of the clouds, precipitation, and sea-level pressure minimum associated with the surface cyclone. Conversely, the convergence aloft upstream of the trough axis (or, equivalently, downstream of the ridge axis) must be accompanied by surface divergence and downward vertical motion in the intervening column. This downward vertical motion is responsible for the generally clear skies and sea-level pressure maximum associated with the surface anticyclone.

4.5 The Relationship between Trajectories and Streamlines

Throughout this discussion we have been making reference to the radius of curvature (R) in formulating the centrifugal force (V^2/R). In considering the gradient wind balance we are, of course, considering the balance of forces acting on an individual

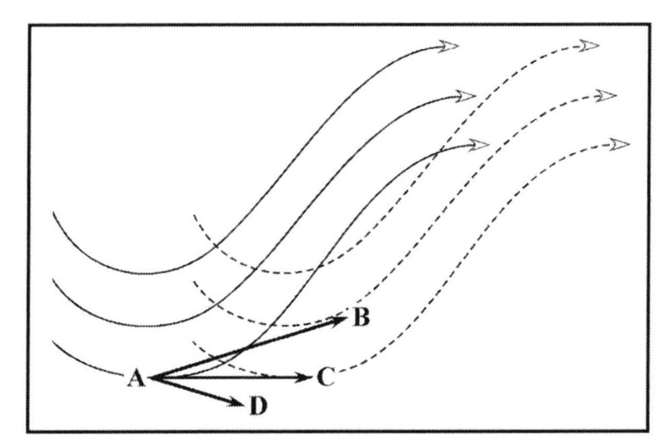

Figure 4.22 Streamlines and trajectories in an eastward-moving upper trough. Thin solid lines represent streamlines of the flow at some initial time while the dashed lines represent streamlines at some later time. The bold arrows (AD, AC, and AB) represent the trajectories of air parcels moving slower than the wave, at the same speed as the wave, and faster than the wave, respectively

parcel of air. Therefore, our expression for R has represented the radius of curvature of the path of an individual air parcel, a parcel *trajectory*. The geopotential height lines on a typical 500 hPa analysis are related to the gradient wind direction but are, in fact, *streamlines* of that flow – not trajectories. It is therefore important to delineate clearly the physical distinction between streamlines and trajectories. We can do so by first defining each term. Streamlines will be considered lines that are everywhere parallel to the *instantaneous* wind velocity. Trajectories will be considered lines that describe the *actual path* of an individual air parcel through space and time. It is clear that both streamlines and trajectories will have a radius of curvature. The difference between the two is illustrated in Figure 4.22. Assuming that the shape of the wave in Figure 4.22 does not change as the wave progresses eastward, then the radius of curvature of the streamlines, R_s, is constant. Note, however, that the radius of curvature of the trajectory (R_t, what we have been referring to as R) for a parcel of air originally located at the base of the trough depends upon the speed of the parcel with respect to the speed of the wave through which it is moving. It is evident that this dependence can give the parcel not only different magnitudes of R_t but even different *signs* as compared to R_s! The distinction between R_s and R_t can be described mathematically as well with the aid of Figure 4.23. From the picture it is clear that

$$\delta s = R\delta\beta \text{ or } \frac{\delta\beta}{\delta s} = \frac{1}{R} \tag{4.47a}$$

where β is the angular direction of the wind, s is a measure of the distance along the path, and R is the radius of curvature. Now, a trajectory represents the rate of change

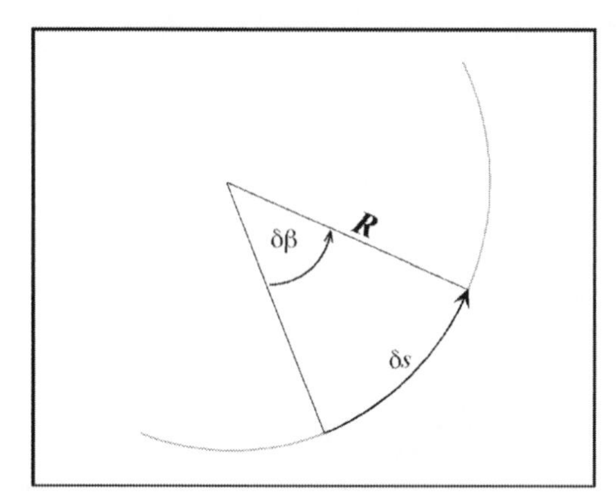

Figure 4.23 Illustration of the relationship between the change in the direction of the wind, $\delta\beta$, and the radius of curvature, R

of wind direction following the parcel so for a trajectory the Lagrangian derivative is appropriate and

$$\frac{d\beta}{ds} = \frac{1}{R_t}. \tag{4.47b}$$

A streamline represents the *local* rate of change of the wind direction so that for a streamline

$$\frac{\partial\beta}{\partial s} = \frac{1}{R_s}. \tag{4.47c}$$

Since $V = ds/dt$ by definition, we have

$$\frac{d\beta}{dt} = \frac{d\beta}{ds}\frac{ds}{dt} = \frac{V}{R_t} \tag{4.47d}$$

representing the change in direction following a parcel. This Lagrangian change in direction can also be expressed as

$$\frac{d\beta}{dt} = \frac{\partial\beta}{\partial t} + V\frac{\partial\beta}{\partial s} = \frac{\partial\beta}{\partial t} + \frac{V}{R_s}. \tag{4.47e}$$

Equating these two expressions for $d\beta/dt$ yields

$$\frac{\partial\beta}{\partial t} = V\left(\frac{1}{R_t} - \frac{1}{R_s}\right) \tag{4.48}$$

which describes the local rate of change of direction. The practical use of (4.48) is limited because a small time increment between observations is necessary to make a reasonable estimate of $\partial\beta/\partial t$. Despite this operational limitation, (4.48) does confirm the intuitive suspicion that when the local rate of change of the wind direction is zero, then trajectories and streamlines coincide. In other words, in a steady-state flow the trajectories and streamlines of the flow are the same thing, but this is a very special case. In summary, it is important to bear in mind that in any application of (4.38b) that includes reference to the centrifugal force term, R represents R_t, not R_s.

Selected References

Bluestein, *Synoptic-Dynamic Meteorology in Midlatitudes, Volume I*, provides discussion on balanced flows.
Hess, *Introduction to Theoretical Meteorology*, offers further discussion on balanced flows and isobaric coordinates.
Palmén and Newton, *Atmospheric Circulation Systems*, is another fine reference for this material.
Holton, *An Introduction to Dynamic Meteorology*, discusses balanced flows and natural coordinates.
Sutcliffe and Godart (1942) is the seminal reference on isobaric analysis.
Montgomery (1937) derives the isentropic streamfunction that bears his name.
Carlson, *Mid-Latitude Weather Systems*, provides an in-depth discussion of the use of isentropic coordinates.

Problems

4.1. Imagine that a hockey puck is given an initial horizontal impulse on an infinite, flat, frictionless ice surface at 45°N.

(a) What force(s) act(s) on the puck after the initial impulse? Explain.
(b) Draw a picture that illustrates the path of the puck under these conditions. Explain your drawing.
(c) Under the influence of the force(s) mentioned in (a), does the speed of the puck ever change during this path? Explain your answer.
(d) Using the x and y equations of motion *appropriate* for the given situation, derive a functional expression for the position (x, y) of the puck. (Hint: recall that $u = dx/dt$ and $v = dy/dt$.)
(e) What is the period of oscillation of this motion assuming the puck does not stray far from 45° N? Show your work.

4.2. Hurricanes are axisymmetric disturbances (i.e. their structure is symmetric about the center). An Atlantic hurricane is located over the Caribbean Sea (latitude 26°N). Balanced wind speeds of 60 m s^{-1} are found, at 950 mb, at a station 200 km from the center.

(a) What is the Rossby number of the flow associated with this storm? Explain your reasoning.
(b) Based upon your answer to (a), what terms in the $\hat{n}$ natural coordinate equation of motion are balanced in this flow? Explain.
(c) If the 950 mb geopotential height at the station is 367 m, what is the sea-level pressure at the eye of the hurricane? Show your work.
(d) Derive an expression for the ageostrophic wind in such a storm in terms of f, the pressure gradient force, and the radius of curvature (R).
(e) Calculate the ageostrophic wind speed at the station. How large would the radius of the hurricane have to be before the flow around it were in approximate geostrophic balance? Explain.

4.3. (a) Prove (both *graphically* and *mathematically*) that the following statement is true:

The advection of the 1000–500 mb thickness by the 1000 mb geostrophic wind is exactly equal to the advection of the 1000–500 mb thickness by the 500 mb geostrophic wind.

(b) Derive an expression for the column-averaged geostrophic temperature advection in an isobaric layer stretching from P_1 to P_2(where $P_2 < P_1$) in terms of the geopotential at P_1 and P_2.

4.4. It is a sunny afternoon at Station X. The pressure decreases to the northeast at a rate of 5 hPa/100 km. The temperature increases to the west at a rate of 5°C/100 km. The local temperature change is -0.5°C/ day^{-1}. In which hemisphere is Station X located? Defend your answer. (Hint: you may assume small Rossby number at Station X.)

4.5. Suppose that a vertical column of the atmosphere at 43°N is initially isothermal from 900 to 500 hPa. The geostrophic wind is 10 m s^{-1} from the south at 900 hPa, 10 m s^{-1} from the west at 700 hPa, and 10 m s^{-1} from the south at 500 hPa. Calculate the mean horizontal temperature gradients in the two layers 900–700 hPa and 700–500 hPa. Compute the rate of advective temperature change in each layer. If the thickness between 600 and 800 hPa remains constant at 2.25 km, how long would this advection pattern have to persist in order to establish a dry adiabatic lapse rate in the 600–800 hPa layer?

4.6. (a) With reference to the thermal wind relationship and some *basic* properties of the Earth's radiation budget, explain why the mean tropospheric vertical wind shear is westerly on Earth.

(b) Using a simple illustration, explain why the direction of this vertical shear is the same in both the northern and southern hemispheres.

4.7. The planet Uranus is characterized by low-Rossby-number flow and has a rotation rate slightly faster than Earth's. It also rotates on its axis at 83° to the plane of the ecliptic. (This means that for one-half of the Uranian 'year' the North Pole is pointed directly at the Sun.)

(a) Based on the above information, what is the predominant force balance for the winds in the Uranian troposphere?

(b) If you could fly a jetplane in the Uranian troposphere during this half of the year, would it be faster to travel from west to east or from east to west? Explain your choice.

4.8. (a) Physically explain (with reference to horizontal forces) why the wind speed is subgeostrophic in a trough and supergeostrophic in a ridge.

(b) Using the $\hat{n}$ equation of motion

$$\frac{V^2}{R} + fV = -\frac{\partial \phi}{\partial n}$$

show that the relationship described in (a) is independent of hemisphere.

4.9. A surface cyclone and a surface anticyclone are shown schematically in Figure 4.1A. Also shown (in dashed lines) are the 1000–850 mb column-averaged isotherms associated with each feature. Explain why the cyclone becomes more intense with height and the anticyclone does not.

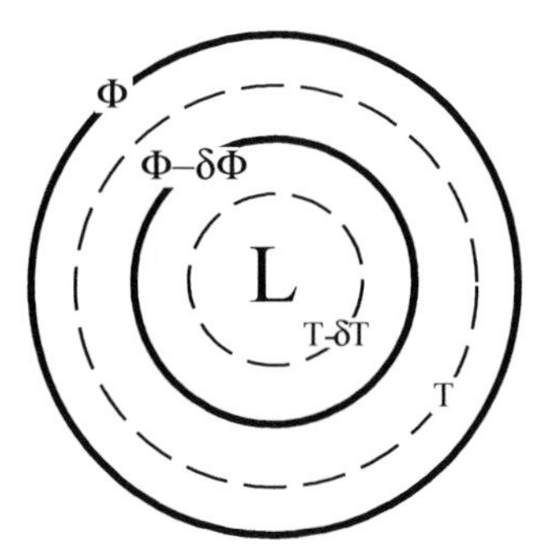

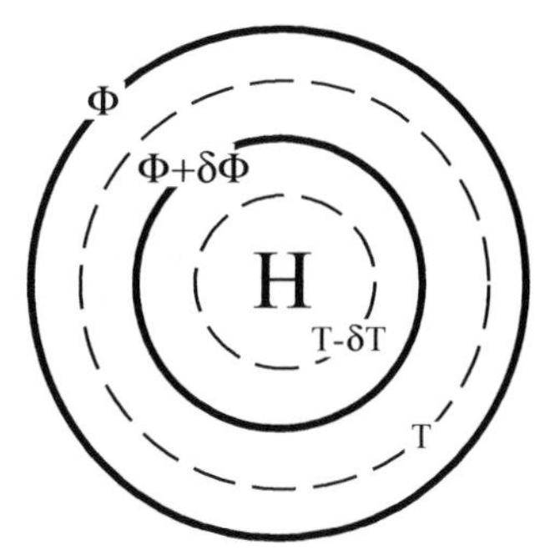

Figure 4.1A

4.10. You have the misfortune of being shipwrecked on a small island in the central Pacific Ocean (latitude 40°N). On a certain day you observe the clouds at three distinct levels in the atmosphere. The lowest-level clouds are moving from north to south. The middle-level clouds are moving from west to east. The upper-level clouds are also moving from north to south. Assume all the clouds are propelled by geostrophic winds. How will the lapse rate over the island change as the day progresses? Explain your answer.

4.11. A vertical coordinate system known as sigma coordinates is used in many numerical prediction models including the Pennsylvania State University/NCAR MM5. Sigma is defined as

$$\sigma = (p/p_s)$$

where p is the pressure and p_s is the surface pressure. Derive an expression for the horizontal pressure gradient force in σ coordinates.

4.12. On a certain day in Madison, WI (latitude 41°N) it is observed that the horizontal pressure gradient is the same at sea level (1005 hPa) as it is at 850 hPa. The thickness in the 1005–850 hPa layer is 1367 m and the temperature at 1005 hPa is 11°C.

(a) If the geostrophic wind speed at 850 hPa is 35 m s^{-1}, what is the geostrophic wind speed at 1005 hPa?
(b) If the kinetic energy generation per unit mass at sea level is 2×10^{-2} J s^{-1}, what is the cross-isobar angle at that level?
(c) What is the magnitude of the frictional drag force at sea level?
(d) How far would one have to travel from Madison to find a sea-level pressure that is 10 hPa lower than it is at Madison?

4.13. (a) Imagine a sunny day on which the flow is steady and without vertical motions and the temperature is constant. If the heating rate is 730 J kg^{-1} h^{-1}, what are the sign and magnitude of the horizontal temperature advection?
(b) If the isotherms are oriented east–west and it is 3°C colder 100 km north of you, what is the wind speed if it is blowing from the WNW (from 290°)?

4.14. The temperature at 700 hPa in Auckland, New Zealand is −5°C. If the local lapse rate near the 700 hPa level is 5 K km^{-1}, the temperature is increasing at a rate of 1.5 K h^{-1}, the wind is northeasterly at 15 m s^{-1}, and the temperature increases toward the north by 4 K for every 150 km, compute the vertical velocity at 700 hPa assuming the flow is adiabatic.

4.15. The geostrophic wind has a magnitude of 35 m s^{-1} near Sapporo, Japan (latitude 45° *N*) on a given day. If the Rossby number is 0.4 and the wind is changing direction locally at a rate of 10° h^{-1}, what is the radius of curvature of the streamlines of the flow?

4.16. Recall from Problem 3.2 that the divergence of the geostrophic wind is not zero if one accounts for the variation of the Coriolis parameter with latitude. For a typical mid-latitude trough at 300 hPa, sketch the distribution of upward and downward vertical motions that would arise from considering just the divergence of the geostrophic wind. Explain your answer based upon the continuity equation. Does this distribution of vertical motions conform with or contradict what is characteristically observed in nature? Explain.

4.17. The following wind data (direction and speed) were received from 50 km to the east, north, west, and south of a station, respectively: 90°, 10 m s^{-1}; 120°, 4 m s^{-1}; 90°, 8 m s^{-1}; 60°, 4 m s^{-1}.

(a) Calculate the approximate horizontal divergence at the station.
(b) Suppose the given wind speeds were all in error by ±10%. What would be the percentage error in the calculated horizontal divergence in the worst case?
(c) Does the answer in (b) suggest why horizontal divergence is not usually a *measured* quantity? What is the reasoning behind your answer?

Solutions

4.1. (e) 16.92 h

4.2. (a) 4.69 (c) 940.25 hPa (d) $V_{ag} = \frac{1}{f}\frac{\partial \phi}{\partial n} - \frac{fR}{2} + (\frac{f^2R^2}{4} - R\frac{\partial \phi}{\partial n})^{0.5}$
(e) 341.41 m s^{-1}

4.5. 25 hours, 21 minutes, 7 seconds

4.11. $PGF_\sigma = -RT\nabla \ln p_s - \nabla\phi$

4.12. (a) 26.19 m s^{-1} (b) 14.81° (c) 323.625 km

4.13. (a) −0.727 K h^{-1} (b) 19.68 m s^{-1}

4.14. 0.347 m s^{-1}

4.15. −3457.16 km

4.17. (a) $2 \times 10^{-5} s^{-1}$ (b) 110%

5

Circulation, Vorticity, and Divergence

Objectives

The atmosphere is characterized by the ubiquitous presence of a variety of swirling fluid eddies. Indeed, this observational fact prompted us in the previous chapter to consider the effects of flow curvature on some simple force balances. In the mid-latitude atmosphere the most important of these many eddies is the large-scale cyclone, also known as the synoptic-scale cyclone. The surface analysis illustrated in Figure 5.1(a) clearly demonstrates that in the northern hemisphere the winds circulate around the center of lowest pressure in a counterclockwise fashion over an enormous geographical area. Conversely, the surface winds circulate clockwise over an equally large area around the center of the northern hemisphere surface high-pressure system depicted in Figure 5.1(b). The ubiquity of large-scale, rotating disturbances in the mid-latitude atmosphere compels us to understand better the nature of fluid rotation and the resulting circulation of these eddies. In this chapter we will investigate the physical quantities known as **circulation** and **vorticity**, both of which serve to quantify the fluid rotation. Along the way we will encounter the intriguing quantity known as **potential vorticity**, the subject of more detailed inquiry in Chapter 9. By formally deriving a vorticity equation, we will find that changes in the vorticity (which, of course, relate to changes in the intensity of the fluid rotation) are inextricably linked to divergence in the fluid. By continuity, the presence of divergence in the fluid implies vertical motions as well. Thus, we will find that understanding the distribution and time tendency of the vorticity is fundamental to *understanding the mid-latitude cyclone and its life cycle.* Finally, given that the mid-latitude atmosphere is in approximate thermal wind balance, we will consider approximations to the vorticity and thermodynamic energy equations appropriate for low-Rossby-number flow. In so doing, we will encounter a system of equations known as the **quasi-geostrophic system**. This important set of relationships will serve as the foundation for inquiry in the second half of the book. We begin by defining the circulation.

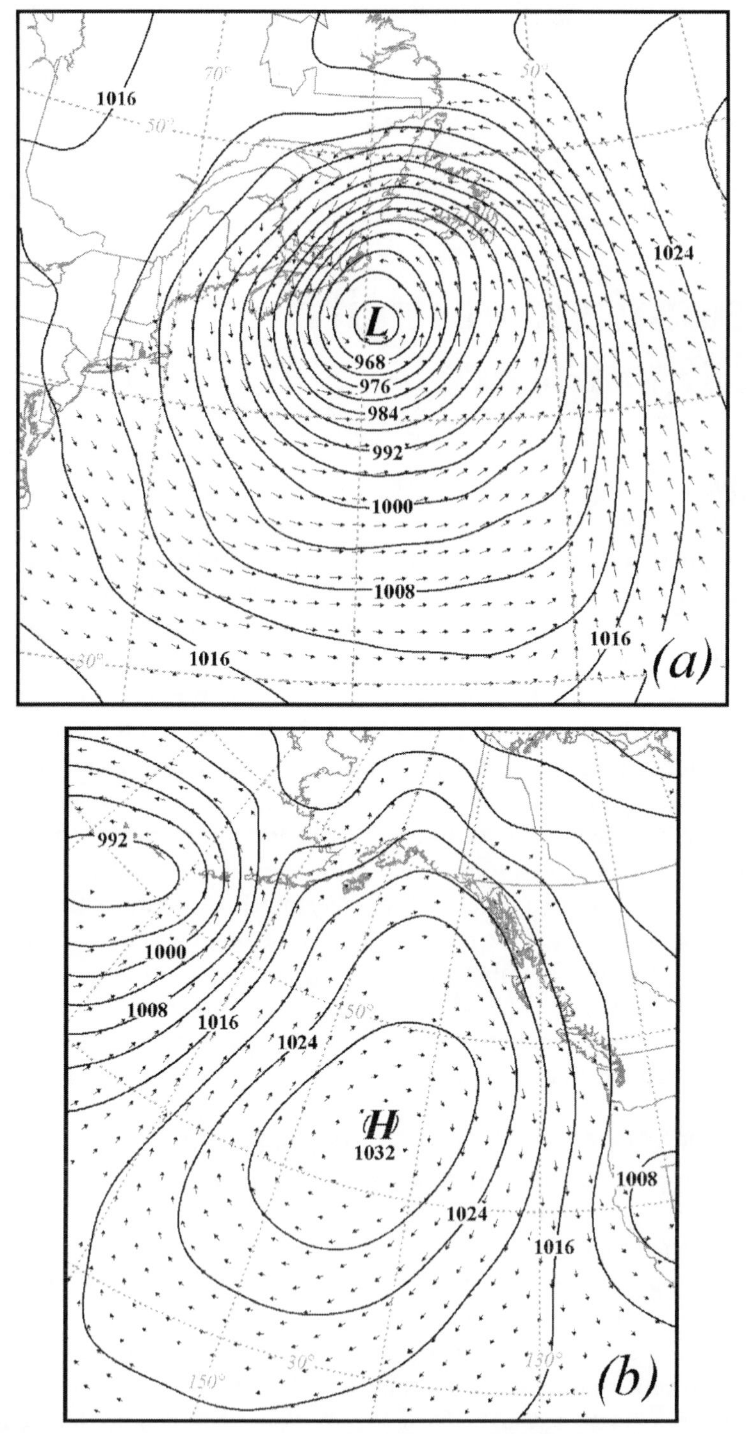

Figure 5.1 (a) Sea-level pressure analysis at 0000 UTC 19 February 2004 from NCEP's Eta model. Solid lines are isobars labeled in hPa and contoured every 4 hPa. Arrows are near surface winds greater than $10\,\mathrm{m\,s^{-1}}$. Note the obvious counterclockwise rotation of the winds about the sea-level pressure minimum. (b) As in (a) but for 0000 UTC 8 July 2004. Arrows are near surface winds greater than $5\,\mathrm{m\,s^{-1}}$. Note the obvious clockwise rotation of the winds about the sea-level pressure maximum

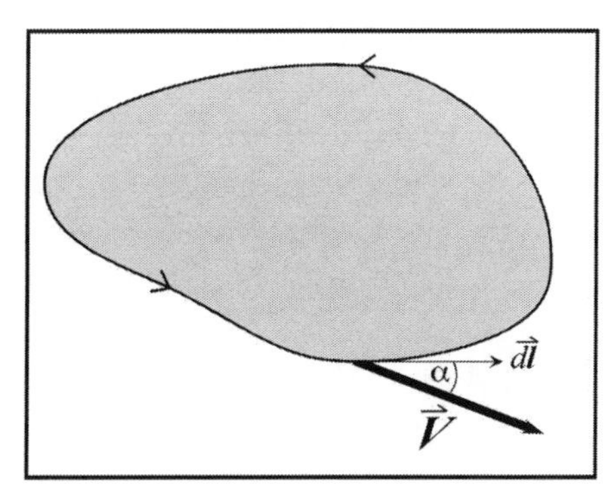

Figure 5.2 Illustration of the means by which circulation around a closed fluid contour is calculated. Symbols are referenced in the text

5.1 The Circulation Theorem and its Physical Interpretation

Imagine a tank of water with a small radius, R, in which the water is initially perfectly still. If some tangential velocity is imposed along the outside edge of the tank (i.e. a torque is applied to the fluid as it is stirred) then the water will clearly acquire some rotation and its angular momentum will increase. Imagine further that a stirring machine is able to move a paddle along the outside edge of the tank at a constant velocity, V. If this stirring paddle is dragged through only one-fourth of the circumference of the tank before being removed a certain amount of circulation of the water results. If the paddle is dragged through half of the circumference, a greater amount of circulation results. A still greater amount of circulation would be observed in the water if the paddle were dragged through the entire circumference of the tank. This simple thought experiment exposes a physical relationship that lies at the heart of the notion of fluid circulation: that is, the circulation imparted to a fluid is related not only to the tangential velocity imparted to the fluid, but also to the distance through which that velocity acts. More formally, we define the circulation, C, around a fluid element with finite area as the line integral, around the fluid element, of the tangential velocity. In formal mathematical terms circulation is expressed as

$$C = \oint \vec{V} \cdot \vec{dl} = \oint |V| \cos\alpha \cdot \vec{dl} \tag{5.1}$$

where $\vec{dl}$ represents the displacement vector along the edge of the fluid element as illustrated with the aid of Figure 5.2. By convention, the integration in (5.1) proceeds in a counterclockwise fashion so that $C > 0$ ($C < 0$) corresponds to cyclonic (anticyclonic) rotation in the northern hemisphere with the situation opposite in the southern hemisphere. Naturally, (5.1) can be rewritten in (x,y) Cartesian coordinates as

$$C = \oint (U dx + V dy) \tag{5.2}$$

where it is understood that U and V represent the tangential velocities around a fluid element in the x and y directions, respectively.

As mentioned previously, it is *changes* in the circulation that we are most interested in when studying the atmosphere since these changes ultimately manifest themselves as intensifications of the high- and low-pressure systems that deliver the sensible weather. Therefore, it is useful to consider the Lagrangian derivative of (5.1) as

$$\frac{dC}{dt} = \frac{d}{dt}\left(\oint \vec{V} \cdot d\vec{l}\right) = \oint \left[\frac{d}{dt}(\vec{V} \cdot d\vec{l})\right]. \tag{5.3a}$$

By the chain rule, $(d/dt)(\vec{V} \cdot d\vec{l})$ can be expressed as

$$\frac{d}{dt}(\vec{V} \cdot d\vec{l}) = \frac{d\vec{V}}{dt} \cdot d\vec{l} + \vec{V} \cdot \frac{d(d\vec{l})}{dt} \tag{5.3b}$$

and since $d\vec{l}$ is a displacement vector, $d(d\vec{l})/dt = d\vec{V}$ so that

$$\frac{d}{dt}(\vec{V} \cdot d\vec{l}) = \frac{d\vec{V}}{dt} \cdot d\vec{l} + \vec{V} \cdot d\vec{V}. \tag{5.3c}$$

Substituting (5.3c) into (5.3a), the rate of change of circulation becomes

$$\frac{dC}{dt} = \oint \frac{d\vec{V}}{dt} \cdot d\vec{l} + \oint \vec{V} \cdot d\vec{V}. \tag{5.4}$$

The second term on the RHS of (5.4) can be written as

$$\oint \vec{V} \cdot d\vec{V} = \frac{1}{2}\oint d(\vec{V}^2) = 0$$

since the integration is around a closed fluid element. Now, if we employ the **absolute acceleration**, $d\vec{V}_a/dt$, in (5.4), then

$$\frac{dC_a}{dt} = \oint \frac{d\vec{V}_a}{dt} \cdot d\vec{l}.$$

We can take advantage of the fact that only the pressure gradient force and the gravitational force influence *absolute* acceleration so that

$$\frac{dC_a}{dt} = \oint \left(-\frac{1}{\rho}\nabla p - \nabla\Phi\right) \cdot d\vec{l} = -\oint \frac{\nabla p}{\rho} \cdot d\vec{l} - \oint \nabla\Phi \cdot d\vec{l} \tag{5.5}$$

where $\nabla\Phi = -g\hat{k}$ on a constant height surface and so represents the gravitational force. Noting that the vertical component of the displacement vector, $d\vec{l}$, is simply dz (i.e. $d\vec{l} \cdot \hat{k} = dz$), then

$$\nabla\Phi \cdot d\vec{l} = -gdz = -d\Phi$$

and

$$-\oint \nabla\Phi \cdot d\vec{l} = \oint d\Phi = 0$$

since $d\Phi$ is a perfect differential and therefore no net work against gravity is done, regardless of the path taken, if a particle ends up where it began. Consequently, (5.5) can be expressed as

$$\frac{dC_a}{dt} = -\oint \frac{\nabla p}{\rho} \cdot \vec{dl} = -\oint \frac{dp}{\rho} \tag{5.6}$$

since $\nabla p \cdot \vec{dl} = dp$. The term on the LHS of (5.6) describes the Lagrangian rate of change of the fluid's rotation so that (5.6) represents the fluid analog of angular acceleration in solid bodies. In the dynamics of solid bodies only torques can produce angular acceleration. Consequently, the term on the RHS of (5.6), known as the **solenoid term**, is the fluid equivalent of a torque. In general, the solenoid term is not zero. In certain environments, however, the density (ρ) is a function only of pressure (i.e. $\rho = \rho(p)$). In such cases, the isobars and isosteres (lines of constant density) are coincident everywhere. This condition is known as **barotropy** and such a fluid is a **barotropic** fluid. In a barotropic fluid, the ideal gas law implies that

$$dp = RTd\rho$$

so that

$$-\oint \frac{dp}{\rho} = \oint \frac{RTd\rho}{\rho} = \oint RTd \ln \rho = 0 \tag{5.7}$$

since, in that case, the solenoid term becomes the closed line integral of the exact differential, $d \ln \rho$. Thus, in a barotropic fluid, the absolute circulation is conserved following the parcel, a result known as **Kelvin's circulation theorem.**

The mid-latitude atmosphere, however, nearly always is best characterized as a **baroclinic** fluid, one in which horizontal density (temperature) contrasts can exist on isobaric surfaces (i.e. density is *not* solely a function of pressure). Thus, most often the isobaric and isopycnal (constant density) surfaces intersect each other as illustrated in Figure 5.3, a schematic vertical cross-section of the pressure/density distribution in the vicinity of a land/sea boundary where the density is higher over the sea than over the land. Since the thickness is lower in the colder column over the sea, even if the pressure is the same near the surface, the isobaric surfaces slope downward toward the sea at higher levels while the isopycnals slope downward toward the warmer land. The intersection of the isobars and isopycnals in Figure 5.3 forms a series of parallelograms which are called solenoids, hence the name for the term under investigation. For the environment depicted in Figure 5.3, we can determine the Lagrangian rate of change of the absolute circulation by evaluating $-dp/\rho$ around the indicated closed path. Movement from A to B occurs on an isobaric surface so there is no contribution to the closed line integral from that portion of the path. Moving from B to C, dp is negative so the contribution to the integral is positive. Another isobaric surface is encountered from C to D with a corresponding lack of contribution to the circulation change. Finally, from D to A

dp is positive so that a negative contribution is made to the circulation change. Since the average density from B to C is less than it is from D to A, and density appears in the denominator of the RHS of (5.6), the positive contribution (B to C) is larger than the negative contribution (D to A) resulting in a net increase in the circulation around the indicated counterclockwise path. Thus, in this example, a circulation will develop in which lighter fluid is made to rise and heavier fluid is made to sink. The effect of this circulation will be to tilt the isopycnals into an orientation in which they are more nearly parallel with the isobars – that is, toward the barotropic state in which subsequent circulation change would be zero. Such a circulation also lowers the center of mass of the entire fluid system and thus reduces the potential energy of that system. The motions that actually accomplish this reduction of potential energy represent a certain amount of kinetic energy. Thus, the resulting circulation is one which converts the potential energy present in the horizontal density contrast into the kinetic energy of the fluid motions involved in the redistribution of mass. It is important to note that the circulation theorem describes only the Lagrangian rate of change of the circulation, not the circulation itself. Thus, our result in Figure 5.3 conveys only that the solenoid term contributes to an increase in the circulation. If a background circulation were already present in the environment portrayed in Figure 5.3, then the solenoid contribution would have to be added to that circulation in order to determine the net circulation in that environment. Any circulation that moves in the same (opposite) direction, as it compelled to do by the solenoid term, is known as a direct (indirect) solenoidal circulation.

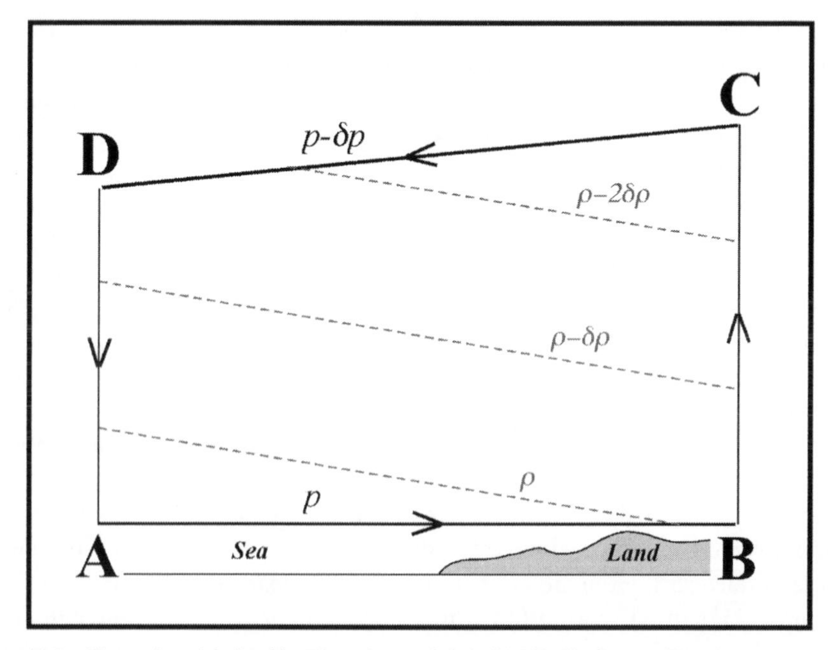

Figure 5.3 Illustration of the land/sea boundary and the solenoid of isobars and isosteres in its vicinity. Discussion of the circulation tendency around the closed loop ABCD is given in the text

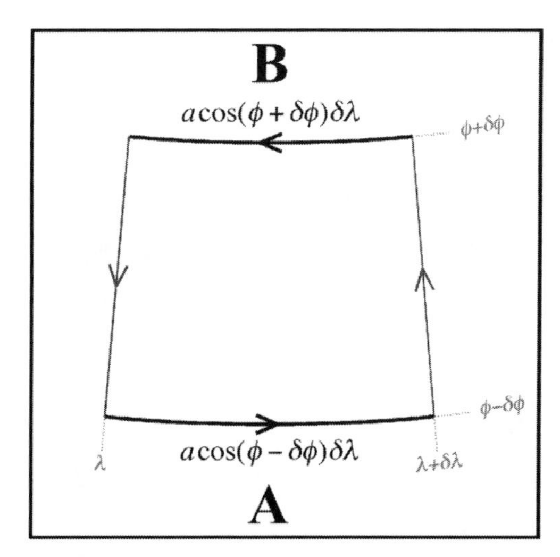

Figure 5.4 Latitude–longitude box around which the Earth's circulation is calculated in the text. The lengths of sides A and B are indicated. The rotation of the Earth is from λ to $\lambda + \delta\lambda$

Thus far we have only discussed the nature of the absolute circulation. Naturally, for the purposes of understanding fluids on Earth, it is much more relevant to consider the *relative* circulation. In order to do so, we must first calculate the circulation that results from the rotation of the Earth and then subtract it from the RHS of (5.6). The velocity around a latitude circle on the spherical Earth is equal to

$$\vec{V} = \vec{\Omega} R$$

where $R = a\cos\phi$. Thus, the westerly motion resulting from rotation of the Earth is

$$U = \Omega a \cos\phi. \tag{5.8}$$

Now we calculate the circulation around the latitude–longitude box shown in Figure 5.4. In spherical coordinates, a length element in the zonal (east–west) direction is given by $dx = a\cos\phi\delta\lambda$ where λ is longitude. Since the rotation of the Earth contributes no meridional (north–south) motion to the box, only sides A and B figure into the calculation of the circulation. Thus, the circulation resulting from the Earth's rotation can be written as

$$\begin{aligned} C_e = \oint U \cdot d\vec{l} &= U_A(dx)_A - U_B(dx)_B \\ &= [\Omega a \cos(\phi - \delta\phi)][a\cos(\phi - \delta\phi)]\delta\lambda \\ &\quad - [\Omega a \cos(\phi + \delta\phi)][a\cos(\phi + \delta\phi)]\delta\lambda \\ &= \Omega a^2[\cos(\phi - \delta\phi)]^2\delta\lambda - \Omega a^2[\cos(\phi + \delta\phi)]^2\delta\lambda. \end{aligned} \tag{5.9}$$

Using the trigonometric identities

$$\begin{aligned} \cos(a - b) &= \cos a \cos b + \sin a \sin b \text{ and } \cos(a + b) \\ &= \cos a \cos b - \sin a \sin b \end{aligned}$$

(5.9) can be rewritten as

$$C_e = \Omega a^2[(\cos\phi\cos\delta\phi + \sin\phi\sin\delta\phi)^2]\delta\lambda - \Omega a^2[(\cos\phi\cos\delta\phi - \sin\phi\sin\delta\phi)^2]\delta\lambda. \quad (5.10)$$

Carrying out the quadratic operations and adding like terms results in

$$C_e = 4\Omega a^2\cos\phi\sin\phi\cos\delta\phi\sin\delta\phi\delta\lambda. \quad (5.11)$$

Now, the area of the box in Figure 5.4 is given by

$$A = (a\cos\phi\delta\lambda) \times (2a\delta\phi) = 2a^2\cos\phi\delta\phi\delta\lambda. \quad (5.12a)$$

If we consider an infinitesimal box in which $\delta\phi$ and $\delta\lambda$ approach zero, then

$$\lim_{\delta\phi\to 0}\cos\delta\phi = 1 \text{and} \lim_{\delta\phi\to 0}\sin\delta\phi = \delta\phi$$

so that (5.12a) can be expressed as

$$A = 2a^2\cos\phi\sin\delta\phi\delta\lambda. \quad (5.12b)$$

Combining (5.11) and (5.12b) we see that the circulation resulting from the Earth's rotation, C_e, is given by

$$C_e = 2\Omega\sin\phi \times A \quad (5.13)$$

so that

$$\frac{dC_e}{dt} = 2\Omega\sin\phi\frac{dA}{dt}. \quad (5.14)$$

Combining (5.14) with (5.6) we arrive at an expression for the Lagrangian rate of change of the relative circulation, C_{rel}:

$$\frac{dC_{rel}}{dt} = \frac{dC_a}{dt} - \frac{dC_e}{dt} = -\oint\frac{dp}{\rho} - 2\Omega\sin\phi\frac{dA}{dt}. \quad (5.15)$$

This expression is known as the **Bjerknes circulation theorem** and it can be applied to realistic flows quite readily.

5.2 Vorticity and Potential Vorticity

Circulation is an important characteristic of fluids as demonstrated in the prior section. Unfortunately, it is not particularly amenable to simple measurement as one needs to sum tangential velocities around the outer edge of a discrete collection of fluid elements in order to arrive at this macroscopic measure of the rotation of the fluid. In Section 1.3 we considered the vorticity as a kinematic property of fluids. As it turns out, the vorticity physically represents a microscopic measure of the rotation in a fluid and, as you may recall from that discussion, is particularly easy to formulate. In this section we explore the relationship between vorticity and circulation in an attempt to interrogate further the nature of rotation in fluids.

The vorticity is a vector quantity defined as the curl (cross-product) of the velocity vector. The absolute vorticity, therefore, is given by $\vec{\upsilon}_a = \nabla \times \vec{V}_a$ while the relative vorticity is given by

$$\vec{\upsilon} = \nabla \times \vec{V} = \left(\frac{\partial w}{\partial y} - \frac{\partial v}{\partial z}\right)\hat{i} + \left(\frac{\partial u}{\partial z} - \frac{\partial w}{\partial x}\right)\hat{j} + \left(\frac{\partial v}{\partial x} - \frac{\partial u}{\partial y}\right)\hat{k} \quad (5.16)$$

in Cartesian coordinates. A large fraction of the rotating fluid systems with which we are interested exhibit rotation in the horizontal plane (i.e. mid-latitude cyclones, hurricanes, tornadoes). Consequently, dynamic meteorology is most often, though not exclusively, interested in the vertical component of the absolute and relative vorticities. These are generally expressed as

$$\eta = \hat{k} \cdot \vec{\upsilon}_a = \hat{k} \cdot \nabla \times \vec{V}_a \quad (5.17a)$$

and

$$\zeta = \hat{k} \cdot \vec{\upsilon} = \hat{k} \cdot \nabla \times \vec{V} = \frac{\partial v}{\partial x} - \frac{\partial u}{\partial y} \quad (5.17b)$$

for the absolute and relative vorticities, respectively. Clearly, the relative vorticity is a much simpler quantity to measure than the circulation as it involves just derivatives of the observed horizontal wind field. But is there a physical relationship between circulation and vorticity?

Consider the tiny fluid element depicted in Figure 5.5. The velocity on side A is given by u while the velocity on side D is given by v. By expanding u and v in a Taylor series, we can get expressions for the velocities on sides C and B as well. With expressions for the velocities on each side of the fluid element, we are in a position to calculate the circulation about that element. Since, by convention, we integrate

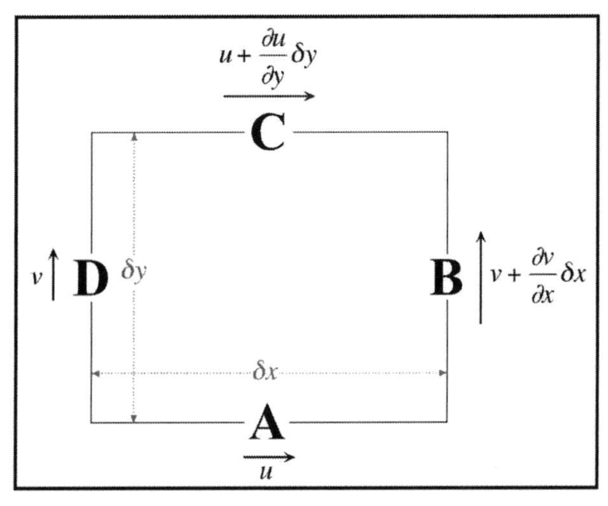

Figure 5.5 Schematic illustrating the calculation of vorticity around an infinitesimal fluid element with Area $= \delta x \delta y$. See text for explanation

around the element in a counterclockwise fashion we find that the circulation is given by

$$C = \oint u dx + v dy = (u)\delta x + \left(v + \frac{\partial v}{\partial x}\delta x\right)\delta y - \left(u + \frac{\partial u}{\partial y}\delta y\right)\delta x - (v)\delta y$$
$$= \left(\frac{\partial v}{\partial x} - \frac{\partial u}{\partial y}\right)\delta x \delta y. \qquad (5.18)$$

Since the area of the fluid element is $\delta x \delta y$, we find that, in the limit as $\delta x \delta y \to 0$, the relative vorticity is simply the relative circulation divided by the area of the fluid element. Therefore, recalling Figure 5.4 and (5.13), the *Earth's* vorticity is given by

$$Vorticity_{Earth} = \frac{C_e}{A} = \frac{2\Omega \sin\phi(A)}{A} = 2\Omega \sin\phi = f \qquad (5.19)$$

so that the vertical component of the absolute vorticity (the sum of the relative vorticity and the Earth's vorticity) is given by

$$\eta = \left(\frac{\partial v}{\partial x} - \frac{\partial u}{\partial y}\right) + f. \qquad (5.20)$$

We can further examine the physical relationship between circulation and vorticity with the aid of Figure 5.6 which portrays a closed fluid element of finite area. If we subdivide the area into a large number of tiny squares (like squares A and B), and then calculate the circulation around each tiny square, an interesting result arises. The side common to A and B contributes the same incremental circulation to both squares but with opposite sign since the integration must proceed in opposite directions. Thus, only the circulation about the outer edges of the two adjoining squares need be considered when summing the total circulation of the squares. If a larger number of even smaller squares are drawn within the closed fluid element in Figure 5.6, cancellation of the shared segments among those squares will still occur. Therefore, only the sides of the squares that run along the outside contour of the fluid element will contribute to the circulation. If we shrink the area of the squares to zero

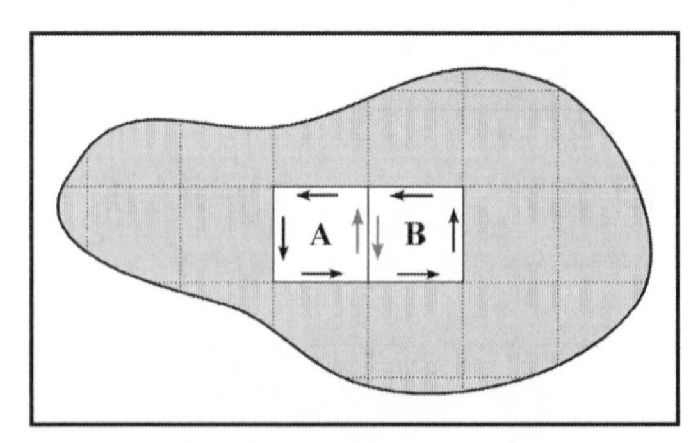

Figure 5.6 Circulation around the shaded fluid element can be represented by summing the circulations around a series of squares such as A and B. Note that the circulation around the combined rectangle AB is comprised of only the contributions from the periphery of AB. The gray arrows, on the side common to both A and B, contribute oppositely to the circulation and therefore cancel

(i.e. $\delta x \delta y \rightarrow 0$), then we can exactly fit any closed fluid element with a collection of such tiny squares. In that case, the total circulation around the fluid element is the sum of the circulations of each infinitesimally tiny square (point vortices, measured by the vorticity) added up over the area of the fluid element. This relationship can be represented by the 2-D form of **Stokes' Theorem**,

$$\oint (udx + vdy) = \iint_{Area} \left(\frac{\partial v}{\partial x} - \frac{\partial u}{\partial y} \right) \partial x \partial y \tag{5.21}$$

where the LHS is the circulation around the fluid element and the RHS is the integral of vorticity over the area enclosed by the curve. A slightly more general way to express this equality is

$$\oint \vec{V} \cdot d\vec{l} = \iint_{A} (\nabla \times \vec{V}) \cdot \hat{n} dA \tag{5.22}$$

where $\hat{n}$ signifies that the summation involves only the component of the vorticity normal to the fluid element.

Though a moving fluid may conform to a nearly infinite set of possible flow configurations, only two broad types are actually associated with vorticity. Defense of this statement is difficult to mount using the Cartesian expression for vorticity. However, consideration of the vertical component of vorticity in natural coordinates offers compelling proof of this simplifying assertion while lending additional insight into the nature of vorticity. Consider the flow parallel to the streamlines in Figure 5.7. In order to develop an expression for the vertical component of vorticity, we need to calculate the circulation around the indicated box and divide by its area. As we have seen before, only the sides of the box aligned along streamlines will contribute

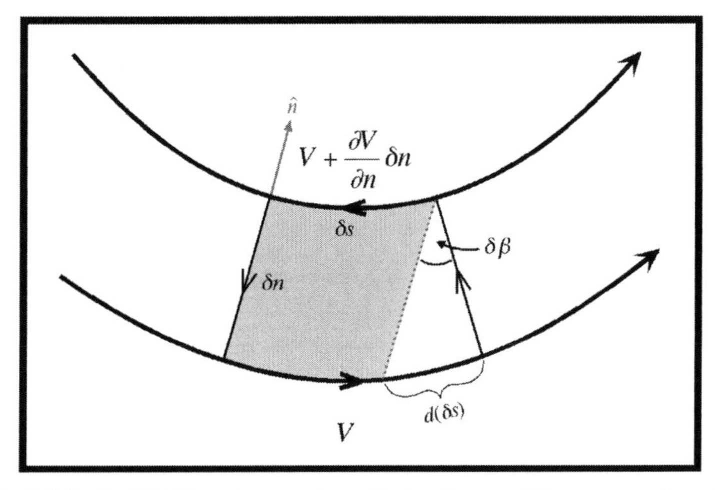

Figure 5.7 Infinitesimal fluid loop in natural coordinates. Heavy solid arrows are streamlines of the flow. The shaded parallelogram has Area $= \delta s \delta n$. The angular change of the flow direction over a distance δs, is given by $\delta\beta$

to the circulation. If the velocity of the flow on the bottom edge of the box is V, then a Taylor series expansion of V across the streamline channel yields $V + (\partial V/\partial n)\delta n$ for the velocity on the top edge of the box. Note that the velocities on the top and bottom edge of the box are in the same direction. The length element on the bottom half of the box is given by $\delta s + d(\delta s)$ where $d(\delta s)$ is the change in δs that results from the fact that the flow in Figure 5.7 is curved. This change in the length element is directly related to the change in direction of the flow as $d(\delta s) = \delta\beta\delta n$ so that the length element on the bottom of the box is given by $\delta s + \delta\beta\delta n$. The contribution to circulation from the bottom half of the box is, therefore, $V(\delta s + \delta\beta\delta n)$. The length element on the top side of the box is δs so the contribution to the circulation on that side of the box is

$$-\left(V + \frac{\partial V}{\partial n}\delta n\right)\delta s$$

where the negative sign is compelled by the fact that the integration must proceed in the counterclockwise direction. Summing these two contributions to the circulation we obtain

$$\begin{aligned} C = \oint \vec{V}\cdot d\vec{l} &= V(\delta s + \delta\beta\delta n) - \left(V + \frac{\partial V}{\partial n}\delta n\right)\delta s \\ &= V\delta s + V\delta\beta\delta n - V\delta s - \frac{\partial V}{\partial n}\delta n\delta s \\ &= V\delta\beta\delta n - \frac{\partial V}{\partial n}\delta n\delta s. \end{aligned} \tag{5.23}$$

Given that vorticity is the circulation divided by the area, as the area tends to zero we have

$$\zeta = \lim_{\delta n\delta s\to 0}\left(\frac{C}{\delta n\delta s}\right) = \frac{V\delta\beta\delta n}{\delta n\delta s} - \frac{\partial V}{\partial n}\frac{\delta n\delta s}{\delta n\delta s} = V\frac{\partial\beta}{\partial s} - \frac{\partial V}{\partial n}. \tag{5.24}$$

Recall from our prior investigation of the relationship between trajectories and streamlines that the angular change in the streamflow as measured along the flow is proportional to the radius of curvature of the streamlines themselves (i.e. $\partial\beta/\partial s = 1/R_s$). Therefore, we can rewrite (5.24) as

$$\zeta = \frac{V}{R_s} - \frac{\partial V}{\partial n} \tag{5.25}$$

which suggests that the vertical vorticity is the sum of two components: (1) the variation of the flow speed normal to the direction of the flow, $-\partial V/\partial n$, known as **shear vorticity**, and (2) variation of the flow direction along a streamline, V/R_s, known as **curvature vorticity**. The mid-latitude atmosphere is full of examples of both types of vorticity as shown in Figure 5.8. A straight jet streak (Figure 5.8a) is the

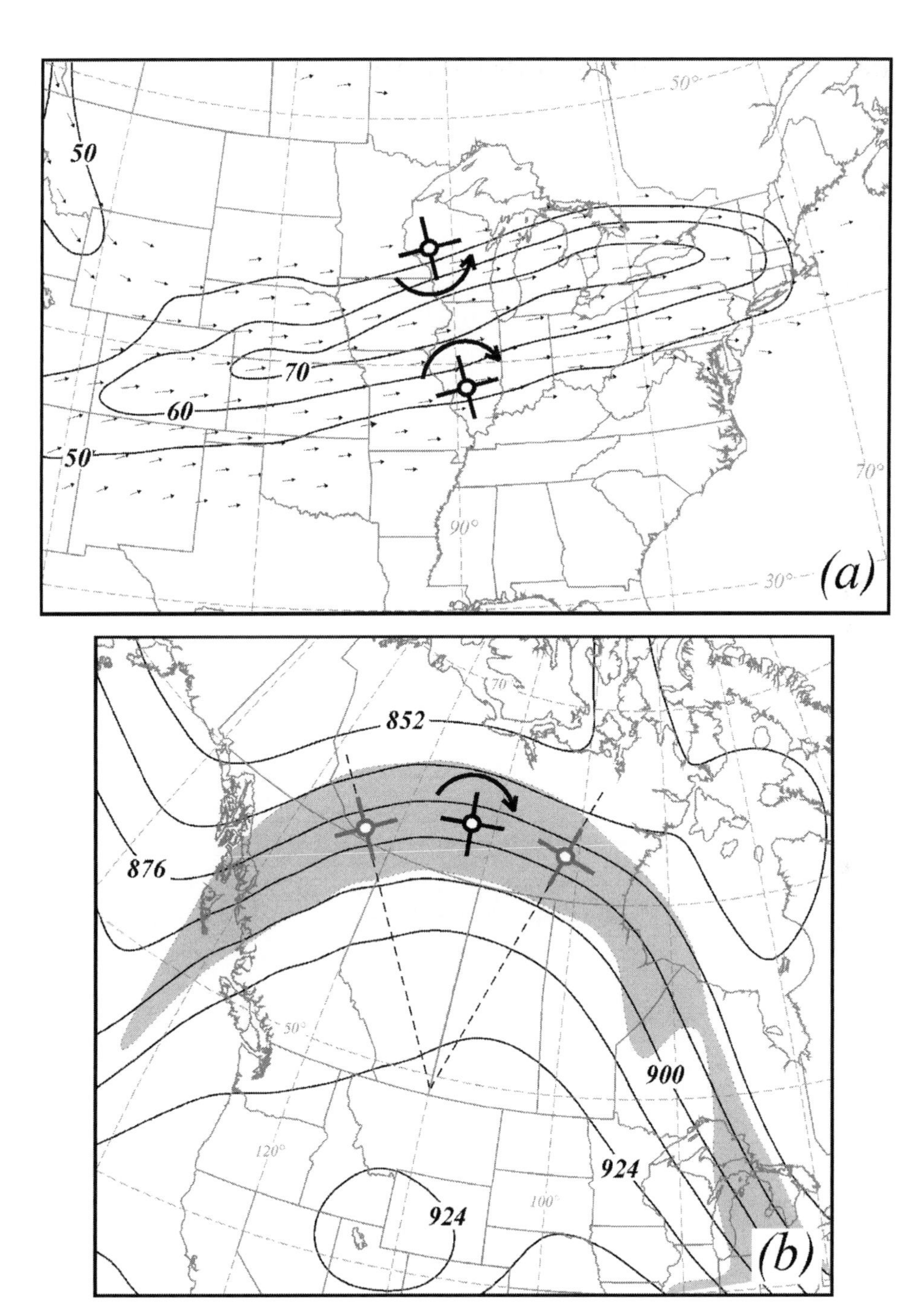

Figure 5.8 (a) The 300 hPa isotachs (solid lines) and wind vectors associated with a straight jet at 0000 UTC 12 November 2003 from NCEP's Eta model analysis. Isotachs are labeled in m s^{-1} and contoured every 10 m s^{-1} starting at 50 m s^{-1}. Olny wind vectors greater than 40 m s^{-1} are shown. Paddlewheels indicate the sense of the vorticity on the north and south sides of the jet. (b) The 300 hPa geopotential heights (solid lines) and 50 m s^{-1} isotach (shaded) associated with a region of anticyclonic curvature at 0000 UTC 14 November 2003 from NCEP's Eta model analysis. Geopotential heights are labeled in dam and contoured every 12 dam. Gray paddlewheels indicate the turning of the wind direction associated with curvature in the flow resulting in anticyclonic vorticity (indicated by the dark paddlewheel)

canonical example of a feature characterized by shear vorticity. It is easy to imagine that a fluid element on the north side of the jet flow in Figure 5.8(a) (represented by the paddlewheel) will be compelled to spin counterclockwise as the flow speed increases to the south of the element. Conversely, a fluid element placed on the south side of that jet will be compelled to flow clockwise as the flow to the north increases in speed. Flow of nearly constant speed through a region of curvature is illustrated in Figure 5.8(b). As parcels progress through the ridge, the upper end of the paddlewheel is forced to traverse a greater distance than the lower end (just as a runner on the outside lane of a track runs a greater distance around the bend in the track). Thus, the paddlewheel is forced to spin clockwise consistent with negative curvature vorticity.

If, for a moment, we consider flow on isentropic surfaces, a rather remarkable relationship between the rotation of a fluid column and its depth presents itself. Recall that the Poisson equation defines potential temperature as $\theta = T(p_0/p)^{R/c_p}$ where $p_0 = 1000$ hPa. If we substitute for T from the ideal gas law we obtain

$$\theta = \frac{p}{\rho R}\left(\frac{p_0}{p}\right)^{R/c_p} \quad \text{or} \quad \rho R\theta = p_0^{(R/c_p)}\, p^{1-(R/c_p)}.$$

Since $c_p - R = c_v$, the foregoing expression can be written as

$$\rho = \frac{p_0^{(R/c_p)}\, p^{(c_v/c_p)}}{R\theta} \tag{5.26}$$

showing that for flow on isentropic surfaces (where θ is constant), density is a function of pressure only, so the flow is barotropic. Therefore, on isentropic surfaces there are no solenoids so the circulation theorem (5.15) becomes

$$\frac{dC}{dt} = -2\Omega \sin\phi \frac{dA}{dt} \quad \text{or} \quad \frac{d}{dt}(C + 2\Omega \sin\phi\, A) = 0. \tag{5.27}$$

Since $\zeta = C/A$, then (5.27) can be written as

$$\frac{d}{dt}[(\zeta_\theta + f)A] = 0. \tag{5.28}$$

Thus, the product $(\zeta_\theta + f)A$ is constant in adiabatic flow on isentropic surfaces, where ζ_θ is the relative vorticity measured on an isentropic surface. Now, recall that the amount of mass in an air column contained between two θ surfaces is directly related to the isobaric depth of the column, $-\delta p$, since $-\delta p = F/A = (\delta M)g/A$. Mass continuity demands that this amount of mass be conserved. Consequently, we can express the cross-sectional area, A, of the column as

$$A = -g\frac{\delta M}{\delta p}.$$

But this is not amenable to measurement since it is not practical to have to measure the mass in the column. We can, however, rewrite this expression equivalently as

$$A = -g\left(\frac{\delta M}{\delta\theta}\frac{\delta\theta}{\delta p}\right).$$

Since both δM and $\delta\theta$ are conserved in adiabatic flow, then their ratio is constant so that

$$A = -g\left(\frac{\delta\theta}{\delta p}\right) \times \mathit{Constant}.$$

Taking the limit as $\delta\theta, \delta p \to 0$ and using (5.28) we obtain

$$(\zeta_\theta + f)\left(-g\frac{\partial\theta}{\partial p}\right) = \mathit{Constant}. \tag{5.29}$$

The above quantity is known as **potential vorticity** and it is a conserved quantity for parcels in adiabatic flow. Just as **potential temperature** refers to the fact that the temperature of a parcel of air can be changed upon adiabatic expansion or compression, the potential vorticity refers to the fact that a parcel's vorticity can be changed by (1) changing latitude (f), and/or (2) changing the static stability ($-\partial\theta/\partial p$). In a so-called homogeneous fluid (one in which density is constant), such as a shallow tank of water, the expression for potential vorticity is even simpler. In such a fluid, there are no pressure–density solenoids and the cross-sectional area of a column of the fluid is given by

$$A = \frac{\delta Mg}{p} = \frac{\delta Mg}{\rho g\delta z} = \frac{\mathit{Constant}}{\delta z}$$

since the density is constant and hydrostatic balance applies. Thus, for a shallow body of water (5.28) suggests that

$$\frac{(\zeta + f)}{\delta z} = \mathit{Constant} = PV_{\mathit{shallow\ water}} \tag{5.30}$$

where ζ is the relative vorticity measured on a surface of constant geometric height. The above relation provides clear physical insight into the nature of this potential vorticity (PV): it is the ratio of the absolute vorticity to the depth of the vortex. If the depth of a rotating column is increased (decreased) by vertical stretching (compression), then the absolute vorticity of the column must also increase (decrease) in order that PV be conserved. Considering (5.29) and a column of air confined between two isentropes, we see that this is also the case for the isentropic flow wherein the depth of the vortex is measured in terms of the static stability, the isobaric separation between adjacent isentropes. Though it seems physically rather curious that the ratio of the vortex strength to vortex depth ought to be a conservative property of fluids, the conservation of PV provides powerful constraints on the behavior of our fluid atmosphere. Consider, for example, a column of air moved adiabatically eastward across the crest of the Rocky Mountains of western North America as illustrated in

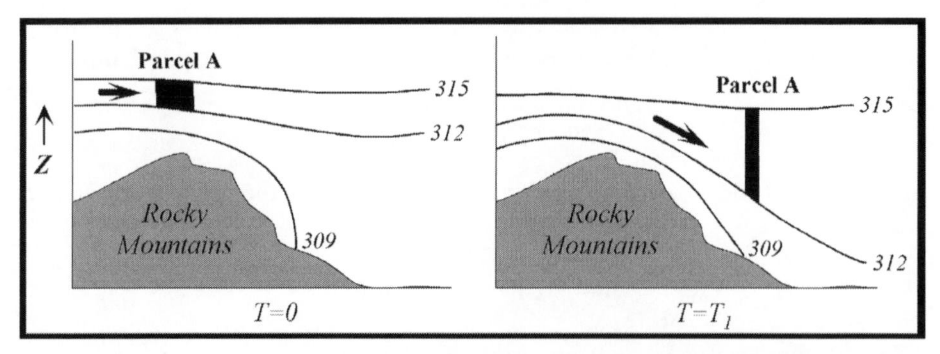

Figure 5.9 Schematic illustration of a fluid column crossing the Rocky Mountains of North America. Parcel A, confined between the 312 and 315 isentropes, is carried across the crest of the mountains by the flow, indicated by the bold arrow, at time $T = 0$. Upon crossing the ridge, the flow subsides and forces a separation between the bounding isentropes ($T = T_1$). Subsidence warming and vorticity increases in the lower troposphere result in the lee of the barrier

Figure 5.9. As the column heads eastward over the Great Plains, the isobaric depth of the column increases. The consequent decrease in static stability requires an attendant increase in the absolute vorticity in order that PV be conserved. The increase in absolute vorticity is manifest in the characteristic leeside trough that forms in the lee of the Rockies under episodes of strong cross-mountain flow (Figure 5.10a). A vertical cross-section through a Rocky Mountain leeside trough (Figure 5.10b) clearly demonstrates that the static stability is greatly reduced in the lee as a consequence of column stretching. Characteristically, the leeside trough axis is coincident with the axis of warmest air near the surface, a direct result of the adiabatic warming produced by the subsidence manifest in the vertical separation of isentropes. We will further investigate both PV and cyclogenesis in Chapter 9.

5.3 The Relationship between Vorticity and Divergence

As is the case with most quantities in meteorology, we are interested in the **time tendency** of the vorticity. Such a relationship is obtained by manipulating the equations of motion. Though the effects of friction could be included, for simplicity we will begin with the frictionless equations of motion in height coordinates:

$$\frac{\partial u}{\partial t} + u\frac{\partial u}{\partial x} + v\frac{\partial u}{\partial y} + w\frac{\partial u}{\partial z} - fv = -\frac{1}{\rho}\frac{\partial p}{\partial x} \tag{5.31a}$$

and

$$\frac{\partial v}{\partial t} + u\frac{\partial v}{\partial x} + v\frac{\partial v}{\partial y} + w\frac{\partial v}{\partial z} + fu = -\frac{1}{\rho}\frac{\partial p}{\partial y}. \tag{5.31b}$$

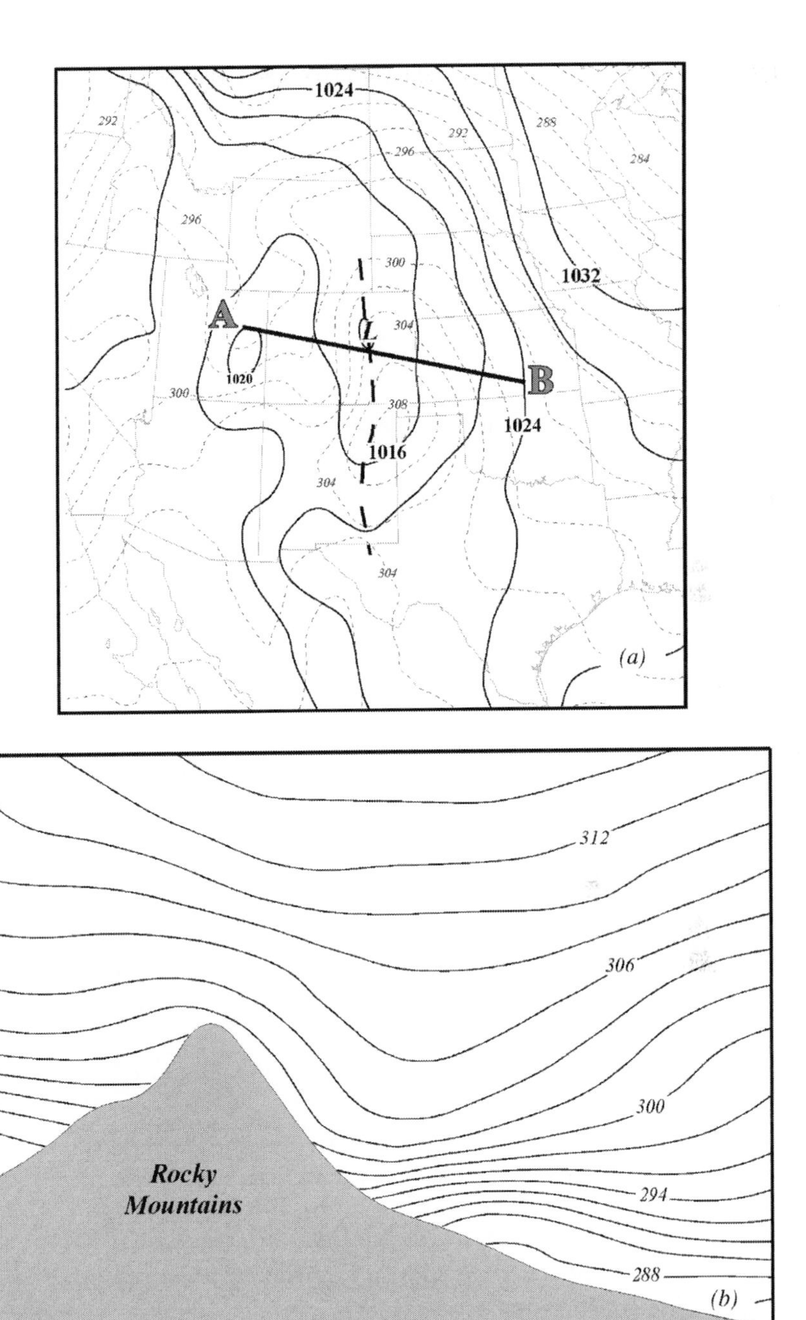

Figure 5.10 (a) Sea-level isobars and 750 hPa potential temperature in the central United States at 0000 UTC 27 December 2004. Thick solid lines are isobars labeled in hPa and contoured every 4 hPa. Dashed lines are isentropes labeled in K and contoured every 2 K. Thick dashed line indicates the axis of the leeside pressure trough. Vertical cross-section along A–B is shown in (b). (b) Vertical cross-section (along line A–B in (a)) of potential temperature. Solid lines are isentropes labeled in K and contoured every 2 K. Note the region of weak stratification in the immediate lee of the Rocky Mountains

Next, we take $\partial/\partial y$ of (5.31a) and subtract that from $\partial/\partial x$ of (5.31b). Using the definition of the vertical component of the relative vorticity,

$$\zeta = \left(\frac{\partial v}{\partial x} - \frac{\partial u}{\partial y}\right),$$

the result becomes

$$\frac{\partial \zeta}{\partial t} + u\frac{\partial \zeta}{\partial x} + v\frac{\partial \zeta}{\partial y} + w\frac{\partial \zeta}{\partial z} + (\zeta + f)\left(\frac{\partial u}{\partial x} + \frac{\partial v}{\partial y}\right)$$
$$+ \left(\frac{\partial w}{\partial x}\frac{\partial v}{\partial z} - \frac{\partial w}{\partial y}\frac{\partial u}{\partial z}\right) + u\frac{df}{dx} + v\frac{\partial f}{\partial y} = \frac{1}{\rho^2}\left(\frac{\partial \rho}{\partial x}\frac{\partial p}{\partial y} - \frac{\partial \rho}{\partial y}\frac{\partial p}{\partial x}\right). \tag{5.32}$$

Now, since

$$\frac{df}{dt} = \frac{\partial f}{\partial t} + u\frac{\partial f}{\partial x} + v\frac{\partial f}{\partial y} + w\frac{\partial f}{\partial z}$$

and f varies only with y, then $df/dt = v\,\partial f/\partial y$ and (5.32) can be rewritten as

$$\frac{d(\zeta + f)}{dt} = -(\zeta + f)\left(\frac{\partial u}{\partial x} + \frac{\partial v}{\partial y}\right) - \left(\frac{\partial w}{\partial x}\frac{\partial v}{\partial z} - \frac{\partial w}{\partial y}\frac{\partial u}{\partial z}\right)$$
$$+ \frac{1}{\rho^2}\left(\frac{\partial \rho}{\partial x}\frac{\partial p}{\partial y} - \frac{\partial \rho}{\partial y}\frac{\partial p}{\partial x}\right) \tag{5.33}$$

which is the **vorticity equation** in height coordinates. This relationship states that the rate of change of the absolute vorticity is given by the sum of the three terms on the RHS of (5.33): (1) the divergence term, (2) the tilting term, and (3) the solenoid term. We now investigate each of these terms independent of the others, beginning with the divergence term.

If only horizontal divergence acts in the fluid, then the vorticity equation becomes

$$\frac{d(\zeta + f)}{dt} = -(\zeta + f)\left(\frac{\partial u}{\partial x} + \frac{\partial v}{\partial y}\right).$$

As illustrated in Figure 5.11(a), when divergence occurs in a fluid (i.e. $(\partial u/\partial x + \partial v/\partial y) > 0$) then the area enclosed by a fluid ring increases with time. Consequently, the absolute vorticity becomes more anticyclonic with time, provided that it was originally cyclonic (as is almost always the case on the synoptic scale). Physically, this result is related to the fact that the circulation of the fluid element must be conserved (in the absence of solenoids). Since vorticity is the ratio of circulation to area, if the area increases through horizontal divergence, then the vorticity must decrease.

The converse is true for a fluid characterized by convergence as illustrated in Figure 5.11(b). Horizontal convergence (i.e. $(\partial u/\partial x + \partial v/\partial y) < 0$) will produce a cyclonic tendency in the vorticity consistent with the shrinking of the area enclosed by the fluid element and the consequent increase in the ratio of circulation to area. The effect of the divergence term is the fluid analog of a figure skater whose rate

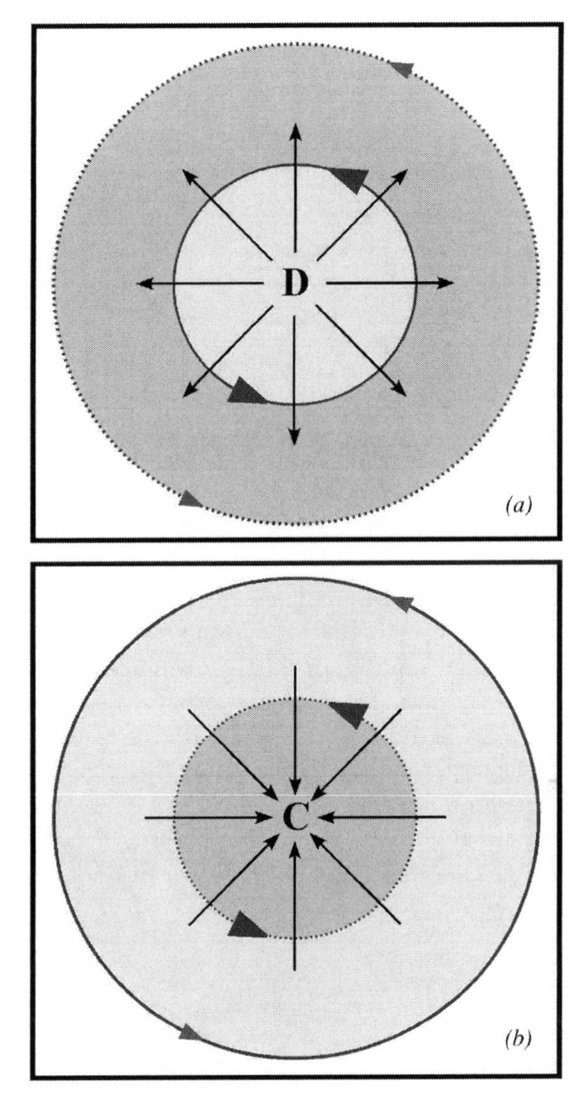

Figure 5.11 (a) Illustration of the effect of divergence (D) on vorticity. Original fluid ring is lightly shaded and bordered by the large circulation arrows. Fluid ring at some later time is darkly shaded and bordered by smaller circulation arrows. (b) As for (a) but for conditions of horizontal convergence (C)

of rotation increases (decreases) when she pulls (extends) her arms to (from) her side, decreasing (increasing) the radius of rotation. Since no torques are applied to the skater, the smaller (larger) radius of rotation requires a larger (smaller) angular velocity. So, in the end, we can say that

Convergence spins up cyclonic vorticity.
Divergence spins up anticyclonic vorticity.

There are profound implications to these statements with regard to mid-latitude weather systems. Surface low-pressure centers are characterized by convergence and thus tend to be foci for the production of low-level cyclonic vorticity. Just the opposite is true for surface anticyclones.

We can investigate the nature of the tilting term by imagining that no divergence or solenoids exist. In such a hypothetical case, (5.33) becomes

$$\frac{d(\zeta + f)}{dt} = -\left(\frac{\partial w}{\partial x}\frac{\partial v}{\partial z} - \frac{\partial w}{\partial y}\frac{\partial u}{\partial z}\right) = \left(\frac{\partial w}{\partial y}\frac{\partial u}{\partial z} - \frac{\partial w}{\partial x}\frac{\partial v}{\partial z}\right).$$

Recall that $(\zeta + f)$ describes only the vertical component of the absolute vorticity vector. Thus, there are rotations about the other two Cartesian axes. For instance, consider the effect of the westerly vertical wind shear that characterizes the middle latitudes on Earth. As illustrated in Figure 5.12, such vertical shear compels the paddlewheel to spin counterclockwise as viewed from the positive y-axis. Every parcel of air aligned along the y-axis will be compelled to spin in the same direction so that a *tube* of air aligned along the y-axis will possess this positive y-component vorticity. Now, if there is a gradient of vertical motion in the y direction, such that $\partial w/\partial y > 0$, then the northern end of the rotating tube will ascend while the southern end will subside. Consequently, a component of the counterclockwise rotation of the tube will be projected onto the z-axis. In other words, the vertical component of the vorticity, originally zero, will gradually acquire positive values through the action of tilting the horizontal vortex tube into the vertical. If you mimic the rotation and vertical motion distribution illustrated in Figure 5.12 by spinning your pencil and lifting and depressing the appropriate ends, you can easily demonstrate the effect of the tilting term for yourself. Mathematically, it is clear that both $\partial u/\partial z$ and $\partial w/\partial y$

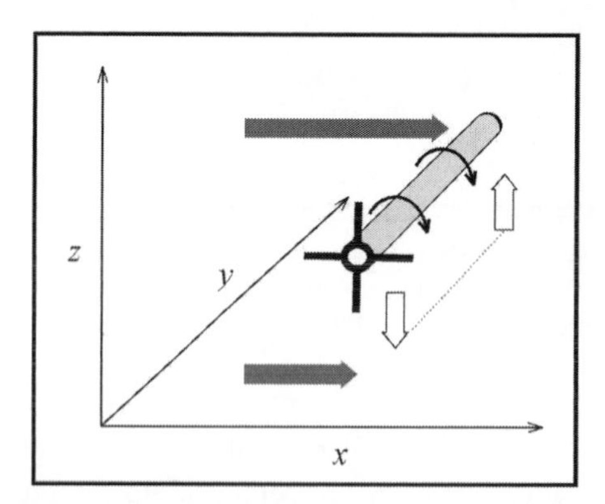

Figure 5.12 Illustration of the effect of vertical shear on horizontal vorticity. Gray shaded arrows represent the westerly winds at two different levels. The paddlewheel is compelled to turn in the indicated direction as is the light gray shaded tube of air aligned along the y-axis. Open arrows represent differential upward and downward vertical motions along the y-axis. The combined effect of the vertical gradient of u and the horizontal gradient of w on the vertical vorticity (via the tilting term) is explained in the text

are positive in Figure 5.12; thus the tilting term is positive as well, leading to an increase in the vertical component of the absolute vorticity.

Finally, it may be unsurprising, given the physical connection between circulation and vorticity, that a solenoid term arises out of the vorticity equation. Solenoids act to rearrange the mass in a fluid into the lowest potential energy state. As it turns out, the solenoid term in (5.33) is the microscopic equivalent of the solenoid term in the circulation theorem. This is easily proven by applying Stokes' Theorem to the solenoid term in the expression for circulation,

$$-\oint \frac{dp}{\rho} = -\oint \alpha dp = -\oint \alpha \nabla p \cdot d\vec{l} = -\iint_A \nabla \times (\alpha \nabla p) \cdot \hat{k}\, dA. \tag{5.34a}$$

Since $\nabla \times (\alpha \nabla p) = \nabla\alpha \times \nabla p$ and the $\hat{k}$ component of $\nabla\alpha \times \nabla p$ is

$$(\nabla\alpha \times \nabla p) \cdot \hat{k} = \frac{1}{\rho^2}\left(\frac{\partial\rho}{\partial x}\frac{\partial p}{\partial y} - \frac{\partial\rho}{\partial y}\frac{\partial p}{\partial x}\right) \tag{5.34b}$$

we can see that the solenoid term in the vorticity equation is equal to the solenoid term in the circulation theorem (5.34a) divided by the area of the fluid element, consistent with the fact that $C/A = \zeta$. The solenoid term shows that given appropriate horizontal configurations of p and ρ, vorticity can be produced. Let us consider the configuration of p and ρ that would characterize cold air advection by the zonal geostrophic wind in the northern hemisphere (Figure 5.13). Imagining that no divergence or tilting is occurring, then the stated conditions render (5.33) as

$$\frac{d(\zeta + f)}{dt} = \frac{1}{\rho^2}\left(\frac{\partial\rho}{\partial x}\frac{\partial p}{\partial y}\right).$$

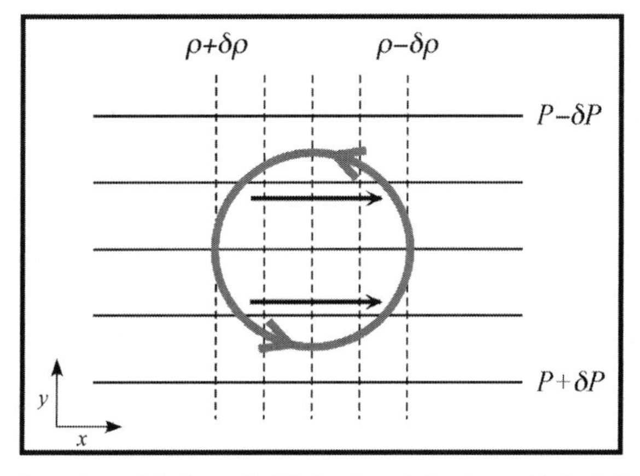

Figure 5.13 Configuration of isobars (solid lines) and isosteres (dashed lines) characterizing geostrophic cold air advection in the northern hemisphere. Bold arrows represent the geostrophic wind. The induced solenoidal circulation is represented by the thick gray circle with arrows

It is clear that $\partial\rho/\partial x < 0$ and $\partial p/\partial y < 0$ for this case. Thus, the solenoid term will generate cyclonic vorticity. The cyclonic vorticity will tend to rotate the isosteres until they are parallel with the isobars in a configuration in which high pressure corresponds to high density and vice versa. Such a configuration of isobars and isosteres represents the state of this fluid system in which ∇T is minimized, representing its lowest potential energy state. Though the physical nature of the solenoid term is tenable, it is not particularly amenable to measurement since it involves knowledge of the density. In prior applications we have adopted pressure as a vertical coordinate and seen that reference to density disappears from some of the basic equations. Use of pressure coordinates to formulate the vorticity equation will eliminate the troublesome solenoid term since $dp = 0$ on isobaric surfaces. Thus, we next formulate the vorticity equation in pressure coordinates in the hope of obtaining a simpler relationship.

In order to derive the vorticity equation in isobaric coordinates we proceed exactly as we did when first deriving it in height coordinates. The frictionless equations of motion in isobaric coordinates are

$$\frac{\partial u}{\partial t} + u\frac{\partial u}{\partial x} + v\frac{\partial u}{\partial y} + \omega\frac{\partial u}{\partial p} - fv = -\frac{\partial \phi}{\partial x} \tag{5.35a}$$

and

$$\frac{\partial v}{\partial t} + u\frac{\partial v}{\partial x} + v\frac{\partial v}{\partial y} + \omega\frac{\partial v}{\partial p} + fu = -\frac{\partial \phi}{\partial y} \tag{5.35b}$$

since the vertical wind is ω, the vertical derivative is $\partial/\partial p$, and the pressure gradient force is represented as $-\nabla_p\phi$ in isobaric coordinates. As before, we take $\partial/\partial y$ of (5.35a) and subtract it from $\partial/\partial x$ of (5.35b). After some algebra and substituting $\zeta = (\partial v/\partial x - \partial u/\partial y)$, we get

$$\frac{\partial \zeta}{\partial t} = -\vec{V}\cdot\nabla(\zeta + f) - \omega\frac{\partial \zeta}{\partial p} - (\zeta + f)(\nabla\cdot\vec{V}) + \hat{k}\cdot\left(\frac{\partial \vec{V}}{\partial p}\times\nabla\omega\right) \tag{5.36}$$

which ascribes local changes in the absolute vorticity (since $\partial f/\partial t = 0$) to (1) horizontal advection, (2) vertical advection, (3) the divergence term, and (4) tilting. As you might guess, the physical interpretation of the last two terms on the RHS of (5.36) is identical to the interpretation given to those encountered in (5.33). The two leading terms on the RHS of (5.36) have a familiar physical interpretation as they both describe advection.

Earlier we found that applying scale analysis to the complicated equations of motion lent us insight into the fundamental balances that characterize the atmospheric flow at middle latitudes. Equations (5.33) and (5.36) are complicated expressions as well and, therefore, we might find it useful to apply a scale analysis to them. We will do so by applying the same basic scalings as we used in Chapter 3 (Section 3.2.2).

Upon doing so we find that the relative vorticity scales as

$$\zeta = \left(\frac{\partial v}{\partial x} - \frac{\partial u}{\partial y}\right) \approx \frac{U}{L} \sim 10^{-5} s^{-1}$$

and, since $f_0 = 10^{-4}\,\text{s}^{-1}$, then $\zeta / f_0 \approx U / f_0 L \equiv R_o = 10^{-1}$ suggesting that for synoptic-scale motions the relative vorticity is an order of magnitude smaller than the planetary vorticity. This implies, in turn, that for mid-latitude synoptic-scale motions, the divergence term can be approximated as

$$-(\zeta + f)(\nabla \cdot \vec{V}) \approx -f(\nabla \cdot \vec{V}).$$

Given these initial scalings, the terms in (5.32) can be scaled as

$$\frac{\partial \zeta}{\partial t}, u\frac{\partial \zeta}{\partial x}, v\frac{\partial \zeta}{\partial y} \approx \frac{U^2}{L^2} \approx 10^{-10} s^{-1}$$

$$w\frac{\partial \zeta}{\partial z} \approx \frac{WU}{HL} \approx 10^{-11} s^{-1}$$

$$v\frac{\partial f}{\partial y} \approx U\beta \approx 10^{-10} s^{-1}$$

$$f(\nabla \cdot \vec{V}) \approx \frac{f_0 U}{L} \approx 10^{-9} s^{-2}$$

$$\left(\frac{\partial w}{\partial x}\frac{\partial v}{\partial z} - \frac{\partial w}{\partial y}\frac{\partial u}{\partial z}\right) \approx 10^{-11} s^{-1}$$

$$\frac{1}{\rho^2}\left(\frac{\partial \rho}{\partial x}\frac{\partial p}{\partial y} - \frac{\partial \rho}{\partial y}\frac{\partial p}{\partial x}\right) \approx 10^{-11} s^{-1}.$$

It should be noted that in the last three terms, which involve the sum of horizontal derivatives (or the sum of their products), it is quite possible that the two parts of the expression might partially cancel leaving the magnitude of the sum smaller than the typical scaling might suggest. For instance, both

$$\frac{\partial w}{\partial x}\frac{\partial v}{\partial z} \quad \text{and} \quad -\frac{\partial w}{\partial y}\frac{\partial u}{\partial z}$$

scale exactly the same way. Their actual values, however, might be such that their sum in the tilting term is smaller than that typical scale. Given the foregoing list, in which vorticity advections, local vorticity tendency, and divergence are the leading order terms, we find that it simply must be the case that the divergence be smaller than its scale in order that the leading terms can satisfy an approximate balance. In other words, since the advection terms are of order $10^{-10}\,\text{s}^{-1}$, and $f_0 = 10^{-4}\,\text{s}^{-1}$, the divergence term must scale as

$$(\nabla \cdot \vec{V}) \approx 10^{-6}\,\mathrm{s}^{-1}.$$

Therefore, for synoptic-scale, mid-latitude motions in the atmosphere, the foregoing scale analysis suggests that the vorticity equation can be approximated as

$$\frac{\partial \zeta}{\partial t} + u\frac{\partial \zeta}{\partial x} + v\frac{\partial \zeta}{\partial y} = \frac{d_h \zeta}{dt} = -f\left(\frac{\partial u}{\partial x} + \frac{\partial v}{\partial y}\right) \tag{5.37}$$

where the operator d_h/dt is given by

$$\frac{d_h}{dt} = \frac{\partial}{\partial t} + u\frac{\partial}{\partial x} + v\frac{\partial}{\partial y}.$$

Equation (5.37) suggests that the Lagrangian rate of change of the absolute vorticity following the horizontal motion on the synoptic scale is largely a consequence of the generation or destruction of vorticity through horizontal divergence. It is important to note that near the centers of typical mid-latitude cyclones, in which the relative vorticity often far exceeds the planetary vorticity, the fundamental scaling that led to (5.37) does *not* apply. Therefore, the problem of mid-latitude cyclogenesis will require consideration of additional physical processes in order to provide useful diagnostic and prognostic information.

The presence and centrality of horizontal divergence in (5.37) suggests that an important set of relationships exists in fluids. The rotation of a fluid depends upon the presence of divergence in that fluid. The presence of divergence in that fluid requires, by continuity, that the fluid also possesses regions of upward and downward motions. In the atmospheric fluid, those upward and downward motions, and the attendant adiabatic warming and cooling that goes along with them, are associated with phase changes of the water substance and the delivery of our sensible weather. In the next section of this chapter we will use a scale analysis similar to that applied in the present discussion to construct a set of approximate relationships that will be used to exploit the physical interrelation of temperature, vorticity, and divergence. The resulting set of equations will provide the basis for the second half of the book in which a detailed understanding of mid-latitude weather systems is developed.

5.4 The Quasi-Geostrophic System of Equations

We have thus far derived expressions for Newton's second law, mass continuity, and energy conservation as they pertain to the motions of the atmosphere in terms of (1) the equations of motion, (2) the continuity equation, and (3) the thermodynamic energy equation, respectively. In this section we will make appropriate simplifications of these relationships in order to develop a fairly simple system of equations which we can exploit to gain physical insight into the nature of mid-latitude weather systems and the mid-latitude synoptic-scale flow. We will pursue the development of these equations using an isobaric coordinate system because such a system, as we have already seen, eliminates the need to consider density. The underlying simplifying assumption that will guide this development is the fact that the fundamental balances constraining the behavior of the mid-latitude atmosphere on Earth are geostrophic balance (in the horizontal) and hydrostatic balance (in the vertical). As we have already seen, these two separate balances are combined in the thermal wind balance, which, consequently, represents the essential balance constraining motions in the mid-latitude atmosphere. If we further assume, consistent with the scale analysis we performed in Section 3.2, that the friction force is not a primary consideration for synoptic-scale flows, then the approximate equations of motion are

$$\frac{du}{dt} = -\frac{\partial \phi}{\partial x} + fv \quad \text{and} \quad \frac{dv}{dt} = -\frac{\partial \phi}{\partial y} - fu, \tag{5.38}$$

the hydrostatic equation is

$$\frac{\partial \phi}{\partial p} = -\alpha = -\frac{RT}{p}, \tag{5.39}$$

the continuity equation is

$$\nabla \cdot \vec{V}_h + \frac{\partial \omega}{\partial p} = 0, \tag{5.40}$$

and the thermodynamic energy equation is

$$\frac{\partial T}{\partial t} + \vec{V}_h \cdot \nabla T - S_p \omega = \frac{\dot{Q}}{c_p} \tag{5.41}$$

where $S_p = -T \partial \ln \theta / \partial p$. We will 'condense' these five expressions (some of which are already approximated) into a set of two that satisfy a scale analysis appropriate for the mid-latitude synoptic-scale flow. We begin by recognizing that the behavior of the horizontal flow, represented by the set (5.38), can also be represented by the isobaric vorticity equation, (5.36), repeated here for convenience:

$$\frac{\partial \zeta}{\partial t} = -\vec{V} \cdot \nabla(\zeta + f) - \omega \frac{\partial \zeta}{\partial p} - (\zeta + f)(\nabla \cdot \vec{V}) + \hat{k} \cdot \left(\frac{\partial \vec{V}}{\partial p} \times \nabla \omega \right).$$

We have already shown through scale analysis that the vertical advection and twisting terms can be neglected for the motions we are considering here. We found that we can additionally neglect ζ compared to f in the divergence term. We can also simplify the advection of planetary vorticity by considering a Taylor series expansion of the Coriolis parameter about a fixed latitude, ϕ_0. Since there is only y-direction variation in f, we need only consider df/dy in this expansion. Letting $\beta = df/dy = 2\Omega \cos \phi_0$, the expression for the Coriolis parameter can be written as

$$f = 2\Omega \sin \phi_0 + 2\Omega \cos \phi_0 y = f_0 + \beta y \tag{5.42}$$

where $y = 0$ at the latitude ϕ_0. There is useful simplifying information in (5.42) since the ratio of the two terms in the expansion scales as

$$\frac{\beta L}{f_0} \approx \frac{\cos\phi_0 L}{\sin\phi_0 a} \approx \frac{L}{a}.$$

Thus, when the latitudinal scale of the motions (represented by L) is much smaller than the radius of the Earth – a condition that is nearly always met for synoptic-scale, mid-latitude motions – then we can assign the constant value f_0 to the Coriolis parameter except where it is differentiated in the advection term in (5.36), in which case we can assign the constant value β to df/dy. Physically, the rationale for this simplification is that if the motions are constrained to a small enough latitudinal range ($L \ll a$), then variation of the Coriolis parameter is negligible. The motion can be thought of as occurring on a hypothetical plane, tangent to the Earth at latitude ϕ_0, where no variation of the Coriolis parameter occurs. For this reason, adoption of the assumptions arising from (5.42) is known as the **beta-plane approximation.**

Returning now to the approximated vorticity equation (5.37),

$$\frac{\partial \zeta}{\partial t} + u\frac{\partial \zeta}{\partial x} + v\frac{\partial \zeta}{\partial y} = \frac{d_h \zeta}{dt} = -f\left(\frac{\partial u}{\partial x} + \frac{\partial v}{\partial y}\right),$$

if we further simplify it by (1) assuming that the horizontal advection is accomplished by the geostrophic winds, and (2) that the relative vorticity can be described by the geostrophic relative vorticity, we have

$$\frac{\partial \zeta_g}{\partial t} + u_g\frac{\partial \zeta_g}{\partial x} + v_g\frac{\partial \zeta_g}{\partial y} = \frac{d_g \zeta}{dt} = -f_0\left(\frac{\partial u}{\partial x} + \frac{\partial v}{\partial y}\right) \tag{5.43}$$

where the operator

$$\frac{d_g}{dt} = \frac{\partial}{\partial t} + u_g\frac{\partial}{\partial x} + v_g\frac{\partial}{\partial y}.$$

Since the geostrophic relative vorticity, ζ_g, can be written as

$$\zeta_g = \frac{\partial v_g}{\partial x} - \frac{\partial u_g}{\partial y} = \frac{\partial}{\partial x}\left(\frac{1}{f_0}\frac{\partial \phi}{\partial x}\right) - \frac{\partial}{\partial y}\left(-\frac{1}{f_0}\frac{\partial \phi}{\partial y}\right) = \frac{1}{f_0}\nabla^2\phi, \tag{5.44}$$

the entire LHS of (5.43) can be expressed in terms of the geopotential. Note that we did not replace the horizontal velocities on the RHS of (5.43) with their geostrophic counterparts since, for constant Coriolis parameter, the geostrophic wind is non-divergent. Using (5.40), we see that

$$-f_0\left(\frac{\partial u}{\partial x} + \frac{\partial v}{\partial y}\right) = f_0\frac{\partial \omega}{\partial p}.$$

Thus, (5.43) can be written as

$$\frac{\partial \zeta_g}{\partial t} = -\vec{V}_g \cdot \nabla\zeta_g + f_0\frac{\partial \omega}{\partial p} \tag{5.45}$$

which is known as the **quasi-geostrophic vorticity equation.**

From (5.39) we see that

$$T = -\frac{p}{R}\frac{\partial\phi}{\partial p}$$

so that (5.41) can be rewritten as

$$\frac{p}{R}\left[\frac{\partial}{\partial t}\left(-\frac{\partial\phi}{\partial p}\right) + \vec{V}_h \cdot \nabla\left(-\frac{\partial\phi}{\partial p}\right)\right] - S_p\omega = \frac{\dot{Q}}{c_p}. \tag{5.46}$$

Recalling that $ds/dt = \dot{Q}/T$ describes the rate of change of entropy, (5.46) can be written as

$$\frac{\partial}{\partial t}\left(-\frac{\partial\phi}{\partial p}\right) + \vec{V}_h \cdot \nabla\left(-\frac{\partial\phi}{\partial p}\right) - \sigma\omega = \frac{\alpha}{c_p}\frac{ds}{dt} \tag{5.47}$$

where

$$\sigma = \frac{RS_p}{p} = -\alpha\frac{\partial \ln\theta}{\partial p}.$$

We will refer to σ as the static stability parameter, noting that in a statically stable atmosphere $\partial\theta/\partial p < 0$ so that $\sigma > 0$. As we did with the vorticity equation, we will approximate the horizontal wind as geostrophic and also assume that the diabatic heating is negligible. Under those restrictions, the thermodynamic energy equation becomes

$$\frac{\partial}{\partial t}\left(-\frac{\partial\phi}{\partial p}\right) = -\vec{V}_g \cdot \nabla\left(-\frac{\partial\phi}{\partial p}\right) + \sigma\omega. \tag{5.48}$$

Note that this expression states that the rate of change of temperature (since $-\partial\phi/\partial p$ is directly related to temperature on isobaric surfaces) is the difference between the advective tendency and the adiabatic warming or cooling produced by vertical motions. Our neglect of diabatic heating is only valid so long as the most significant temperature changes occur as a result of either of the two processes represented on the RHS of (5.48). This is never actually the case but it is not a paralyzingly incorrect assumption for synoptic-scale motions in the mid-latitudes. In the context of extratropical cyclone development, in which latent heating plays a significant role, the appropriateness of this assumption erodes considerably.

The combination of (5.45) and (5.48) represents the two simplified expressions which we sought at the beginning of this section. Since the geostrophic wind, the geostrophic vorticity, and (though it may not seem obvious) σ can all be written in terms of the geopotential, (5.45) and (5.48) represent a set of two equations with two unknowns (ϕ and ω). In subsequent chapters we will manipulate this pair of expressions to compute the geopotential tendency ($\partial\phi/\partial t$) and the vertical motion (ω) given only the instantaneous field of geopotential at a variety of levels in the atmosphere. Amazingly, this implies that much of the behavior of the geostrophically and hydrostatically balanced mid-latitude atmosphere can be diagnosed or predicted without direct measurement of the velocity field! It is easy to see why these two expressions, which constitute the quasi-geostrophic system of equations, sit at the very epicenter of modern dynamical meteorology. The lion's share of equation development has now been completed. In the rest of this book, we will use these basic tools to develop physical understanding of the behavior of mid-latitude weather systems. It is a thrilling ride that will amply reward the discipline and patience you have invested in our endeavor to this point!

Selected References

Acheson, *Elementary Fluid Dynamics*, provides a solid introduction to vorticity.
William and Elder, *Fluid Physics for Oceanographers and Physicists*, does the same.
Hess, *Introduction to Theoretical Meteorology*, offers the most accessible introductory discussion of vorticity and its relation to circulation.
Eliassen (1984) provides an elegant overview of the development of the quasi-geostrophic system of equations.
Bleck (1973) discusses potential vorticity at a reasonably introductory level.

Problems

5.1. (a) Derive an expression for the isobaric, geostrophic relative vorticity in terms of ϕ.
(b) Demonstrate that, in the northern hemisphere, this expression describes positive relative vorticity for regions of cyclonic curvature of streamlines and negative relative vorticity for regions of anticyclonic curvature.
(c) An **equivalent barotropic** environment is one in which geopotential height lines are everywhere parallel to isotherms or thickness isopleths. If such an environment is characterized by cyclonic vorticity, is the geostrophic relative vorticity of larger magnitude at 500 hPa or at 900 hPa? Explain your reasoning with a picture and an appeal to the governing equations.

5.2. A cylindrical column of air at 30°N with radius 100 km expands to twice its original radius. The air is initially at rest.

(a) Calculate the circulation about this column after expansion.
(b) What is the relative vorticity after expansion? Explain.
(c) How is the result of this problem consistent with the vorticity equation? What physical process, implicit in this problem, is a powerful mechanism for changing the relative vorticity?

5.3. Recall that we concluded that horizontal divergence was so subject to error when calculated from observations that we could not 'measure' it with any confidence (Problem 4.17). Vertical vorticity is, like divergence, a first derivative of the horizontal wind field. Can you suggest a reason why calculations of relative vorticity based on observed, mid-latitude winds are not as prone to error as calculations of divergence?

5.4. Some people believe the Earth is flat. If it really were, and it still rotated with rotation rate Ω, what value would the Coriolis parameter (i.e. the planetary vorticity) have?

5.5. A square 800 km on each side is embedded in an easterly flow (i.e. westward-moving air) that decreases in magnitude toward the north at a rate of 10 m s^{-1} per 400 km.

(a) What is the circulation about the square? Show your work.
(b) What is the mean relative vorticity in the square? Show *two* methods of arriving at this answer.

5.6. A cylinder (rotating counterclockwise) is filled with different color fluids as shown in Figure 5.1A. The inner radius is 2 m and the outer radius is 4 m. The tangential velocity distribution is given by the function $V = A/r$ where $A = 10$ m^2 s^{-1}.

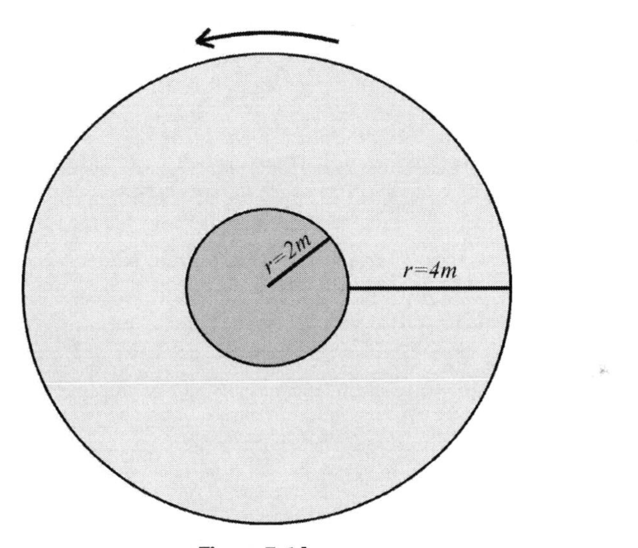

Figure 5.1A

(a) What is the average vorticity in the annulus filled with dark gray fluid? (Show your work).
(b) What is the average vorticity of the light gray fluid? Explain.
(c) Verify this result by using the natural coordinate form of vorticity at $r = 3$ m (the average radius of the annulus).

5.7. For geostrophic, steady-state flow with constant Coriolis parameter (f_0) show that

$$\frac{\partial \omega}{\partial p} = \nabla \cdot \left(\frac{\vec{V}_g \zeta_g}{f_0} \right).$$

5.8. Imagine the first major winter storm of the year has just visited the central United States. The storm delivered heavy snow to northern Nebraska but no accumulation to southern Nebraska. In the wake of the storm, a broad surface anticyclone has settled over the region. In the absence of any significant pressure gradient force over the state, a persistent northerly surface flow develops over southern Nebraska. Why? Will this northerly flow have an appreciable diurnal cycle? Why or why not?

5.9. An atmospheric column initially sitting atop the Chilean Andes (elevation 5000 m, latitude 40°S) stretches to 10 km and has zero relative vorticity. If westerly flow advects that column down the east slopes of the Andes into the Pampa of Argentina (elevation 1000 m) what will its relative vorticity be assuming the flow is barotropic?

5.10. Under what precise conditions can curved, geostrophic flow have zero relative vorticity? Explain. Draw schematic diagrams to illustrate your answer.

5.11. A popular tale suggests that the spin of the 'bathtub vortex' is determined by the Earth's rotation. In order to evaluate this proposition, consider a situation in which a small drain is opened in the center of a large tank of initially still water. After a certain time, t, the tangential velocity 1 centimeter from the drain axis is observed to be 0.5 cm s^{-1}. The tank is located at 45°N.

(a) If only the Earth's rotation was responsible for the observed swirl, and if friction can be neglected, what was the initial radial distance of the fluid ring observed at 1 cm radius at time t?
(b) Suppose that instead of being initially motionless, the water had a small, initial, clockwise swirl. At the initial radial distance calculated in (a), how large would the tangential velocity of this swirl have to have been in order to reverse the spin of the drain vortex?
(c) What can be concluded about the veracity of this popular tale?

5.12. A pair of cyclonic and anticyclonic vortices are observed in the atmosphere at 43°N. Both vortices have the same area-averaged value of relative vorticity ($|\zeta| = 1 \times 10^{-5}\, s^{-1}$). Suppose that a uniform horizontal convergence and divergence associated with the cyclonic and anticyclonic vortices, respectively, persists during an entire day with equal magnitudes ($|\nabla \cdot \vec{V}| = 2 \times 10^{-6}\, s^{-1}$).

(a) Estimate the respective changes in ζ as a consequence of this circumstance.
(b) It is observed that departures from mean sea-level pressure are larger in magnitude for extreme cyclones than for extreme anticyclones. Does the result in (a) suggest a dynamical reason for this asymmetry? Explain.

Solutions

5.1. (a) $\zeta_g = \frac{1}{f}\nabla^2\phi$

5.2. (a) $-6.872 \times 10^6\, m^2\, s^{-1}$ (b) $-5.469 \times 10^{-5}\, s^{-1}$

5.4. 2Ω

5.5. (a) $-1.6 \times 10^7\ \mathrm{m^2\,s^{-1}}$ (b) $-2.5 \times 10^{-5}\ \mathrm{s^{-1}}$

5.6. (a) $5\ \mathrm{s^{-1}}$ (b) 0

5.9. $-7.5 \times 10^{-5}\ \mathrm{s^{-1}}$

5.10. $R_s = \frac{V_g}{\partial V_g/\partial n}$

5.11. (a) 0.985 m (b) $-5.07 \times 10^{-5}\ \mathrm{m\,s^{-1}}$

5.12. (a) $\zeta_t = 3.06 \times 10^{-5}\ \mathrm{s^{-1}}$ for the cyclone and $\zeta_t = -2.29 \times 10^{-5}\ \mathrm{s^{-1}}$ for the anticyclone

6

The Diagnosis of Mid-Latitude Synoptic-Scale Vertical Motions

Objectives

Regions of upward vertical motion are often associated with clouds and precipitation since rising air cools by expansion. This cooling increases the relative humidity of the air which can eventually lead to condensation and cloud formation. Regions of rising air are also often associated with mass divergence in an atmospheric column and, consequently, surface pressure falls and cyclogenesis. Regions of downward vertical motion are often cloud free as air dries and warms upon being compressed as it sinks to higher pressure. Mass convergence into an atmospheric column, characteristic of regions of downward vertical motion, results in surface pressure rises and surface anti-cyclogenesis. As a result of the fundamental nature of these relationships, it is not an exaggeration to say that determination of where, when, and to what degree the air is rising or sinking is of fundamental importance for accurately diagnosing the current weather or predicting its future state. In this chapter we will investigate a number of different methods for diagnosing synoptic-scale vertical motions in typical mid-latitude weather systems.

Some of these diagnostic methods will derive from careful consideration of the ageostrophic wind vector itself. Several others (the Sutcliffe development theorem as well as the traditional and $\vec{Q}$-vector forms of the quasi-geostrophic omega equation) will arise from simultaneously solving the quasi-geostrophic vorticity and thermodynamic energy equations for the vertical motion, ω, and will make reference only to the instantaneous mass distribution. Taken together, the collection of diagnostics to be developed in this chapter will provide us with a formidable set of tools for understanding the synoptic-scale behavior of mid-latitude weather systems. We begin our investigation by considering the ageostrophic wind.

Mid-Latitude Atmospheric Dynamics Jonathan E. Martin

6.1 The Nature of the Ageostrophic Wind: Isolating the Acceleration Vector

Recall that the geostrophic wind is non-divergent on an f plane. In fact, under such conditions only departures from geostrophy contribute to horizontal divergence and, through the continuity of mass, to vertical motions as shown in (4.9). For this reason it is extremely important to examine means by which the ageostrophic motions in the mid-latitude atmosphere might be diagnosed. We begin with the frictionless equation of motion

$$\frac{d\vec{V}}{dt} = -f\hat{k} \times \vec{V} - \nabla\phi, \tag{6.1}$$

and take the vertical cross-product of this expression to obtain

$$\frac{\hat{k}}{f} \times \frac{d\vec{V}}{dt} = \frac{\hat{k}}{f} \times (-f\hat{k} \times \vec{V}) - \frac{\hat{k}}{f} \times \nabla\phi. \tag{6.2a}$$

The right hand rule dictates that $\hat{k} \times \hat{k} \times \vec{A} = -\vec{A}$, and $\vec{V}_g = (\hat{k}/f) \times \nabla\phi$, so

$$\frac{\hat{k}}{f} \times \frac{d\vec{V}}{dt} = \vec{V} - \vec{V}_g = \vec{V}_{ag}. \tag{6.2b}$$

The famous British meteorologist R. C. Sutcliffe[1] reasoned that surface pressure falls resulted from vertical differences in mass divergence in a column. Larger mass divergence aloft than at the surface resulted in surface pressure falls and vice versa for surface pressure rises. Such differences in divergence could be related to differential accelerations at the surface and aloft through application of (6.2b). Thus, Sutcliffe argued that isolation of the acceleration vector could give insights into the sense of the vertical motion in an atmospheric column. Before presenting the elegant theory of Sutcliffe (1939), let us endeavor to isolate the acceleration vector, and its ageostrophic consequences, in two rather simple cases. These cases correspond to the two broad classes of circumstances in which geostrophic balance is violated: the presence of along-flow speed change and curvature in the flow.

The canonical synoptic example of along-flow speed change is the isolated jet streak. Shown in Figure 6.1 is the isotach distribution associated with an isolated wind speed maximum at 300 hPa in the northern hemisphere. The dashed line drawn perpendicular to the jet axis divides the jet into the so-called entrance region to its left and the exit region to its right. A parcel of air located on the western edge of the entrance region (indicated by the solid circle in Figure 6.1) would quite obviously experience an acceleration in the direction of the flow at that location. Hence, the

[1] R. C. Sutcliffe (1904–1991) received his Ph.D. in statistics but found employment with the British Meteorological Office in 1927. He worked with the famous Tor Bergeron while in Malta where he opined that weather forecasting was scarcely worthy of description as a scientific activity. He was among the first, and greatest, atmospheric scientists to insist that weather forecasting and diagnosis should proceed from the equations of motion. This insistence led to what might be considered the first major breakthrough in modern synoptic–dynamic meteorology and his most famous contribution, 'A contribution to the problem of development' (Sutcliffe 1947).

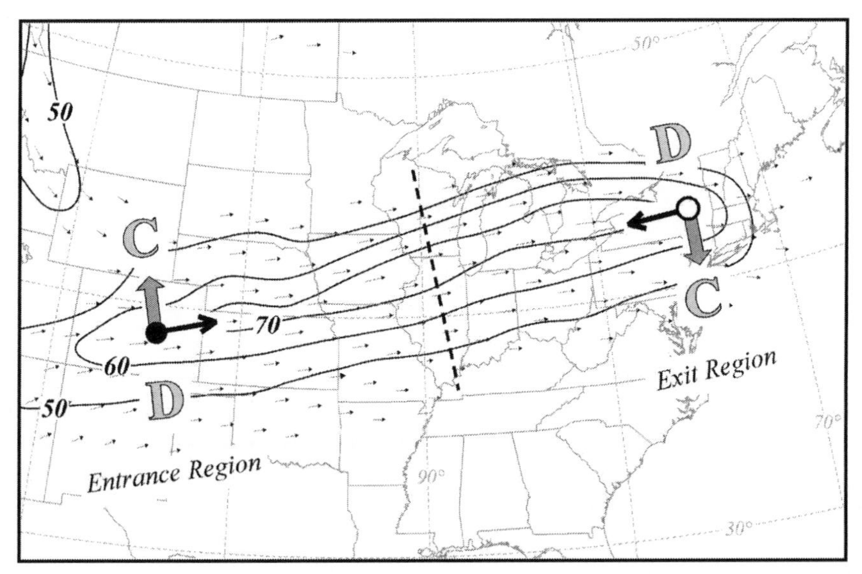

Figure 6.1 The 300 hPa isotachs (solid lines) and wind vectors associated with a straight jet at 0000 UTC 12 November 2003 from NCEP's Eta model analysis. Isotachs are labeled in m s^{-1} and contoured every 10 m s^{-1} starting at 50 m s^{-1}. Only wind vectors greater than 40 m s^{-1} are shown. Thick black arrows indicate the direction of the acceleration vector $d\vec{V}/dt$ at the entrance region (solid black circle) and exit regions (open circle) of the jet. The gray shaded arrow is the resultant ageostrophic wind vector, $\vec{V}_{ag}$, at both locations. C and D represent the locations of 300 hPa ageostrophic convergence and divergence, respectively

vector $d\vec{V}/dt$ points eastward toward the center of the jet streak. Consequently, the ageostrophic wind vector, $\vec{V}_{ag}$, points northward at the indicated point. The result of this distribution of ageostrophic winds in the entrance region of the jet is that there is convergence of air at 300 hPa to the north of the indicated position and divergence of air at 300 hPa to the south of the indicated position. Given that 300 hPa is nearly at the top of the troposphere, upper-level divergence (convergence) is associated with upward (downward) vertical motion in the intervening column and so a thermally direct vertical circulation generally exists in the entrance region of a straight jet streak.

A parcel of air located on the eastern edge of the exit region (indicated by the open circle in Figure 6.1) would quite obviously experience a deceleration in the direction opposite the flow at that location. Hence, the vector $d\vec{V}/dt$ points westward toward the center of the jet streak. Consequently, the ageostrophic wind vector, $\vec{V}_{ag}$, points southward at the indicated point. The result of this distribution of ageostrophic winds in the exit region of the jet is that there is convergence of air at 300 hPa to the south of the indicated position and divergence of air at 300 hPa to the north of the indicated position. Upward vertical motion occurs in the column beneath the upper divergence maxima and, thus, a thermally indirect vertical circulation generally exists in the exit region of a straight jet streak.

Curvature in the flow is also a circumstance that violates the geostrophic assumption. Consider flow through an upper tropospheric trough–ridge couplet where the

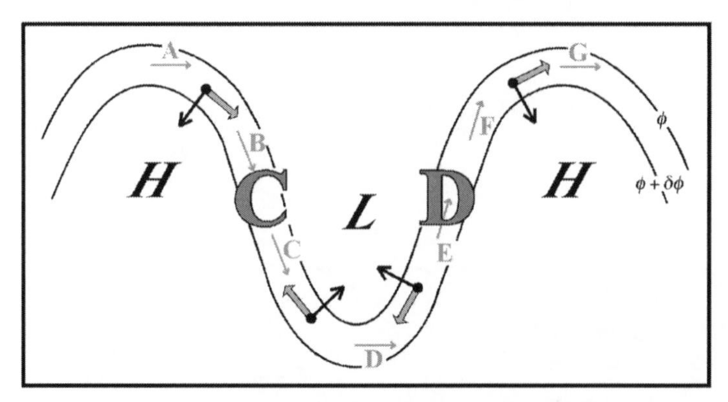

Figure 6.2 Schematic upper tropospheric trough–ridge wave train in which the speed of the flow is the same everywhere. Thick black arrows represent the acceleration vectors, $d\vec{V}/dt$, at the indicated points determined graphically by finite differencing between adjacent wind arrows (in gray and labeled as described in the text). Gray shaded arrows represent resultant ageostrophic winds, $\vec{V}_{ag}$, at the indicated points. Convergence and divergence are indicated by C and D, respectively

wind speed is constant and everywhere parallel to the geopotential height lines as shown in Figure 6.2. Under such circumstances, the acceleration of the wind will be entirely a consequence of directional changes. Thus, between points A and B in Figure 6.2, a southwestward-directed acceleration is required to turn the wind from westerly at point A to northwesterly at point B. There is no direction change between points B and C and, thus, no acceleration vector. A northeastward-directed acceleration is required to turn the northwesterly wind at point C to a westerly direction at point D. In order to turn the westerly at point D to a southwesterly at point E, a northwestward-directed acceleration is required. No change in direction exists between points E and F but a change from southwesterly at F to westerly at point G requires a southeastward-directed acceleration as shown. Given the four acceleration vectors drawn in Figure 6.2, it is simple to draw the ageostrophic winds in this trough–ridge couplet. The ageostrophic winds clearly converge on the western side of the upper trough (on its upstream side) leading to downward vertical motion in the column in that location. Meanwhile, the divergence of the ageostrophic winds on the downstream side of the upper trough is associated with upward vertical motions in the column in that location. This result provides a first insight into the physical reason why inclement weather is often found downstream of upper-level trough axes while clear skies are often found downstream of upper-level ridge axes. This basic relationship lies at the heart of understanding the distribution of sensible weather in the middle latitudes.

6.1.1 Sutcliffe's expression for net ageostrophic divergence in a column

Having examined the distribution of the ageostrophic winds in these canonical synoptic examples, let us now turn our attention to the insightful work of Sutcliffe

(1939). We begin by considering the surface wind $\vec{V}_0$, the wind at some upper tropospheric level, $\vec{V}$, and the vertical shear between the two layers, $\vec{V}_s$. Based upon these simple definitions, it is clear that $\vec{V} = \vec{V}_0 + \vec{V}_s$ and therefore

$$\frac{d\vec{V}}{dt} = \frac{d\vec{V}_0}{dt} + \frac{d\vec{V}_s}{dt} \tag{6.3}$$

where

$$\frac{d}{dt} = \frac{\partial}{\partial t} + \vec{V} \cdot \nabla$$

is the Lagrangian operator used to describe $d\vec{V}/dt$. Given these definitions, (6.3) can be expanded into

$$\frac{d\vec{V}}{dt} = \frac{\partial \vec{V}_0}{\partial t} + (\vec{V}_0 + \vec{V}_s) \cdot \nabla \vec{V}_0 + \frac{d\vec{V}_s}{dt}. \tag{6.4}$$

Alternatively, (6.4) can be written as

$$\frac{d\vec{V}}{dt} = \frac{\partial \vec{V}_0}{\partial t} + \vec{V}_0 \cdot \nabla \vec{V}_0 + \vec{V}_s \cdot \nabla \vec{V}_0 + \frac{d\vec{V}_s}{dt}. \tag{6.5}$$

Recognizing that the first two terms on the RHS of (6.5) describe the acceleration of the wind at the surface, $(d\vec{V}/dt)_0$, an expression for the differential acceleration in the layer arises:

$$\frac{d\vec{V}}{dt} - \left(\frac{d\vec{V}}{dt}\right)_0 = \vec{V}_s \cdot \nabla \vec{V}_0 + \frac{d\vec{V}_s}{dt}. \tag{6.6}$$

This expression relates the fact that if there is shearing over the surface wind ($\vec{V}_s \cdot \nabla \vec{V}_0$) or a change in the shear vector ($d\vec{V}_s/dt$) then there must be a difference in acceleration between the upper tropospheric wind and the surface wind. Based upon (6.2b), this implies that there must be some net divergence in the column and therefore, by continuity, vertical motions. Let us now examine the physical significance of the two terms on the RHS of (6.6). As we will do with every other diagnostic expression, we will consider the effect of each term in isolation, beginning with the shearing over the surface wind.

(a) *Shearing over the surface wind: $d\vec{V}/dt - (d\vec{V}/dt)_0 = \vec{V}_s \cdot \nabla \vec{V}_0$*

Our expression begins by first expanding this term into its full component form given by

$$\vec{V}_s \cdot \nabla \vec{V}_0 = \left(u_s \frac{\partial u_0}{\partial x} + v_s \frac{\partial u_0}{\partial y}\right) \hat{i} + \left(u_s \frac{\partial v_0}{\partial x} + v_s \frac{\partial v_0}{\partial y}\right) \hat{j}. \tag{6.7}$$

Figure 6.3 depicts a schematic sea-level pressure minimum and some 1000–500 hPa thickness contours. Considering the value of this term at the center of the low-pressure

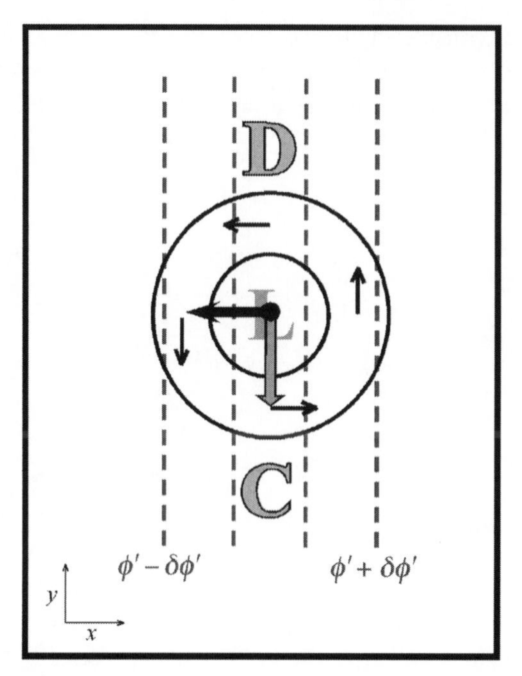

Figure 6.3 Sea-level isobars (solid lines) and 1000–500 hPa thickness (dashed lines) near a developing surface low-pressure center in the northern hemisphere. Thin dark arrows represent the sea-level geostrophic winds. The thick black arrow represents $\vec{V}_s . \nabla \vec{V}_0$. The gray shaded arrow represents the difference between the upper-level and surface ageostrophic wind, $\vec{V}_{ag_U} - \vec{V}_{ag_L}$. Net column convergence and divergence are indicated by C and D, respectively

center greatly reduces the mathematical complexity of applying (6.7). We will assume that the winds are geostrophic everywhere, which dictates that a thermal wind vector be directed along the positive y-axis in the northern hemisphere. At the center of the low, therefore, there is no x-direction vertical shear so that $u_s = 0$. It is also clear that there is no $\partial v_0/\partial y$ at the center of the low-pressure center. Thus, (6.7) reduces to

$$\vec{V}_s \cdot \nabla \vec{V}_0 = v_s \frac{\partial u_0}{\partial y} \hat{i}$$

for the scenario illustrated in Figure 6.3. We have already found that v_s is positive in this case. We now discern that $\partial u_0/\partial y$ is negative so that the product $v_s \partial u_0/\partial y$ is negative. Consequently, the vector $\vec{V}_s \cdot \nabla \vec{V}_0$ points in the negative x direction as indicated. Since $\vec{V}_s \cdot \nabla \vec{V}_0$ represents the acceleration at the top of the column minus the acceleration at the bottom of the column, taking the vertical cross-product of $\vec{V}_s \cdot \nabla \vec{V}_0$ indicates the direction of the column-differential ageostrophic wind. Figure 6.3 shows that, in this case, there is greater ageostrophic divergence (convergence) aloft than at the surface north (south) of the surface low implying ascent (descent) in that location. The surface cyclone will propagate toward the net column mass divergence (i.e. in the direction of the ascending air) as only mass divergence and ascending air will be associated with sustained pressure falls at the surface. Application of similar reasoning to the case of a surface anticyclone (a recommended exercise for the reader) leads us to a general statement: *the sea-level pressure perturbation will propagate in the direction of the thermal wind.*

(b) Rate of change of the shear vector: $d\vec{V}_s/dt$

Figure 6.4(a) shows some 1000–500 hPa thickness isopleths along with the thermal wind vector in the layer in the northern hemisphere at some time $T = 0$. Some time later ($T = T_1$), the horizontal thickness gradient has been increased by some agency in the atmosphere such as confluent horizontal flow. The result of such an increase in the baroclinicity is a larger thermal wind, still directed to the north as shown in Figure 6.4(b). If we assume the winds are everywhere geostrophic, then the difference in the thermal wind vectors in Figures 6.4(a) and 6.4(b) represents a change in the shear vector and can be represented by the expression $d\vec{V}_s/dt$ so long as the change has been measured following an individual air parcel. Thus, $d\vec{V}_s/dt$ is directed in the positive y direction. The vertical cross-product of $d\vec{V}_s/dt$ (which

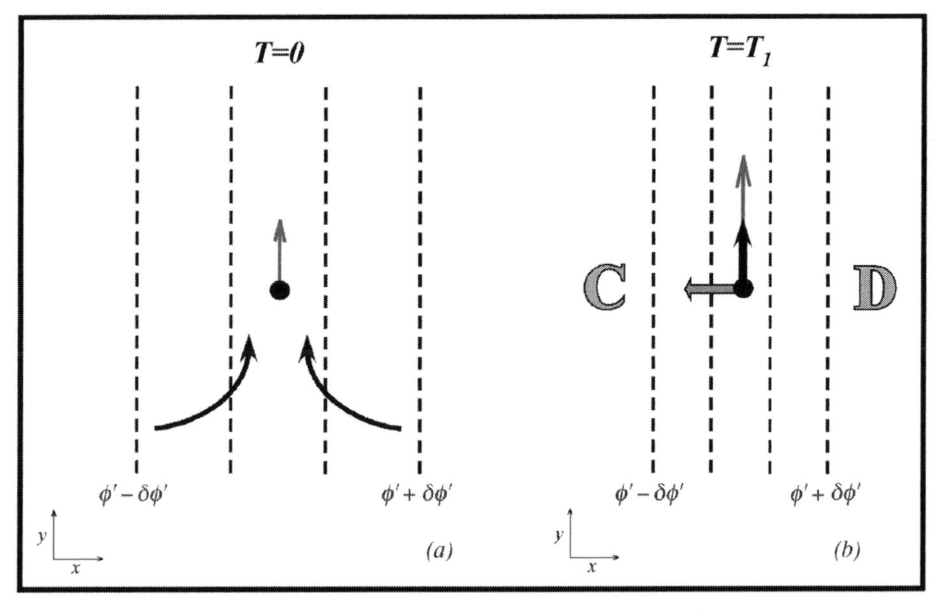

Figure 6.4 Illustration of the effect of the rate of change of shear term, $d\vec{V}_s/dt$, from (6.6). Dashed lines are 1000–500 hPa thickness contours whose gradient increases in magnitude from time $T = 0$ to time $T = T_1$. The increase in effected by horizontal confluence, represented by the thin black arrows in (a). The thin gray arrow represents $\vec{V}_s$, the thermal wind shear. At $T = T_1$, the thermal wind is larger and the bold black arrow in (b) represents the Lagrangian change in the shear, $d\vec{V}_s/dt$. The gray shaded arrow in (b) represents the difference between upper-level and near surface ageostrophic wind, $\vec{V}_{ag_U} - \vec{V}_{ag_L}$. Net column convergence and divergence are indicated by C and D, respectively

points directly toward the low thicknesses in Figure 6.4b) represents the column-differential ageostrophic wind for this example. The column of air on the warm (cold) side of the thickness gradient experiences greater divergence (convergence) aloft than at the surface and so it rises (sinks). Thus we find that anytime the horizontal flow acts to increase the thickness (or temperature) gradient, the response is the development of a thermally direct vertical circulation in which warm air rises and cold air sinks. Conversely, any systematic relaxation of the horizontal gradient of temperature by the action of the horizontal flow induces a thermally indirect vertical circulation. This physical insight, a direct consequence of the fact that the rate of change of the shear vector produces divergence in the column, is central to the dynamics of frontogenesis, a topic we will explore in great detail in Chapter 7.

6.1.2 Another perspective on the ageostrophic wind

We now turn our attention to a more formal expansion of the ageostrophic wind relationship (6.2b) which, recall, stated that

$$\vec{V}_{ag} = \frac{\hat{k}}{f} \times \frac{d\vec{V}}{dt}.$$

The Lagrangian derivative in the preceding expression can be expanded so that

$$\vec{V}_{ag} = \frac{\hat{k}}{f} \times \left(\frac{\partial \vec{V}}{\partial t} + \vec{V} \cdot \nabla \vec{V} + \omega \frac{\partial \vec{V}}{\partial p} \right). \tag{6.8}$$

The three terms on the RHS of (6.8) represent three contributions to the total ageostrophic wind: (1) the local wind tendency component, (2) the inertial advective component, and (3) the convective component. If we substitute $\vec{V}_g$ for $\vec{V}$ everywhere in (6.8) then

$$\vec{V}_{ag} = \frac{\hat{k}}{f} \times \left(\frac{\partial \vec{V}_g}{\partial t} + \vec{V}_g \cdot \nabla \vec{V}_g + \omega \frac{\partial \vec{V}_g}{\partial p} \right). \tag{6.9}$$

Our aim in this development is to diagnose the synoptic-scale vertical motion by first isolating the distribution of the ageostrophic wind. As is clear from (6.9), diagnosis of the convective component of the ageostrophic wind requires a priori knowledge of the vertical motion. Thus, it is not feasible to perform the intended diagnosis on the convective component. For this reason we will consider only the first two terms on the RHS of (6.9) in the foregoing analysis, starting with the local wind tendency component.

The local wind tendency component of the ageostrophic wind ($\vec{V}_{ag_T}$) can be related to geopotential height or pressure changes since

$$\vec{V}_{ag_T} = \frac{\hat{k}}{f} \times \frac{\partial \vec{V}_g}{\partial t} = \frac{\hat{k}}{f} \times \frac{\partial}{\partial t} \left(\frac{\hat{k}}{f} \times \nabla \phi \right) = -\frac{1}{f^2} \nabla \frac{\partial \phi}{\partial t} \tag{6.10a}$$

on pressure surfaces or

$$\frac{\hat{k}}{f} \times \frac{\partial \vec{V}_g}{\partial t} = -\frac{1}{\rho f^2} \nabla \frac{\partial p}{\partial t} \tag{6.10b}$$

on height surfaces. This component of the ageostrophic wind is known as the **isallobaric wind** as a result of its dependence on the gradient of isallobars (lines of constant pressure tendency, $\partial p/\partial t$). Knowledge of the isallobaric wind, like any component of the ageostrophic wind, only tells us about the distribution of vertical motion when we know its divergence. Thus, we are most interested in the divergence of the isallobaric wind, given by

$$\nabla \cdot \vec{V}_{isal} = -\frac{1}{f^2} \nabla^2 \frac{\partial \phi}{\partial t} \tag{6.11a}$$

on pressure surfaces or

$$\nabla \cdot \vec{V}_{isal} = -\frac{1}{\rho f^2} \nabla^2 \frac{\partial p}{\partial t} \tag{6.11b}$$

on height surfaces. It is left as an exercise to show that pressure (or geopotential) falls are associated with convergence of the isallobaric wind while pressure (or geopotential) rises are associated with divergence of the isallobaric wind.

The inertial–advective component ($\vec{V}_{IA}$) of the ageostrophic wind is given by

$$\vec{V}_{IA} = \frac{\hat{k}}{f} \times \left[\left(u_g \frac{\partial u_g}{\partial x} + v_g \frac{\partial u_g}{\partial y} \right) \hat{\imath} + \left(u_g \frac{\partial v_g}{\partial x} + v_g \frac{\partial v_g}{\partial y} \right) \hat{\jmath} \right]. \tag{6.12}$$

A number of different cases can be used to illustrate the effect of this term. Here we will examine two starting with the case of upper tropospheric diffluent flow illustrated in Figure 6.5(a). At the point in question, the expression in (6.12) is considerably simplified by the fact that the value of v_g at the point is identically zero. Additionally, there is no value of $\partial v_g/\partial x$ there. Thus, the expression (6.12) reduces to

$$\vec{V}_{IA} = \frac{\hat{k}}{f} \times u_g \frac{\partial u_g}{\partial x} \hat{\imath}.$$

Noting that $u_g > 0$ and $\partial u_g/\partial x < 0$ at the indicated point, $u_g \partial u_g/\partial x$ points in the negative x direction. Thus, the inertial advective wind, $\vec{V}_{IA}$, points in the negative y direction as indicated in Figure 6.5(a). As a consequence, there is upper-level divergence to the north of the indicated point and upper-level convergence to the south of it. Thus, the air will rise in the column to the north and sink in the column to the south. This pattern is precisely the same thermally indirect vertical circulation at the jet exit region that we diagnosed earlier.

We can also employ (6.12) to consider the effect of flow curvature on ageostrophy and vertical motions. A northern hemispheric 500 hPa ridge is shown schematically in Figure 6.5(b). If we imagine a case in which the magnitude of the geostrophic flow does not vary across the indicated points, then considerable simplification of (6.12) results. At the central point in Figure 6.5(b) it is clear that there is no value of v_g.

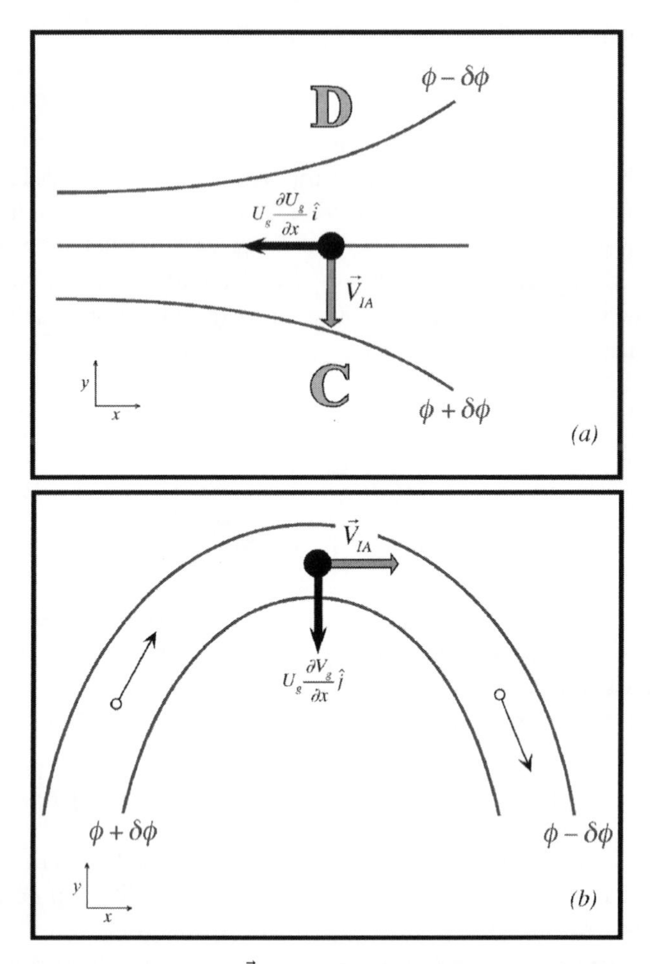

Figure 6.5 (a) Inertial advective wind, $\vec{V}_{IA}$, in diffluent horizontal flow in the northern hemisphere. Solid lines are geopotential height at some upper tropospheric level (such as 300 hPa). Arrows are labeled as described in the text. Upper-level convergence and divergence of the inertial advective wind are represented by C and D, respectively. (b) Inertial advective wind through an upper-level ridge axis in the northern hermisphere. Lines labeled as in (a). Thin arrows represent the geostrophic flow through the ridge. See text for explanation

There is also no value for the derivative $\partial u_g/\partial x$ at that point. Thus, (6.12) reduces to

$$\vec{V}_{IA} = \frac{\hat{k}}{f} \times u_g \frac{\partial v_g}{\partial x} \hat{j}$$

for the situation illustrated in Figure 6.5(b). At the crest of the ridge u_g is positive and $\partial v_g/\partial x$ is negative so that $u_g \partial v_g/\partial x$ points in the negative y direction. Consequently, the inertial advective wind, $\vec{V}_{IA}$, points in the direction of the geostrophic flow through the ridge. This analysis has demonstrated the familiar fact that the flow through a ridge axis is supergeostrophic and that this circumstance arises as a result of the inertial advective component of the ageostrophic wind. The reader should consider the arguments made for confluent and cyclonically curved flow, respectively. Now that we have spent a good deal of effort considering vertical motions that arise from the divergence of the ageostrophic wind (i.e. $\nabla \cdot (\hat{k}/f \times d\vec{V}/dt)$) we will move toward a related expression first derived by Sutcliffe in 1947.

6.2 The Sutcliffe Development Theorem

Sutcliffe made a refinement to his earlier theory (1939) by adopting the geostrophic assumption. Starting with the vector identity $\vec{A} \cdot \vec{B} \times \vec{C} = -\vec{B} \cdot \vec{A} \times \vec{C}$, Sutcliffe reasoned that the divergence of the ageostrophic wind was closely related to changes in the vertical component of vorticity. In mathematical terms,

$$-\nabla \cdot \hat{k} \times \frac{d\vec{V}}{dt} = \hat{k} \cdot \nabla \times \frac{d\vec{V}}{dt}. \tag{6.13}$$

Using the frictionless equations of motion in the x and y directions, the RHS of (6.13) becomes

$$\frac{\partial}{\partial x}\left(\frac{dv}{dt} = -\frac{\partial \phi}{\partial y} - fu\right) - \frac{\partial}{\partial y}\left(\frac{du}{dt} = -\frac{\partial \phi}{\partial x} + fv\right) \tag{6.14}$$

Letting

$$\frac{d}{dt} = \frac{\partial}{\partial t} + u\frac{\partial}{\partial x} + v\frac{\partial}{\partial y}, \quad \zeta = \frac{\partial v}{\partial x} - \frac{\partial u}{\partial y}, \quad \text{and} \quad \eta = \zeta + f,$$

(6.14) is a form of the vorticity equation[2]

$$\frac{d\eta}{dt} = \frac{d(\zeta + f)}{dt} = -(\zeta + f)\nabla \cdot \vec{V} \tag{6.15}$$

showing that vorticity changes are a result of divergence in the fluid. Next, Sutcliffe expanded this expression into its components

$$\frac{\partial(\zeta + f)}{\partial t} + \vec{V} \cdot \nabla(\zeta + f) + \omega\frac{\partial(\zeta + f)}{\partial p} = -(\zeta + f)\nabla \cdot \vec{V} \tag{6.16}$$

and assumed that (1) the vorticity and horizontal winds are geostrophic, (2) the vertical advection of vorticity is negligible, and (3) the relative vorticity can be neglected in the divergence term.[3] This yielded a simplified form of (6.16)

$$\frac{\partial(\zeta_g + f)}{\partial t} + \vec{V}_g \cdot \nabla(\zeta_g + f) = -f_0\nabla \cdot \vec{V} \tag{6.17a}$$

[2] It is derived in the manner of (5.36) but since we have neglected vertical advection ($\omega\partial/\partial p$) in the expression for the Lagrangian derivative, (6.15) contains no tilting term.

[3] These assumptions are precisely those that led to the quasi-geostrophic vorticity equation (5.43).

which can be rewritten as

$$\frac{1}{f}\nabla^2\frac{\partial\phi}{\partial t} + \vec{V}_g \cdot \nabla(\zeta_g + f) = -f_0\nabla \cdot \vec{V} \tag{6.17b}$$

since $\zeta_g = (1/f)\nabla^2\phi$. Now if we consider the differences in divergence between the top and bottom of an air column, we can rewrite (6.17b) as

$$f_0(\nabla \cdot \vec{V} - \nabla \cdot \vec{V}_0) = -\vec{V}_g \cdot \nabla(\zeta_g + f) + \vec{V}_{g_0} \cdot \nabla(\zeta_{g_0} + f) - \frac{1}{f}\nabla^2\frac{\partial\phi'}{\partial t} \tag{6.18}$$

where

$$\frac{\partial\phi'}{\partial t} = \frac{\partial\phi}{\partial t} - \frac{\partial\phi_0}{\partial t}$$

represents the rate of change of thickness in the column. Thus, (6.18) demonstrates that vertical motion is related to (1) the change in the vertical distribution of vorticity by advection, and (2) the Laplacian of variation in the temperature field.

Considering the thickness tendency term first we find that since

$$\frac{\partial\phi'}{\partial t} = \underset{A}{\frac{d\phi'}{dt}} - \underset{B}{\vec{V} \cdot \nabla\phi'} - \underset{C}{\omega\frac{\partial\phi'}{\partial p}}$$

there are three physical processes that can lead to a local change in thickness: (1) diabatic heating (A), (2) horizontal advection (B), and (3) vertical advection (adiabatic temperature changes) (C). If only diabatic heating were acting in the column, then $\partial\phi'/\partial t > 0$ so that $-\nabla^2(\partial\phi'/\partial t) > 0$ and upward vertical motion would result according to (6.18). Conversely, diabatic cooling is associated with downward vertical motion. Adiabatic effects result from the very vertical motions we are trying to diagnose so the vertical advection term is difficult to interpret in this simplified framework and will therefore be neglected as it was in the vorticity equation (6.17a).

Let us consider horizontal advection as a means of producing the local thickness tendency. In such a case, the last term on the RHS of (6.18) could be written as

$$-\frac{1}{f}\nabla^2\frac{\partial\phi'}{\partial t} = \frac{1}{f}\nabla^2\left(\overline{u}_g\frac{\partial\phi'}{\partial x} + \overline{v}_g\frac{\partial\phi'}{\partial y}\right) \tag{6.19}$$

where the overbars denote column-averaged geostrophic winds. Substituting from the thermal wind equation ($fv'_g = \partial\phi'/\partial x$ and $-fu'_g = \partial\phi'/\partial y$) yields

$$-\frac{1}{f}\nabla^2\frac{\partial\phi'}{\partial t} = \nabla^2(\overline{u}_g v'_g - \overline{v}_g u'_g). \tag{6.20a}$$

This expression can be rearranged to yield

$$
\begin{aligned}
-\frac{1}{f}\nabla^2\frac{\partial\phi'}{\partial t}
&= \left(\bar{u}_g\frac{\partial}{\partial x}+\bar{v}_g\frac{\partial}{\partial y}\right)\left(\frac{\partial v'_g}{\partial x}-\frac{\partial u'_g}{\partial y}\right)+\left(\bar{u}_g\frac{\partial}{\partial y}-\bar{v}_g\frac{\partial}{\partial x}\right)\left(\frac{\partial u'_g}{\partial x}+\frac{\partial v'_g}{\partial y}\right)\\
&\quad -\left(u'_g\frac{\partial}{\partial x}+v'_g\frac{\partial}{\partial y}\right)\left(\frac{\partial \bar{v}_g}{\partial x}-\frac{\partial \bar{u}_g}{\partial y}\right)-\left(u'_g\frac{\partial}{\partial y}-v'_g\frac{\partial}{\partial x}\right)\left(\frac{\partial \bar{u}_g}{\partial x}+\frac{\partial \bar{v}_g}{\partial y}\right)
\end{aligned} \tag{6.20b}
$$

so long as certain *products* of derivatives, i.e. terms such as

$$\frac{\partial \bar{u}_g}{\partial x}\frac{\partial v'_g}{\partial x},$$

are neglected.[4] Such terms are known as **deformation terms** and we will consider the consequences of their neglect later in this chapter. The second and fourth terms on the RHS of (6.20b) contain odd mixed derivatives of the divergence of the thermal wind (second term) and the divergence of the layer mean geostrophic wind (fourth term). Both of these quantities are zero and so (6.20b) can be further reduced to

$$-\frac{1}{f}\nabla^2\left(\frac{\partial\phi'}{\partial t}\right) = (\bar{\vec{V}}_g\cdot\nabla)\zeta'_g - (\vec{V'}_g\cdot\nabla)\bar{\zeta}_g. \tag{6.21}$$

Additional simplification arises by employing our definitions of the mean geostrophic wind in the layer ($\bar{\vec{V}}_g = (\vec{V}_g + \vec{V}_{g_0}/2)$), thermal wind in the layer ($\vec{V'}_g = \vec{V}_g - \vec{V}_{g_0}$), along with similar vertical average and vertical difference terms for the geostrophic vorticity ($\bar{\zeta}_g = (\zeta_g + \zeta_{g_0})/2$) and $\zeta'_g = \zeta_g - \zeta_{g_0}$). In this case, (6.21) becomes

$$-\frac{1}{f}\nabla^2\left(\frac{\partial\phi'}{\partial t}\right) = \vec{V}_{g_0}\cdot\nabla\zeta_g - \vec{V}_g\cdot\nabla\zeta_{g_0}. \tag{6.22}$$

This last expression describes only the contribution to net column divergence (via thickness tendencies) made by horizontal advection. So, we must substitute (6.22) into (6.18) to get

$$
\begin{aligned}
f_0(\nabla\cdot\vec{V}-\nabla\cdot\vec{V}_0) &= -\vec{V}_g\cdot\nabla(\zeta_g+f)+\vec{V}_{g_0}\cdot\nabla(\zeta_{g_0}+f)\\
&\quad +\vec{V}_{g_0}\cdot\nabla\zeta_g-\vec{V}_g\cdot\nabla\zeta_{g_0}.
\end{aligned}
$$

[4] The full expansion of $\nabla^2(\bar{u}_g v'_g)$, for instance, involves expanding $\partial^2(\bar{u}_g v'_g)/\partial x^2$ which, by the chain rule, is equal to

$$\frac{\partial}{\partial x}\left[\frac{\partial}{\partial x}(\bar{u}_g v'_g)\right] = \frac{\partial}{\partial x}\left[\bar{u}_g\frac{\partial v'_g}{\partial x}+v'_g\frac{\partial \bar{u}_g}{\partial x}\right] = \bar{u}_g\frac{\partial^2 v'_g}{\partial x^2}+2\frac{\partial \bar{u}_g}{\partial x}\frac{\partial v'_g}{\partial x}+v'_g\frac{\partial^2 \bar{u}_g}{\partial x^2}.$$

The first and third terms of this expression are found in (6.20b), but the product of derivatives term is not.

This expression can be reduced to

$$f_0(\nabla \cdot \vec{V} - \nabla \cdot \vec{V}_0) = -(\vec{V}_g - \vec{V}_{g_0}) \cdot \nabla(\zeta_{g_0} + \zeta_g + f),$$

or finally,

$$f_0(\nabla \cdot \vec{V} - \nabla \cdot \vec{V}_0) = -\vec{V}' \cdot \nabla(\zeta_{g_0} + \zeta_g + f) \tag{6.23}$$

which states that synoptic-scale upward (downward) vertical motions, the result of greater divergence (convergence) aloft than near the surface in any air column, are forced by cyclonic (anticyclonic) vorticity advection by the thermal wind! This is a remarkable result and represented one of the first operationally applicable theoretical results in the history of synoptic–dynamic meteorology. Consider the fact that given geopotential height analyses at two different isobaric levels, say 1000 and 500 hPa, it is easy to calculate graphically the distribution of thickness isopleths, parallel to which flows the thermal wind, $\vec{V}'$. Since $\zeta_g = (1/f)\nabla^2\phi$, it is equally simple to acquire a quick sense of the geostrophic vorticity at both levels. Thus, with just the geopotential height distribution at two levels, a theoretically solid basis for estimating the synoptic-scale vertical motion is offered by (6.23). Figure 6.6 illustrates the utility and ease of application of the Sutcliffe development theorem. Of course, the actual vertical motion distribution (Figure 6.6c) in any given storm is considerably more complicated than what might be expected from the Sutcliffe development term (Figure 6.6b) but the gross features are nicely captured by the simple approximated expression in (6.23).

6.3 The Quasi-Geostrophic Omega Equation

An alternative path to a diagnostic equation for synoptic-scale vertical motions arises from considering the quasi-geostrophic vorticity and thermodynamic energy equations. Recall that these expressions are given by

$$\frac{\partial \zeta_g}{\partial t} = -\vec{V}_g \cdot \nabla(\zeta_g + f) + f_0 \frac{\partial \omega}{\partial p} \tag{6.24a}$$

and

$$\frac{\partial}{\partial t}\left(-\frac{\partial \phi}{\partial p}\right) = -\vec{V}_g \cdot \nabla\left(-\frac{\partial \phi}{\partial p}\right) + \sigma\omega, \tag{6.25a}$$

respectively. Since the geostrophic relative vorticity can be expressed as the Laplacian of geopotential, this pair of equations can be rewritten as

$$\frac{1}{f_0}\nabla^2\left(\frac{\partial \phi}{\partial t}\right) = -\vec{V}_g \cdot \nabla(\zeta_g + f) + f_0 \frac{\partial \omega}{\partial p} \tag{6.24b}$$

$$-\frac{\partial}{\partial p}\left(\frac{\partial \phi}{\partial t}\right) = -\vec{V}_g \cdot \nabla\left(-\frac{\partial \phi}{\partial p}\right) + \sigma\omega. \tag{6.25b}$$

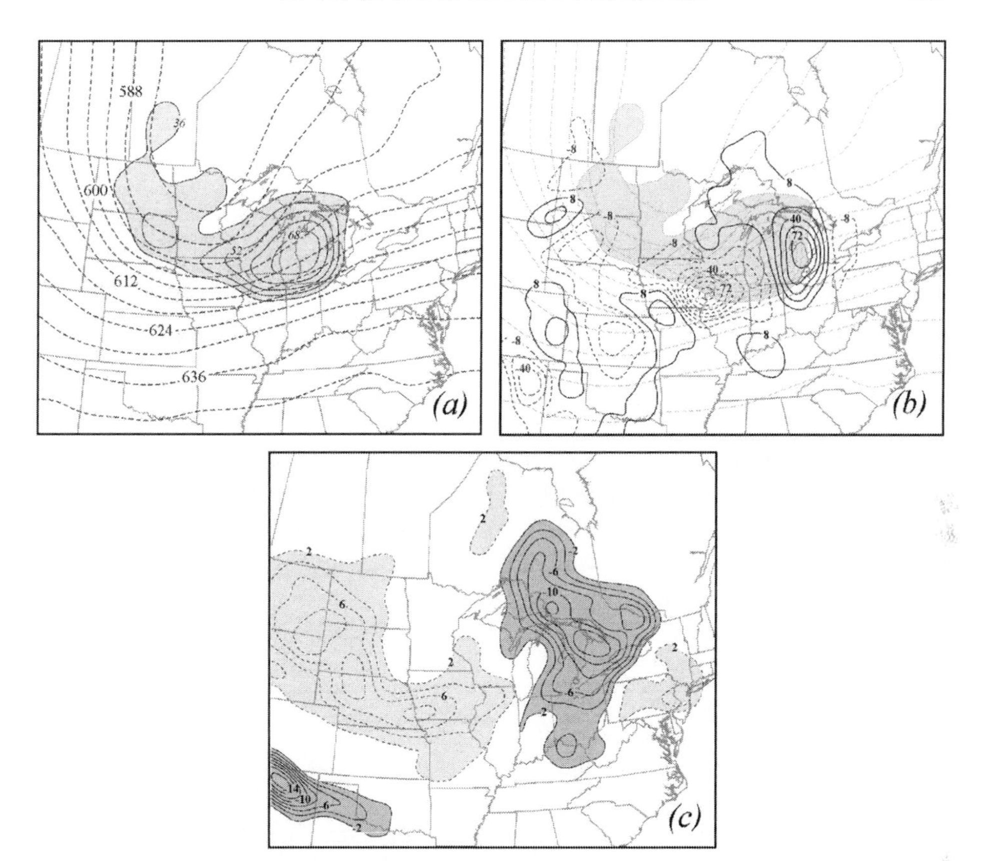

Figure 6.6 (a) The 300–700 hPa thickness (dashed lines) and the sum $\zeta_{g300} + \zeta_{g700} + f$ (shaded) at 0000 UTC 13 November 2003. Thickness is labeled in dam and contoured every 6 dam. Geostrophic vorticity is labeled in 10^{-5} s^{-1} and contoured every 8×10^{-5} s^{-1} starting at 36×10^{-5} s^{-1}. (b) Advection of the sum $\zeta_{g300} + \zeta_{g700} + f$ by the 300–700 hPa thermal wind labeled in 10^{-9} m kg^{-1} and contoured every 16×10^{-9} m kg^{-1} starting at 8×10^{-9} m kg^{-1}. Vorticity and thickness from (a) are shown lightly in the background. (c) The 500 hPa vertical motion at 0000 UTC 13 November 2003. Vertical motion is labeled in μbar s^{-1} (dPa s^{-1}) and contoured every 2 μbar s^{-1} with dark (light) shading corresponding to upward (downward) vertical motion

In order to eliminate the time derivatives in both expressions, we take $f_0\partial/\partial p$ of (6.24b) and ∇^2 of (6.25b) to get

$$\frac{\partial}{\partial p}\nabla^2\left(\frac{\partial\phi}{\partial t}\right) = f_0\frac{\partial}{\partial p}[-\vec{V}_g\cdot\nabla(\zeta_g + f)] + f_0^2\frac{\partial^2\omega}{\partial p^2} \tag{6.24c}$$

$$-\frac{\partial}{\partial p}\nabla^2\left(\frac{\partial\phi}{\partial t}\right) = \nabla^2\left[-\vec{V}_g\cdot\nabla\left(-\frac{\partial\phi}{\partial p}\right)\right] + \sigma\nabla^2\omega. \tag{6.25c}$$

The sum of these two expressions yields

$$0 = f_0\frac{\partial}{\partial p}[-\vec{V}_g\cdot\nabla(\zeta_g + f)] + f_0^2\frac{\partial^2\omega}{\partial p^2} + \nabla^2\left[-\vec{V}_g\cdot\nabla\left(-\frac{\partial\phi}{\partial p}\right)\right] + \sigma\nabla^2\omega,$$

which can be rearranged into

$$\sigma\left(\nabla^2 + \frac{f_0^2}{\sigma}\frac{\partial^2}{\partial p^2}\right)\omega = f_0\frac{\partial}{\partial p}[\vec{V}_g \cdot \nabla(\zeta_g + f)] + \nabla^2\left[\vec{V}_g \cdot \nabla\left(-\frac{\partial \phi}{\partial p}\right)\right], \tag{6.26}$$

known as the **quasi-geostrophic omega equation**. What does this expression mean? First, note that only derivatives in space exist in (6.26) so that it is a *diagnostic* equation for ω in terms of the *instantaneous* geopotential height field. The value of such an expression is that we can obtain from it a measure of ω that is not dependent on accurate observations of the wind. It is a rather complicated-looking expression so we will need to consider what physical meaning the mathematics contains.

The term on the LHS of (6.26), despite its complicated-looking nature, is essentially a 3-D Laplacian term. If we assume that the vertical motion field displays a sinusoidal vertical profile (which turns out to be a very solid assumption) then $\partial^2\omega/\partial p^2 \propto -\omega$. Also, since the Laplacian is a second-derivative operator, a local maximum (minimum) in $\nabla^2\omega$ implies a local minimum (maximum) in ω itself. Thus, whenever the RHS of (6.26) is found to be positive (negative), then $\nabla^2\omega$ is positive (negative) implying that ω is negative (positive), corresponding to upward (downward) vertical motion.

The first term on the RHS of (6.26) physically represents the vertical derivative $(-\partial/\partial p)$ of geostrophic vorticity advection $(-\vec{V}_g \cdot \nabla(\zeta_g + f))$. Thus, if an environment is characterized by geostrophic cyclonic vorticity advection increasing (decreasing) with height, then this term is positive (negative) implying that the environment will be characterized by upward (downward) vertical motion. Note the physical similarity between this term and the role of differential geostrophic vorticity advection in the Sutcliffe development theorem (6.18). Consider the schematic in Figure 6.7. Since geostrophic vorticity near the surface high and low is concentrated at those locations and the circulations are nearly closed at that level, the geostrophic vorticity advection near the surface is rather small. Geostrophic vorticity advection above the surface low, however, is large and positive so that that column experiences upward-increasing cyclonic vorticity advection and, hence, upward vertical motion. Geostrophic vorticity advection above the surface high is large and negative and that column experiences upward-decreasing cyclonic vorticity advection and, hence, downward vertical motion. It is important to note that since geostrophic vorticity is proportional to $\nabla^2\phi$ when the column above the surface low experiences a greater increase in geostrophic vorticity aloft than near the surface, this implies that

$$\frac{\partial}{\partial t}(\nabla^2\phi - \nabla^2\phi_0) > 0$$

or, alternatively,

$$\nabla^2\frac{\partial \phi'}{\partial t} > 0$$

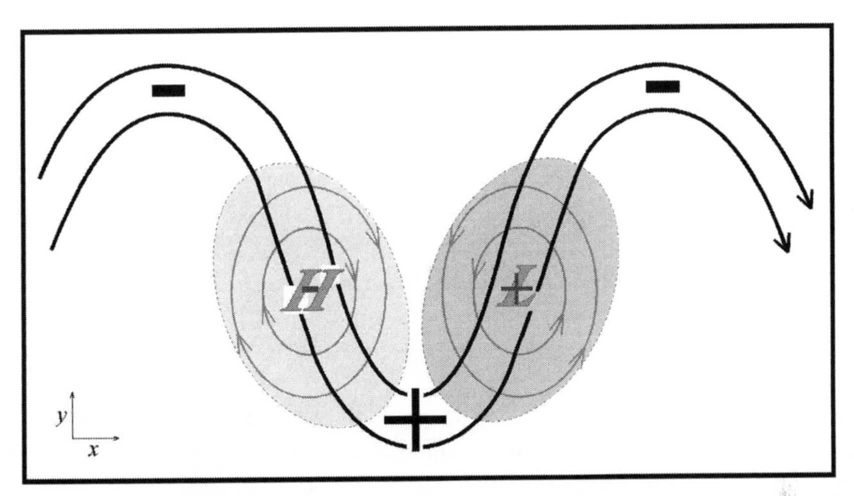

Figure 6.7 The 500 hPa open wave over a surface low and surface high in the northern hemisphere. Thin gray lines are sea-level isobars with accompanying arrows indicating the direction of the near surface geostrophic flow. Thick black lines are the 500 hPa geopotential height with arrows indicating the direction of the 500 hPa geostrophic flow. The '+' and '−' represent the locations of positive and negative relative vorticities, respectively, with the size of the symbol indicating the relative magnitude of the vorticity (i.e. larger at upper levels). Dark (light) gray shaded ovals indicate regions of upward (downward) vertical motions. See text for explantion

which requires that

$$\frac{\partial \phi'}{\partial t} < 0.$$

In order to experience the requisite thickness decrease, the column must cool. The cooling is achieved by adiabatic expansion of the rising air.

The second term on the RHS of (6.26) can be rewritten as

$$\nabla^2\left[\vec{V}_g \cdot \nabla\left(-\frac{\partial \phi}{\partial p}\right)\right] = -\nabla^2\left[-\vec{V}_g \cdot \nabla\left(-\frac{\partial \phi}{\partial p}\right)\right].$$

This alternative form makes it clear that the term in the brackets physically represents horizontal temperature advection. Thus, this entire expression describes the Laplacian of horizontal temperature advection. If an environment is characterized by a *local* maximum in warm (cold) air advection, then this term is positive (negative), corresponding to upward (downward) vertical motion. It is important to note that, according to the quasi-geostrophic omega equation, warm (cold) air advection alone is not enough to diagnose the sense of the vertical motion – it is the Laplacian of the temperature advection that matters. This implies that *only heterogeneity in the thermal advection field is associated with* ω. This is easily demonstrated by considering an alternative form of this term:

$$-\nabla \cdot \nabla\left[-\vec{V}_g \cdot \nabla\left(-\frac{\partial \phi}{\partial p}\right)\right].$$

It is clear that if the *gradient* of horizontal temperature advection is zero (i.e. there is uniform horizontal temperature advection) then the whole term will be zero and no vertical motion will be forced.

The vertical motion fields described by the quasi-geostrophic omega equation are precisely those vertical motions that are required to keep the thermal and mass fields in hydrostatic and geostrophic balance. These diagnosed vertical motions also tend to be an accurate description of the large-scale vertical motions observed in the mid-latitude atmosphere. We will investigate why vertical motions are required to maintain thermal wind balance presently. First, it is enlightening to examine a simplified form of (6.26) that renders a result similar to that described by the Sutcliffe development theorem.

Trenberth (1978) argued that carrying out all the derivatives on the RHS of (6.26) could simplify the forcing function of the quasi-geostrophic omega equation. Expanding the terms in square brackets on the RHS of (6.27) yields

$$\sigma\left(\nabla^2 + \frac{f_0^2}{\sigma}\frac{\partial^2}{\partial p^2}\right)\omega = f_0\frac{\partial}{\partial p}\left[u_g\frac{\partial(\zeta_g + f)}{\partial x} + v_g\frac{\partial(\zeta_g + f)}{\partial y}\right] + \nabla^2\left[-u_g\frac{\partial^2\phi}{\partial x\partial p} - v_g\frac{\partial^2\phi}{\partial y\partial p}\right]. \qquad (6.27)$$

Employing the geostrophic wind relationships

$$u_g = -\frac{1}{f}\frac{\partial\phi}{\partial y} \quad \text{and} \quad v_g = \frac{1}{f}\frac{\partial\phi}{\partial x}$$

along with the definition of the geostrophic relative vorticity ($\zeta_g = (1/f_0)\nabla^2\phi$) gives

$$\sigma\left(\nabla^2 + \frac{f_0^2}{\sigma}\frac{\partial^2}{\partial p^2}\right)\omega = f_0\frac{\partial}{\partial p}\left[-\frac{1}{f}\frac{\partial\phi}{\partial y}\frac{\partial}{\partial x}\left(\frac{1}{f_0}\nabla^2\phi + f\right) + \frac{1}{f}\frac{\partial\phi}{\partial x}\frac{\partial}{\partial y}\left(\frac{1}{f_0}\nabla^2\phi + f\right)\right] - \frac{1}{f}\nabla^2\left[-\frac{\partial\phi}{\partial y}\frac{\partial^2\phi}{\partial x\partial p} + \frac{\partial\phi}{\partial x}\frac{\partial^2\phi}{\partial y\partial p}\right]. \qquad (6.28)$$

Using the *Jacobian* operator, $J(A, B)$, where

$$J(A, B) = \left(\frac{\partial A}{\partial x}\frac{\partial B}{\partial y} - \frac{\partial A}{\partial y}\frac{\partial B}{\partial x}\right),$$

(6.28) can be rewritten as

$$\sigma\left(\nabla^2 + \frac{f_0^2}{\sigma}\frac{\partial^2}{\partial p^2}\right)\omega = \frac{1}{f}[F_1 + F_2]$$

where $F_1$and F_2 are given by

$$F_1 = -J\left(\nabla^2\phi, \frac{\partial\phi}{\partial p}\right) - J\left(\phi, \nabla^2\frac{\partial\phi}{\partial p}\right) - 2\left[J\left(\frac{\partial\phi}{\partial x}, \frac{\partial^2\phi}{\partial x\partial p}\right) + J\left(\frac{\partial\phi}{\partial y}, \frac{\partial^2\phi}{\partial y\partial p}\right)\right] \tag{6.29}$$

$$F_2 = J\left(\frac{\partial\phi}{\partial p}, \nabla^2\phi\right) + J\left(\frac{\partial\phi}{\partial p}, ff_0\right) + J\left(\phi, \nabla^2\frac{\partial\phi}{\partial p}\right), \tag{6.30}$$

respectively. Notice that the second term on the RHS of (6.29) is the additive inverse of the third term on the RHS of (6.30). Also, F_1 contains the deformation terms (the square bracketed term on the RHS of (6.29)) which we neglected in the Sutcliffe development theorem. Upon neglecting them here again, we find that the RHS of (6.28) can be approximated as

$$\frac{1}{f}[F_1 + F_2] \approx \frac{1}{f}\left[2J\left(\frac{\partial\phi}{\partial p}, \nabla^2\phi\right) + J\left(\frac{\partial\phi}{\partial p}, ff_0\right)\right] \tag{6.31a}$$

which can be approximated further, without much error and to facilitate application, as

$$\frac{1}{f}[F_1 + F_2] \approx \frac{2}{f}\left[J\left(\frac{\partial\phi}{\partial p}, \nabla^2\phi\right) + J\left(\frac{\partial\phi}{\partial p}, ff_0\right)\right]. \tag{6.31b}$$

Upon expansion of (6.31b), we find that an approximate form of the RHS of (6.26), the quasi-geostrophic omega equation, is given by

$$\sigma\left(\nabla^2 + \frac{f_0^2}{\sigma}\frac{\partial^2}{\partial p^2}\right)\omega \approx 2\left[f_0\frac{\partial\vec{V}_g}{\partial p}\cdot\nabla(\zeta_g + f)\right]. \tag{6.32}$$

Thus, even when proceeding from the classical quasi-geostrophic omega equation, the fundamental physical insight achieved by Sutcliffe is confirmed: that is, large-scale mid-latitude vertical motions are forced by thermal wind advection of absolute geostrophic vorticity.

This result is both reassuring and convenient in the sense that it compresses a rather complicated expression (6.26) into a single forcing term that is easy to employ qualitatively using standard observations. Some nagging questions undoubtedly persist in the reader's mind at this point in our discussion, however. Perhaps chief among them is: *Is it reasonable that we continually neglect the so-called deformation terms in deriving diagnostic expressions for the large-scale, mid-latitude vertical motions?* Also, perhaps lurking deeper in the mind of the reader: *How do these quasi-geostrophic vertical motions serve to maintain the thermal wind balance?* Before we consider the second question, let us first consider the nature of the neglected deformation terms. The deformation terms appeared explicitly in (6.29) as

$$DEF = -\frac{2}{f}\left[J\left(\frac{\partial\phi}{\partial x}, \frac{\partial^2\phi}{\partial x\partial p}\right) + J\left(\frac{\partial\phi}{\partial y}, \frac{\partial^2\phi}{\partial y\partial p}\right)\right]. \tag{6.33a}$$

Employing the geostrophic ($fv_g = \partial\phi/\partial x$ and $-fu_g = \partial\phi/\partial y$) and the hydrostatic ($\partial\phi/\partial p = -RT/p$) relationships, this can be expressed as

$$DEF = -\frac{2}{f}\left[J\left(fv_g, -\frac{R}{p}\frac{\partial T}{\partial x}\right) + J\left(-fu_g, -\frac{R}{p}\frac{\partial T}{\partial y}\right)\right]$$

or

$$DEF = \frac{2R}{p}\left[J\left(v_g, \frac{\partial T}{\partial x}\right) - J\left(u_g, \frac{\partial T}{\partial y}\right)\right]. \qquad (6.33b)$$

Carrying out the indicated derivatives and then grouping like terms together yields

$$DEF = \frac{2R}{p}\left[\left(\frac{\partial v_g}{\partial x} + \frac{\partial u_g}{\partial y}\right)\frac{\partial^2 T}{\partial x \partial y} - \frac{\partial v_g}{\partial y}\frac{\partial^2 T}{\partial x^2} - \frac{\partial u_g}{\partial x}\frac{\partial^2 T}{\partial y^2}\right]. \qquad (6.33c)$$

Denoting the geostrophic shearing deformation, $(\partial v_g/\partial x + \partial u_g/\partial y)$, as SH, the geostrophic stretching deformation, $(\partial u_g/\partial x - \partial v_g/\partial y)$, as ST, and employing the non-divergence of the geostrophic wind, (6.33c) can be rewritten as

$$DEF = \frac{2R}{P}\left[(SH)\frac{\partial^2 T}{\partial x \partial y} + \frac{(ST)}{2}\left(\frac{\partial^2 T}{\partial x^2} - \frac{\partial^2 T}{\partial y^2}\right)\right] \qquad (6.33d)$$

which illustrates that the deformation terms will be significant where second derivatives of temperature are coincident with deformation (i.e. first derivatives) in the geostrophic wind field. Mid-latitude frontal regions, as we will see, are defined by such conditions. This fact has led to the historical assumption that the deformation term is only large in frontal regions. As it turns out, a number of other recurrent but non-frontal thermal structures in mid-latitude cyclones are also characterized by these conditions – most notably the large-scale thermal ridge often associated with occluded cyclones. From this perspective, neglect of the deformation term is liable to lead to significant misdiagnosis in many canonical mid-latitude cyclone environments (as we will show later). Next we will derive an alternative expression for the forcing for quasi-geostrophic vertical motions that includes these terms, lends additional insight into the nature of the mid-latitude atmosphere, and is amenable to simple graphical evaluation.

6.4 The $\vec{Q}$-Vector

The remainder of this chapter will be devoted to examining the so-called $\vec{Q}$-vector form of the quasi-geostrophic omega equation introduced by Hoskins *et al.* (1978). Consideration of the $\vec{Q}$-vector reveals an unexpected and intriguing characteristic of the thermal wind balance that will serve as a cornerstone in the development of a deeper conceptual understanding of the nature of quasi-geostrophic vertical motions. We begin by investigating the geostrophic paradox.

6.4.1 The geostrophic paradox and its resolution

Consider the jet entrance region depicted in Figure 6.8. The confluent, geostrophic wind field depicted there acts to tighten the horizontal temperature gradient at C. Any such increase in the magnitude of the temperature gradient forces an increase in the geostrophic vertical shear via the thermal wind relationship. Simultaneously, the geostrophic wind advects lower geostrophic momentum (quantified by the isotachs of the y-direction geostrophic wind) into the jet core. The momentum advection tends to decrease the wind speed at C and, thus, contributes to a decrease in the vertical shear of the geostrophic wind in that column. Thus, the very same geostrophic flow that serves to increase the magnitude of the horizontal temperature gradient at C also serves to decrease the vertical shear of the geostrophic wind at C via negative geostrophic momentum advection. This set of circumstances presents a paradox: that is, on the one hand, geostrophic temperature advection should increase the thermal wind at C and, on the other, geostrophic momentum advection should decrease it at C. So, the geostrophic wind actually destroys thermal wind balance by affecting opposite signed changes to the two components of that balance. Since the thermal wind balance is a form of the geostrophic balance, it can therefore be said that the geostrophic wind destroys itself! We will refer to this property of the geostrophic flow as the geostrophic paradox.

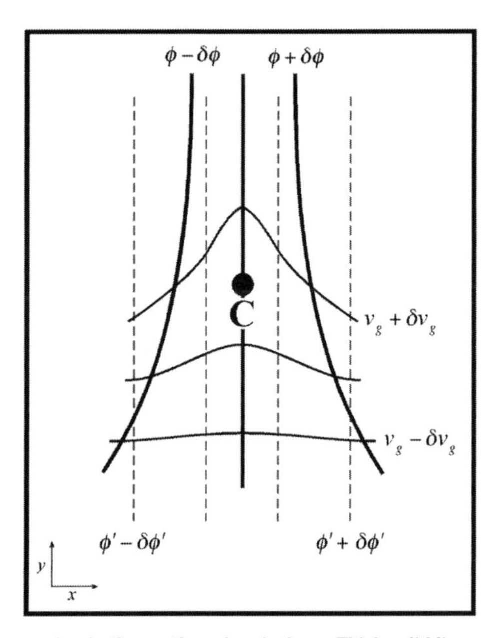

Figure 6.8 Jet entrance region in the northern hemisphere. Thick solid lines are 500 hPa geopotential height, dashed lines are 1000–500 hPa thickness, and thin solid lines are isotachs of the y-direction geostrophic wind. Point C is mentioned in the explanation given in the text

Interestingly, however, observations suggest that the synoptic-scale flow in the middle latitudes is very nearly in geostrophic balance at all times. How can this be in the face of what we have just described? There must be another portion of the flow that acts to maintain the geostrophic balance in the face of its self-destructive tendency. That portion of the flow is the forced, ageostrophic, secondary circulation.[5] Since the geostrophic flow tends to create thermal wind imbalance, the forced secondary circulation must bring the flow back toward a state of geostrophic balance. This may be accomplished if the secondary circulation counteracts the tendencies induced by the geostrophic wind itself. Therefore, the secondary, ageostrophic circulation operating in the vicinity of the jet entrance region depicted in Figure 6.8 must simultaneously (1) decrease the magnitude of the horizontal temperature gradient, and (2) increase the vertical shear. We now examine a derivation that quantifies the geostrophic paradox and in so doing leads to a description of the forced, secondary circulation that resolves it.

We begin by considering both the thermodynamic energy equation and the y equation of motion at the level of quasi-geostrophic theory:

$$\left(\frac{\partial}{\partial t} + \vec{V}_g \cdot \nabla\right) v_g + f_0 u_{ag} = 0 \quad \text{and} \quad \left(\frac{\partial}{\partial t} + \vec{V}_g \cdot \nabla\right)\left(-\frac{\partial \phi}{\partial p}\right) - \sigma\omega = 0.$$

Neglecting the ageostrophy for the moment, these expressions can be rewritten as

$$\left(\frac{\partial}{\partial t} + \vec{V}_g \cdot \nabla\right) v_g = 0 \tag{6.34a}$$

and

$$\left(\frac{\partial}{\partial t} + \vec{V}_g \cdot \nabla\right)\left(-\frac{\partial \phi}{\partial p}\right) = 0. \tag{6.35a}$$

Recall that the thermal wind balance for the situation depicted in Figure 6.8 is given by

$$f_0 \frac{\partial v_g}{\partial p} = \frac{\partial^2 \phi}{\partial x \partial p}.$$

Now, $f_0 \partial/\partial p$ of (6.34a) is equal to

$$\begin{aligned} f_0 \frac{\partial}{\partial p}\left[\left(\frac{\partial}{\partial t} + \vec{V}_g \cdot \nabla\right) v_g\right] &= f_0 \frac{\partial}{\partial p}\left[\frac{\partial v_g}{\partial t} + u_g \frac{\partial v_g}{\partial x} + v_g \frac{\partial v_g}{\partial y}\right] \\ &= \left(\frac{\partial}{\partial t} + \vec{V}_g \cdot \nabla\right)\left(f_0 \frac{\partial v_g}{\partial p}\right) \\ &\quad + f_0 \left[\frac{\partial u_g}{\partial p}\frac{\partial v_g}{\partial x} + \frac{\partial v_g}{\partial p}\frac{\partial v_g}{\partial y}\right]. \end{aligned}$$

[5] This flow is referred to as 'secondary' in order to distinguish it from the primary, geostrophic flow.

Employing the thermal wind relationship and the non-divergence of the geostrophic wind, this can be rewritten as

$$f_0 \frac{\partial}{\partial p}\left[\left(\frac{\partial}{\partial t} + \vec{V}_g \cdot \nabla\right) v_g\right] = \left(\frac{\partial}{\partial t} + \vec{V}_g \cdot \nabla\right)\left(f_0 \frac{\partial v_g}{\partial p}\right) + \left[\frac{\partial \vec{V}_g}{\partial x} \cdot \nabla\left(-\frac{\partial \phi}{\partial p}\right)\right]. \tag{6.34b}$$

Interestingly, $-\partial/\partial x$ of (6.35a) is equal to

$$\begin{aligned}
&-\frac{\partial}{\partial x}\left[\left(\frac{\partial}{\partial t} + \vec{V}_g \cdot \nabla\right)\left(-\frac{\partial \phi}{\partial p}\right)\right] \\
&= -\frac{\partial}{\partial x}\left[\frac{\partial}{\partial t}\left(-\frac{\partial \phi}{\partial p}\right) + u_g \frac{\partial}{\partial x}\left(-\frac{\partial \phi}{\partial p}\right) + v_g \frac{\partial}{\partial y}\left(-\frac{\partial \phi}{\partial p}\right)\right] \\
&= \left(\frac{\partial}{\partial t} + \vec{V}_g \cdot \nabla\right)\left(\frac{\partial^2 \phi}{\partial x \partial p}\right) - \left[\frac{\partial \vec{V}_g}{\partial x} \cdot \nabla\left(-\frac{\partial \phi}{\partial p}\right)\right].
\end{aligned} \tag{6.35b}$$

Examination of the last lines of (6.34b) and (6.35b) proves that the geostrophic tendencies of $f_0 \partial v_g/\partial p$ and $\partial^2\phi/\partial x \partial p$ (the two components of the thermal wind balance) have equal magnitude but opposite sign! Thus, the geostrophic wind destroys itself by changing the two parts of the thermal wind balance equally, but in opposite directions. Let us denote the magnitude of this geostrophic tendency as Q_1 so that

$$Q_1 = -\frac{\partial \vec{V}_g}{\partial x} \cdot \nabla\left(-\frac{\partial \phi}{\partial p}\right).$$

If we now reinsert the ageostrophic terms that we previously neglected in developing (6.34a) and (6.35a), we get

$$f_0 \frac{\partial}{\partial p}\left[\left(\frac{\partial}{\partial t} + \vec{V}_g \cdot \nabla\right) v_g + f_0 u_{ag}\right] = \left(\frac{\partial}{\partial t} + \vec{V}_g \cdot \nabla\right)\left(f_0 \frac{\partial v_g}{\partial p}\right) - Q_1 + f_0^2 \frac{\partial u_{ag}}{\partial p} \tag{6.36}$$

and

$$-\frac{\partial}{\partial x}\left[\left(\frac{\partial}{\partial t} + \vec{V}_g \cdot \nabla\right)\left(-\frac{\partial \phi}{\partial p}\right) - \sigma\omega\right] = \left(\frac{\partial}{\partial t} + \vec{V}_g \cdot \nabla\right)\left(\frac{\partial^2 \phi}{\partial x \partial p}\right) + Q_1 + \sigma \frac{\partial \omega}{\partial x}. \tag{6.37}$$

Multiplying (6.37) by –1 and adding it to (6.36) eliminates the time derivatives (since $f_0 \partial v_g / \partial p = \partial^2 \phi / \partial x \partial p$ by the thermal wind) and yields

$$-2Q_1 = \sigma \frac{\partial \omega}{\partial x} - f_0^2 \frac{\partial u_{ag}}{\partial p}. \tag{6.38}$$

The same set of operations can be performed on the x equation of motion and the thermodynamic energy equation resulting in

$$-2Q_2 = \sigma \frac{\partial \omega}{\partial y} - f_0^2 \frac{\partial v_{ag}}{\partial p} \tag{6.39}$$

where

$$Q_2 = -\frac{\partial \vec{V}_g}{\partial y} \cdot \nabla \left(-\frac{\partial \phi}{\partial p} \right).$$

Finally, taking $\partial / \partial x$ of (6.38) and adding it to $\partial / \partial y$ of (6.39) produces

$$-2 \left(\frac{\partial Q_1}{\partial x} + \frac{\partial Q_2}{\partial y} \right) = \sigma \left(\frac{\partial^2 \omega}{\partial x^2} + \frac{\partial^2 \omega}{\partial y^2} \right) - f_0^2 \frac{\partial}{\partial p} \left(\frac{\partial u_{ag}}{\partial x} + \frac{\partial v_{ag}}{\partial y} \right)$$

which becomes, upon substituting from the continuity equation,

$$-2 \left(\frac{\partial Q_1}{\partial x} + \frac{\partial Q_2}{\partial y} \right) = \sigma \left(\frac{\partial^2 \omega}{\partial x^2} + \frac{\partial^2 \omega}{\partial y^2} \right) + f_0^2 \frac{\partial^2 \omega}{\partial p^2} = \sigma \left(\nabla^2 + \frac{f_0^2}{\sigma} \frac{\partial^2}{\partial p^2} \right) \omega. \tag{6.40}$$

The RHS of (6.40) is identically the 3-D Laplacian operator found on the LHS of the classical quasi-geostrophic omega equation (6.26). The forcing function in this form of the quasi-geostrophic omega equation is given by twice the convergence of a 2-D horizontal vector quantity, the $\vec{Q}$-vector, defined as $\vec{Q} = (Q_1, Q_2)$ or

$$\vec{Q} = \left[\left(-\frac{\partial \vec{V}_g}{\partial x} \cdot \nabla \left(-\frac{\partial \phi}{\partial p} \right) \right) \hat{\imath}, \left(-\frac{\partial \vec{V}_g}{\partial y} \cdot \nabla \left(-\frac{\partial \phi}{\partial p} \right) \right) \hat{\jmath} \right]. \tag{6.41}$$

Using the hydrostatic relationship ($\partial \phi / \partial p = -RT/p$) we can rewrite this expression in a more convenient form as

$$\vec{Q} = -\frac{R}{p} \left[\left(\frac{\partial \vec{V}_g}{\partial x} \cdot \nabla T \right) \hat{\imath}, \left(\frac{\partial \vec{V}_g}{\partial y} \cdot \nabla T \right) \hat{\jmath} \right]$$

which is easier to employ with real weather maps. Looking again at (6.40), we see that if $\vec{Q}$ is convergent (divergent) then upward (downward) vertical motion results. Also, note that in deriving (6.40) there was no neglect of the deformation terms as we had been forced to do in prior derivations of an omega equation.

Now let us return to our original example of confluent flow superimposed upon a temperature gradient shown in Figure 6.8. The traditional approximations to the quasi-geostrophic omega equation might not be of much help in diagnosing omega in this environment since vorticity advection is rather difficult to determine here. The completeness of the $\vec{Q}$-vector comes at the price of increased complication, however. Therefore, we examine the full expression of the $\vec{Q}$-vector in order to determine if a simplification, applicable to the example shown in Figure 6.8, is possible. The full expression of $\vec{Q}$ is given by

$$\vec{Q} = -\frac{R}{p}\left[\left(\frac{\partial u_g}{\partial x}\frac{\partial T}{\partial x} + \frac{\partial v_g}{\partial x}\frac{\partial T}{\partial y}\right)\hat{i} + \left(\frac{\partial u_g}{\partial y}\frac{\partial T}{\partial x} + \frac{\partial v_g}{\partial y}\frac{\partial T}{\partial y}\right)\hat{j}\right]. \quad (6.42)$$

But there is no $\partial T/\partial y$ in Figure 6.8 so, again employing the non-divergence of the geostrophic wind, $\vec{Q}$ simplifies to

$$\begin{aligned}\vec{Q} &= -\frac{R}{p}\left[\left(\frac{\partial u_g}{\partial x}\frac{\partial T}{\partial x}\right)\hat{i} + \left(\frac{\partial u_g}{\partial y}\frac{\partial T}{\partial x}\right)\hat{j}\right] = -\frac{R}{p}\left(\frac{\partial T}{\partial x}\right)\left(\frac{-\partial v_g}{\partial y}\hat{i} + \frac{\partial u_g}{\partial y}\hat{j}\right) \\ &= -\frac{R}{p}\left(\frac{\partial T}{\partial x}\right)\left[\hat{k} \times \frac{\partial \vec{V}_g}{\partial y}\right]. \end{aligned} \quad (6.43)$$

So if one measures the change in the geostrophic wind vector along isotherms (i.e. along the y-axis), then the direction of the resulting $\vec{Q}$-vector is determined as the vertical cross-product of that vector change with its magnitude modulated by the intensity of the x-direction temperature gradient.

Figure 6.9(a) shows the $\vec{Q}$-vectors for the confluent jet entrance of Figure 6.8. This configuration of $\vec{Q}$-vectors results in $\vec{Q}$ convergence in the warm air and $\vec{Q}$ divergence in the cold air. Consequently, we have diagnosed a thermally direct, secondary, vertical circulation in which the warm air rises and the cold air sinks (Figure 6.9b). Such a secondary ageostrophic circulation achieves two important modifications of the environment. First, adiabatic cooling of the rising warm air and adiabatic warming of the sinking cold air decrease the magnitude of ∇T. This exactly counteracts the tendency of the geostrophic temperature advection in the confluent flow! Second, under the influence of the Coriolis force, the horizontal branches of this secondary ageostrophic circulation tend to increase the vertical wind shear – exactly counteracting the tendency of the geostrophic momentum advection in the confluent flow! Thus, the secondary ageostrophic circulation diagnosed with the $\vec{Q}$-vectors is precisely that necessary to restore the thermal wind balance in the face of the geostrophic wind's tendency to destroy the balance.

6.4.2 A natural coordinate version of the $\vec{Q}$-Vector

As we have just seen, the $\vec{Q}$-vector is a rather bulky expression but $-2\nabla \cdot \vec{Q}$ represents a complete form of the forcing in the quasi-geostrophic omega equation. Here we consider an expression for the $\vec{Q}$-vector distilled into a natural coordinate version

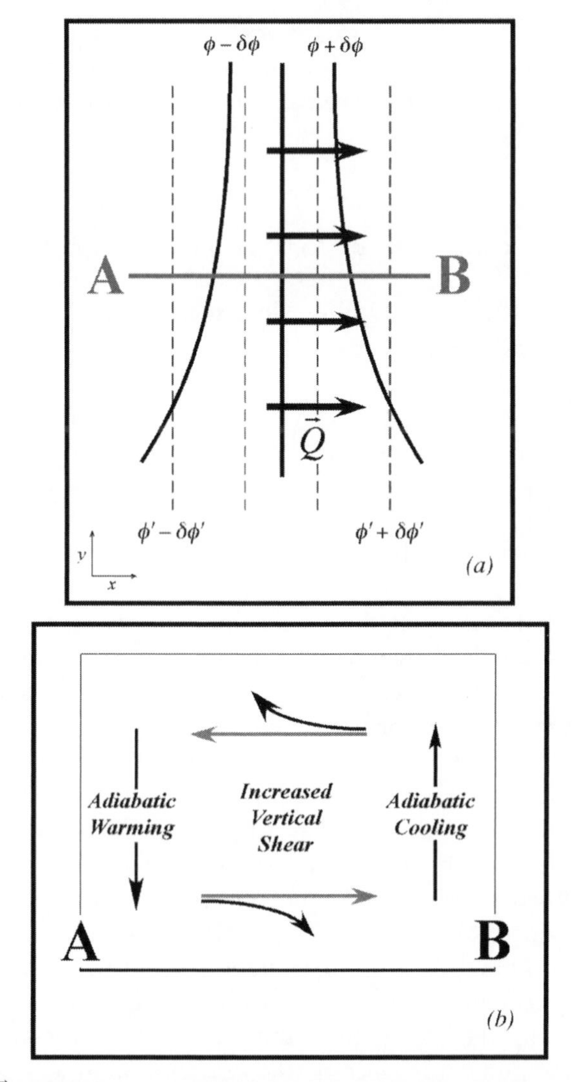

Figure 6.9 (a) $\vec{Q}$-vectors for the confluent jet entrance region depicted in Figure 6.8. Vertical cross-section along line A–B is shown in (b). (b) Vertical cross-section along line A–B in (a). Black arrows represent the vertical and horizontal branches of the secondary, ageostrophic circulation associated with the $\vec{Q}$-vector distribution in (a). Gray arrows represent the direction of the horizontal branch of the forced circulation before the *Coriolis* force turns in to the right. See text for explanation

that is easily applied to weather maps.[6] We begin with (6.42)

$$\vec{Q} = -\frac{R}{p}\left[\left(\frac{\partial u_g}{\partial x}\frac{\partial T}{\partial x} + \frac{\partial v_g}{\partial x}\frac{\partial T}{\partial y}\right)\hat{i} + \left(\frac{\partial u_g}{\partial y}\frac{\partial T}{\partial x} + \frac{\partial v_g}{\partial y}\frac{\partial T}{\partial y}\right)\hat{j}\right]$$

[6] This discussion follows work originally done by Sanders and Hoskins (1990).

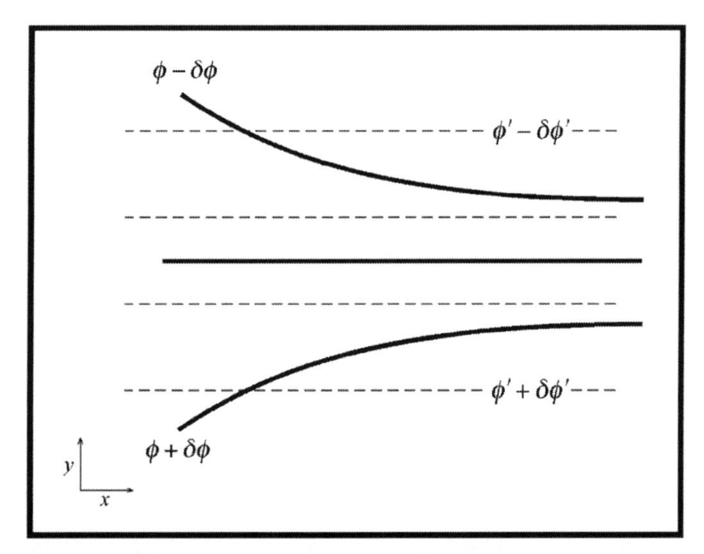

Figure 6.10 Zonally oriented, confluent jet entrance region in the northern hemisphere. Thick solid lines are 500 hPa geopotential height, dashed lines are 1000–500 hPa thickness. Note that for this flow configuration, $\partial T/\partial x = 0$

and consider, independently, two extreme examples in which $\partial T/\partial x = 0$ and $\partial T/\partial y = 0$. For the case of $\partial T/\partial x = 0$ we consider the confluent entrance region of a zonally oriented jet as in Figure 6.10. In such an environment, the above expression reduces to

$$\vec{Q} = -\frac{R}{p}\left(\frac{\partial T}{\partial y}\right)\left[\frac{\partial v_g}{\partial x}\hat{i} + \frac{\partial v_g}{\partial y}\hat{j}\right] = -\frac{R}{p}\left(\frac{\partial T}{\partial y}\right)\left[\frac{\partial v_g}{\partial x}\hat{i} - \frac{\partial u_g}{\partial x}\hat{j}\right]$$
$$= \frac{R}{p}\left(\frac{\partial T}{\partial y}\right)\left[\hat{k} \times \frac{\partial \vec{V}_g}{\partial x}\right]$$

since the geostrophic wind is non-divergent and

$$\frac{\partial v_g}{\partial x}\hat{i} - \frac{\partial u_g}{\partial x}\hat{j} = -\hat{k} \times \frac{\partial \vec{V}_g}{\partial x}.$$

Note that in this example, the x-axis is in the along-flow direction and the y-axis is in the across-flow direction, pointing toward colder air.

For the case of $\partial T/\partial y = 0$, we appeal to the confluent jet entrance in Figure 6.8 used to illustrate the utility of the $\vec{Q}$-vector. In that example, we found that the expression for $\vec{Q}$ reduced to

$$\vec{Q} = -\frac{R}{p}\left(\frac{\partial T}{\partial x}\right)\left[\hat{k} \times \frac{\partial \vec{V}_g}{\partial y}\right]$$

and the y-axis was in the along-flow direction with the x-axis in the across-flow direction pointing toward warmer air.

Let us now adopt natural coordinates $(\hat{s}, \hat{n})$ such that $\hat{s}$ is directed along the isotherms and $\hat{n}$ is directed across the isotherms toward warmer air. For the case of $\partial T/\partial x = 0$ (Figure 6.10) we could say that $\partial T/\partial y = -|\partial T/\partial n|$ (since $\partial T/\partial y < 0$). Analogously, we could say that $\partial \vec{V}_g/\partial x = \partial \vec{V}_g/\partial s$ so that our natural coordinate expression for $\vec{Q}$ would be

$$\vec{Q} = -\frac{R}{p}\left|\frac{\partial T}{\partial n}\right|\left[\hat{k} \times \frac{\partial \vec{V}_g}{\partial s}\right].$$

For the case of $\partial T/\partial y = 0$, we could say that $\partial T/\partial x = |\partial T/\partial n|$ (since $\partial T/\partial x > 0$). Also, we could say that $\partial \vec{V}_g/\partial y = \partial \vec{V}_g/\partial s$ so that our natural coordinate expression for $\vec{Q}$ would be, again,

$$\vec{Q} = -\frac{R}{p}\left|\frac{\partial T}{\partial n}\right|\left[\hat{k} \times \frac{\partial \vec{V}_g}{\partial s}\right], \tag{6.44}$$

demonstrating that this expression serves as the general natural coordinate expression for $\vec{Q}$. In order to apply this expression, we simply denote the vector change in the geostrophic wind along isotherms, take the vertical cross-product of that vector, and flip the resultant direction by 180° (as we must multiply by –1) to determine the direction of $\vec{Q}$. The magnitude is modulated by $|\partial T/\partial n|$.

Now we examine some examples for which the answers should be fairly familiar. First, let us consider a pattern of sea-level isobars and isotherms for an idealized train of cyclones and anticyclones, illustrated in Figure 6.11. Choosing the middle isotherm as our $\hat{s}$-axis, we need only consider the vector change in the geostrophic wind along *that* isotherm. Upon doing so we find that the $\vec{Q}$-vectors converge to the east of the sea-level low-pressure center and diverge to its west. Thus, we have diagnosed ascent to the east of the cyclone and descent to the east of the anticyclone.

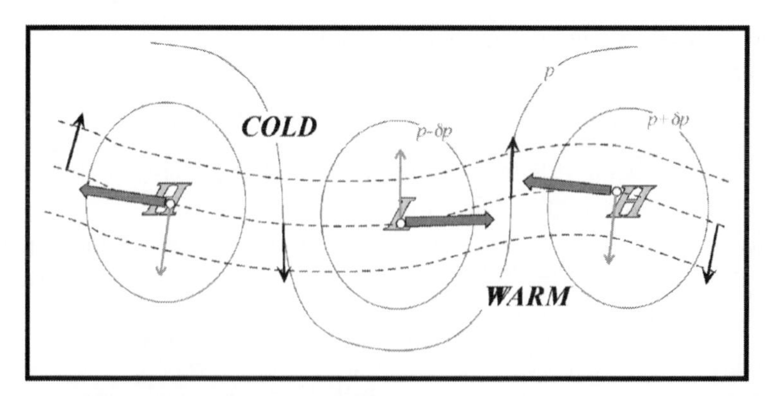

Figure 6.11 Schematic train of lows and highs in the northern hemisphere. Thin solid lines are sea-level isobars, black dashed lines are 1000–500 hPa thickness, black arrows are surface geostrophic winds, light gray arrows represent $\partial \vec{V}_g/\partial s$, and shaded arrows are $\vec{Q}$-vectors. See text for explanation

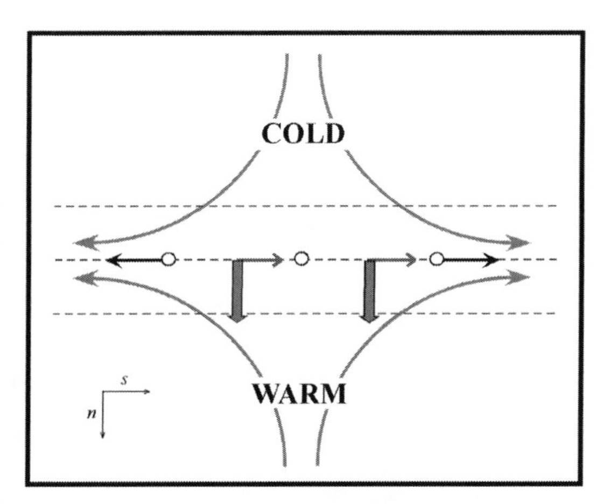

Figure 6.12 Isotherms in a region of geostrophic deformation. Curved gray arrows are geostrophic streamlines, dashed lines are isotherms or thickness isopleths, black arrows are geostrophic winds at the indicated circles. This gray arrows represent $\partial \vec{V}_g / \partial s$, and the shaded gray arrows are the $\vec{Q}$-vectors

In this way, the train of cyclones and anticyclones propagates to the east, in the direction of the thermal wind – a result we noted earlier in the chapter.

Next we consider a zonally oriented bundle of isentropes placed in a region of pure geostrophic deformation as illustrated in Figure 6.12. Clearly, this environment would not be easily diagnosed using the traditional form of the quasi-geostrophic omega equation nor any of the approximations to it that we have examined. Picking the middle isotherm as the $\hat{s}$-axis, we need only consider the geostrophic wind variation along that isotherm. The resulting $\vec{Q}$-vectors are uniformly pointed toward the warm side of the baroclinic zone, indicating rising warm air and sinking cold air – a thermally direct vertical circulation. The differential thermal advection occurring in this deformation zone would tend to bring the isotherms closer together in the horizontal, thereby increasing the thermal wind shear. This same underlying dynamical principle was discussed in reference to Figure 6.4(b). In the next chapter we will more fully discuss the relationship between changes in the temperature gradient and attendant vertical circulations as we discuss frontogenesis. Finally we consider a hypothetical field of uniform geostrophic temperature advection as depicted in Figure 6.13. It is easy to demonstrate that since there is no variation of the geostrophic wind along any isotherm, there is no $\vec{Q}$-vector field and, hence, no quasi-geostrophic vertical motion.

We have said that the $\vec{Q}$-vector form of the quasi-geostrophic omega equation is a complete form of the forcing. This distinguishes it from the Sutcliffe and Trenberth approximations wherein the deformation terms are neglected. Two reasonable questions to ask at this point in our discussion are (1) where are the deformation terms hiding in the $\vec{Q}$–vector forcing, and (2) are they really negligible? The first question is rather academic but the second is crucially important to operational forecasting.

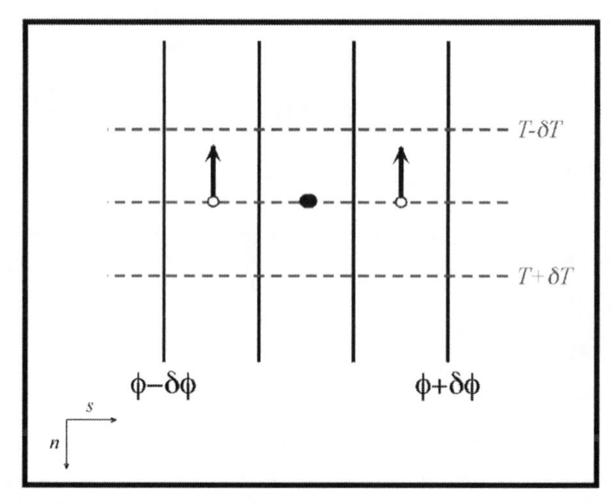

Figure 6.13 Geopotential heights (thick black lines) and isotherms (dashed lines) in a field of uniform geostrophic warm air advection. Arrows are the geostrophic winds at the indicated points. Since the geostrophic flow is uniform, $\partial\vec{V}_g/\partial s$ is zero at the black dot and hence there is no $\vec{Q}$-vector and no $\vec{Q}$-vector divergence

Recall that the forcing for ω in the $\vec{Q}$-vector form of the quasi-geostrophic omega equation is given by

$$Forcing = -2\nabla\cdot\vec{Q} = -2\left(\frac{\partial Q_1}{\partial x}+\frac{\partial Q_2}{\partial y}\right). \tag{6.45}$$

Using (6.42), this can be written as

$$Forcing = -2\frac{R}{p}\left[\frac{\partial}{\partial x}\left(-\frac{\partial\vec{V}_g}{\partial x}\cdot\nabla T\right)+\frac{\partial}{\partial y}\left(-\frac{\partial\vec{V}_g}{\partial y}\cdot\nabla T\right)\right]$$

which expands to four terms after applying the chain rule to yield

$$\begin{aligned} Forcing = -2\frac{R}{p}\Bigg\{&\left[\frac{\partial}{\partial x}\left(-\frac{\partial\vec{V}_g}{\partial x}\right)\cdot\nabla T+\frac{\partial}{\partial y}\left(-\frac{\partial\vec{V}_g}{\partial y}\right)\cdot\nabla T\right]\\ &+\left[-\frac{\partial\vec{V}_g}{\partial x}\cdot\nabla\frac{\partial T}{\partial x}-\frac{\partial\vec{V}_g}{\partial y}\cdot\nabla\frac{\partial T}{\partial y}\right]\Bigg\}. \end{aligned} \tag{6.46}$$

It is left as an exercise to the reader to show that the first square bracketed term on the RHS of (6.46) is exactly equal to the Sutcliffe/Trenberth approximation to the forcing function of the quasi-geostrophic omega equation. Of course, that means that the second square bracketed term on the RHS of (6.46) represents the oft neglected deformation terms. As pointed out previously, these terms will be significant any time a second derivative of temperature is coincident with a first derivative of the geostrophic wind. Frontal zones fit this description but many other characteristic thermal structures observed in mid-latitude cyclones do as well. Figure 6.14

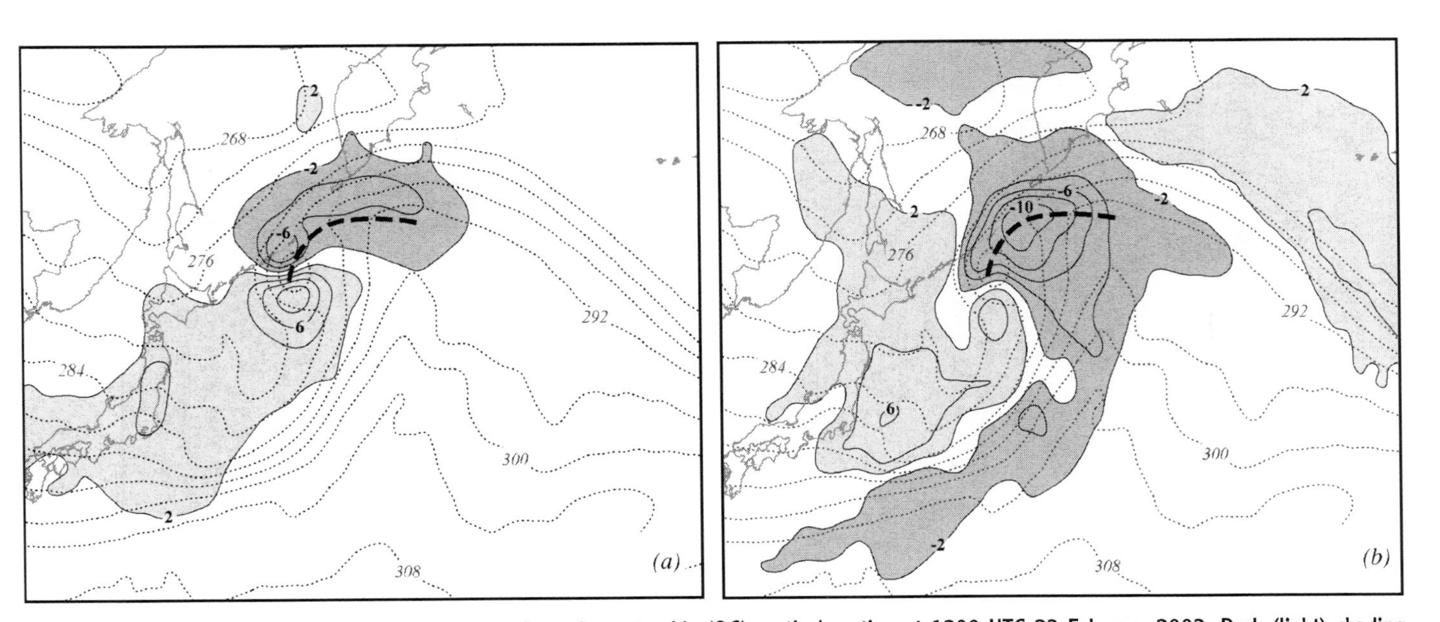

Figure 6.14 The 700 hPa potential temperature and quasi-geostrophic (QG) vertical motion at 1200 UTC 23 February 2003. Dark (light) shading represents upward (downward) vertical motion labeled in μ bar s^{-1} (dPa s^{-1}) and contoured every -2 (2) μ bar s^{-1} starting at -2 (2) μ bar s^{-1}. (a) Sutcliffe/Trenberth approximation to the QG vertical motion. (b) The deformation term contribution to the QG vertical motion. In both panels the thick, dashed line is the axis of the occluded thermal ridge

illustrates the quasi-geostrophic (QG) omega resulting from both the Sutcliffe/Trenberth forcing terms (Figure 6.14a) and the deformation terms (Figure 6.14b) for a modest occluded cyclone. Note that the occluded thermal ridge, a non-frontal thermal structure, is the seat of significant QG vertical motions associated with the deformation terms. This ascent would not be accounted for in the Sutcliffe/Trenberth approximation to the QG omega equation.

6.4.3 The along- and across-isentrope components of $\vec{Q}$

A final word concerning the physical meaning of the $\vec{Q}$-vector is appropriate before we begin to discuss frontogenesis in Chapter 7. This comment begins by rewriting the hydrostatic equation in the form $-\partial\phi/\partial p = f\gamma\theta$ where θ is the potential temperature and γ is a constant on isobaric surfaces, i.e.

$$\gamma = \frac{R}{f p_0}\left(\frac{p_0}{p}\right)^{c_v/c_p},$$

with p_0 usually taken to be 1000 hPa. Employing this form of the hydrostatic equation allows (6.41) to be rewritten as

$$\vec{Q} = f\gamma\left[\left(-\frac{\partial \vec{V}_g}{\partial x}\cdot\nabla\theta\right)\hat{\imath}, \left(-\frac{\partial \vec{V}_g}{\partial y}\cdot\nabla\theta\right)\hat{\jmath}\right]. \tag{6.47}$$

Now let us consider the Lagrangian rate of change of $\nabla\theta$ following the geostrophic flow, in symbols,

$$\frac{d}{dt_g}\nabla\theta = \left(\frac{\partial}{\partial t} + \vec{V}_g\cdot\nabla\right)\nabla\theta = \left(\frac{\partial}{\partial t} + \vec{V}_g\cdot\nabla\right)\left(\frac{\partial\theta}{\partial x}\hat{\imath} + \frac{\partial\theta}{\partial y}\hat{\jmath}\right). \tag{6.48}$$

It is left to the reader to show that, under adiabatic conditions,

$$f\gamma\frac{d}{dt_g}\nabla\theta = \vec{Q}.$$

Thus, a profound physical meaning can be ascribed to the $\vec{Q}$-vector: that is, $\vec{Q}$ *describes the rate of change of* $\nabla\theta$ *following the geostrophic flow.* This property of the $\vec{Q}$-vector will be exploited in our subsequent discussions of both frontogenesis and cyclogenesis. For now, it is enough that we take advantage of this physical fact to develop additional insight from the $\vec{Q}$-vector.

Given that

$$\vec{Q} = f\gamma\frac{d}{dt_g}\nabla\theta,$$

it is useful to consider separately the along- and across-isentrope components of $\vec{Q}$, denoted as $\vec{Q}_s$ and $\vec{Q}_n$ (where $\vec{Q} = \vec{Q}_s + \vec{Q}_n$), respectively, illustrated in schematic form in Figure 6.15. Before deriving mathematical expressions corresponding to $\vec{Q}_s$

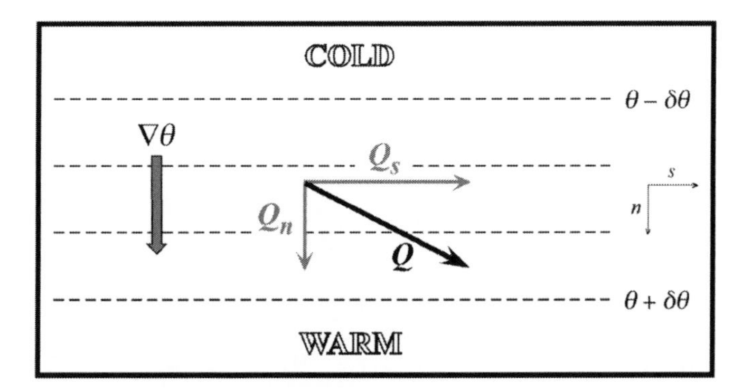

Figure 6.15 Natural coordinate partition of the $\vec{Q}$-vector into its along-isentrope ($\vec{Q}_s$) and across-isentrope ($\vec{Q}_n$) components. See text for explanation

and $\vec{Q}_n$, let us consider their respective physical meanings. Noting that the vector $\nabla\theta$, like all vectors, has both magnitude and direction, it is clear that $\vec{Q}_n$, which is directed along $\nabla\theta$, can only affect changes in the *magnitude* of $\nabla\theta$. Since $\vec{Q}_s$ is directed perpendicularly to $\nabla\theta$ it can only affect changes in the *direction* of $\nabla\theta$. Now, $\vec{Q}_n$ is simply the component of $\vec{Q}$ along the vector $\nabla\theta$ and simple vector calculus yields a mathematical expression for $\vec{Q}_n$ as

$$\vec{Q}_n = \left(\frac{\vec{Q}\cdot\nabla\theta}{|\nabla\theta|}\right)\frac{\nabla\theta}{|\nabla\theta|}. \tag{6.49}$$

Allowing the unit vector in the $\nabla\theta$ direction ($\nabla\theta/|\nabla\theta|$) to be written as $\hat{n}$, and the magnitude of $\vec{Q}_n$ ($\vec{Q}\cdot\nabla\theta/|\nabla\theta|$) to be written as Q_n, (6.49) can be rewritten as $\vec{Q}_n = Q_n\hat{n}$. Similarly, $\vec{Q}_s$ is the component of $\vec{Q}$ along the vector $\hat{k}\times\nabla\theta$ and so can be written as

$$\vec{Q}_s = \left[\frac{\vec{Q}\cdot(\hat{k}\times\nabla\theta)}{|\nabla\theta|}\right]\frac{\hat{k}\times\nabla\theta}{|\nabla\theta|} \tag{6.50}$$

where we have taken advantage of the fact that $\left|\hat{k}\times\nabla\theta\right| = |\nabla\theta|$. Allowing the unit vector in the $\hat{k}\times\nabla\theta$ direction to be denoted as $\hat{s}$ and the magnitude of $\vec{Q}_s$ ($\vec{Q}\cdot(\hat{k}\times\nabla\theta)/|\nabla\theta|$) to be denoted as Q_s, (6.50) can be written as $\vec{Q}_s = Q_s\hat{s}$. Substituting the expressions for both $\vec{Q}_n$ and $\vec{Q}_s$, we can write

$$\vec{Q} = Q_n\hat{n} + Q_s\hat{s}. \tag{6.51}$$

Since the total QG vertical motion is related to $-2\nabla\cdot\vec{Q}$, the foregoing partition allows us to see that total as the sum of two orthogonal parts associated with $-2\nabla\cdot\vec{Q}_n$ and $-2\nabla\cdot\vec{Q}_s$, respectively. Given the orientations of $\vec{Q}_n$ and $\vec{Q}_s$, these components of the total vertical motion will be distributed in couplets across the thermal wind (transverse) and along the thermal wind (shearwise), respectively.

It will be shown in the next chapter that the transverse component of the QG omega is directly related to the dynamics of the frontal zones that characterize the

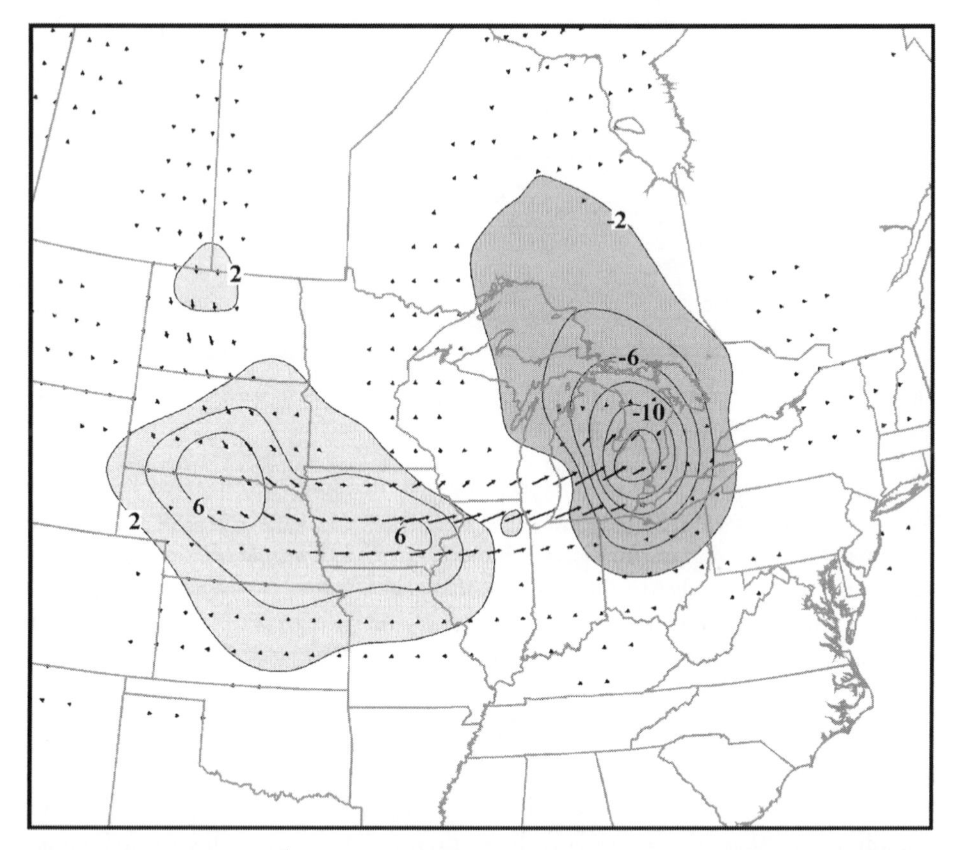

Figure 6.16 The 700 hPa $\vec{Q}_{T_R}$ vectors (black arrows) and associated QG vertical motion at 0000 UTC 13 November 2003. Vertical motion shown in units of μ bar s^{-1} (dPa s^{-1}) contoured every 2μ bar s^{-1} with dark shading showing upward vertical motions and light shading showing downward vertical motion

mid-latitude cyclone. Insight into the nature of the shearwise component arises by considering an alternative form of the Trenberth approximation to the QG omega equation in which the thermal wind advection of geostrophic absolute vorticity was the principal forcing mechanism for vertical motions. Starting with (6.32)

$$\sigma\left(\nabla^2 + \frac{f_0^2}{\sigma}\frac{\partial^2}{\partial p^2}\right)\omega \approx 2\left[f_0\frac{\partial \vec{V}_g}{\partial p}\cdot\nabla(\zeta_g + f)\right]$$

and taking advantage of the non-divergence of the geostrophic wind while neglecting the contribution of the planetary vorticity to the geostrophic absolute vorticity, we note that the RHS can be written in a flux divergence form as

$$\sigma\left(\nabla^2 + \frac{f_0^2}{\sigma}\frac{\partial^2}{\partial p^2}\right)\omega \approx 2\nabla\cdot\left[f_0\frac{\partial \vec{V}_g}{\partial p}\zeta_g\right]. \tag{6.52a}$$

But since

$$\frac{\partial \vec{V}_g}{\partial p} = \frac{\hat{k}}{f} \times \nabla \frac{\partial \phi}{\partial p} = -\gamma(\hat{k} \times \nabla\theta),$$

(6.52a) can be rewritten as

$$\sigma\left(\nabla^2 + \frac{f_0^2}{\sigma}\frac{\partial^2}{\partial p^2}\right)\omega \approx -2\nabla \cdot \vec{Q}_{TR} \tag{6.52b}$$

where $\vec{Q}_{TR} = f_0\gamma\zeta_g(\hat{k} \times \nabla\theta)$. Thus, the approximate Trenberth form of the QG omega equation can be written in a form that is identical to the $\vec{Q}$-vector form of the full omega equation. Note that the vector $\vec{Q}_{TR}$ must be everywhere parallel to isentropes and thus $\vec{Q}_{TR}$ represents at least a portion of $\vec{Q}_s$.[7] An illustration of the distribution of $\vec{Q}_{TR}$ vectors and the associated QG vertical motions from the developing cyclone previously examined in Figure 6.6 are illustrated in Figure 6.16. The distinction between such shearwise and transverse vertical motions will prove valuable when we discuss the process of mid-latitude cyclogenesis in Chapter 8.

Selected References

Sutcliffe (1939) offers an illuminating discussion of the ageostrophic wind and its role in producing vertical motions.

Sutcliffe (1947) describes his famous development theorem.

Trenberth (1978) describes the cancellation among terms in the traditional QG omega equation.

Hoskins *et al.* (1978) provide the seminal derivation and discussion of the $\vec{Q}$-vector.

Martin (1998b) examines the appropriateness of neglecting the deformation terms in the QG omega equation from the perspective of the $\vec{Q}$-vector.

Problems

6.1. (a) For the mid-latitude, upper tropospheric wave train shown in Figure 6.1A, indicate where the regions of ascent and descent are found. Explain your answer.

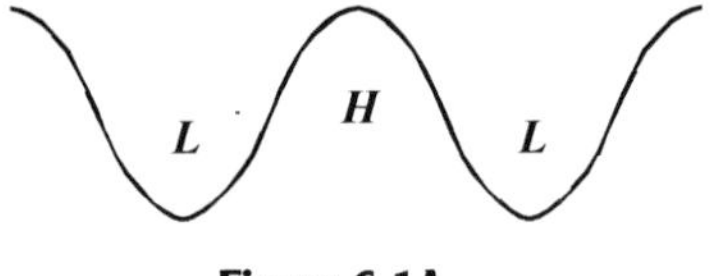

Figure 6.1A

---

[7] A full description of both the $\vec{Q}_n$ and $\vec{Q}_s$ components of the $\vec{Q}$-vector, along with their application to the diagnosis of vertical motions in the occluded quadrant of mid-latitude cyclones, is given in Martin (1999).

(b) The propagating wave train shown in Figure 6.1A will have an associated distribution of height rises and falls as shown in Figure 6.1B. What can be concluded about the relative magnitudes of the isallobaric and inertial advective components of the ageostrophic wind at that level? Explain your answer.

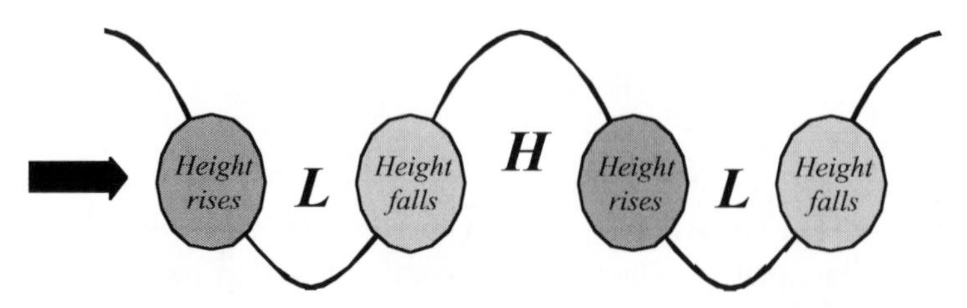

Figure 6.1B

6.2. An expansion of the ageostrophic wind consistent with the assumptions of quasi-geostrophic theory can be written as

$$\frac{\hat{k}}{f} \times \frac{d\vec{V}_g}{dt} = \frac{\hat{k}}{f} \times \left(\frac{\partial \vec{V}_g}{\partial t} + \vec{V}_g \cdot \nabla \vec{V}_g \right) = \vec{V}_{ag}.$$

(a) Given this assumption, show that an expression for the inertial advective component of the ageostrophic wind is given by

$$\vec{V}_{IA} = -\frac{\vec{V}_g \zeta_g}{f}.$$

With simple pictures, show that the distribution of $\vec{V}_{IA}$ explains:

(b) the classic four-quadrant vertical motion distribution associated with a straight jet streak, and

(c) the distribution of vertical motion associated with an upper tropospheric wave train in the geopotential height field (such as is shown in Figure 6.1A).

6.3. This problem refers to the diagnosis of development described by Sutcliffe (1939). The net column ageostrophic wind associated with the Lagrangian rate of change of the shear vector ($d\vec{V}_s/dt$) is always perpendicular to the shear vector itself, which means that any process that changes the shear forces a vertical circulation that is transverse to the shear. Show that the net column ageostrophic wind associated with shearing over the surface wind ($\vec{V}_s \cdot \nabla \vec{V}_0$) is *not* always parallel to the shear vector.

6.4. Consider Figure 6.2A which shows 1000–500 hPa thickness contours (dashed lines) along with isopleths of an unknown variable Q which has the same value at 1000 hPa as it does at 500 hPa.

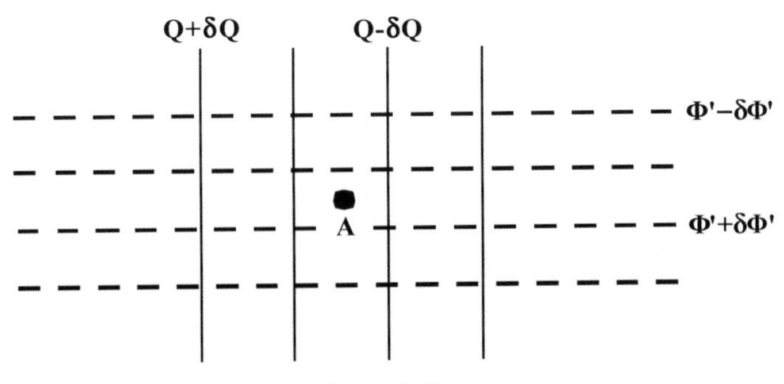

Figure 6.2A

At which level, 1000 or 500 hPa, is the geostrophic advection of Q at Station A larger? Station A is in the northern hemisphere. Explain your answer. (Hint: use the most basic, *physical* definition of the thermal wind to prove your answer.)

6.5. There is partial cancellation between the two separate forcing terms in the traditional quasi-geostrophic omega equation. Describe in words what process is represented by the portion that cancels.

6.6. Show that

$$-\frac{\partial \phi}{\partial p} = f\gamma\theta$$

where

$$\gamma = \frac{R}{fp_0}\left(\frac{p_0}{p}\right)^{\frac{c_v}{c_p}}, \quad R + c_v = c_p, \quad \text{and} \quad \theta = T\left(\frac{p_0}{p}\right)^{\frac{R}{c_p}}.$$

6.7. Do the *entrance/exit* region circulations associated with a straight jet streak in the southern hemisphere mid-latitudes have the same characteristics as those associated with jet streaks in the northern hemisphere? Explain.

6.8. Figure 6.3A illustrates the 700 hPa geopotential height and temperature analysis for a developing mid-latitude cyclone east of the Kamchatka Peninsula in October 2004.

(a) Draw $\vec{Q}$-vectors at the indicated points using the natural coordinate expression for the $\vec{Q}$-vector.
(b) Sketch the areas of convergence and divergence of the $\vec{Q}$-vectors drawn in (a).
(c) Do the areas of convergence and divergence correspond to your intuition about where the air is likely to be rising and sinking in this storm? Explain.

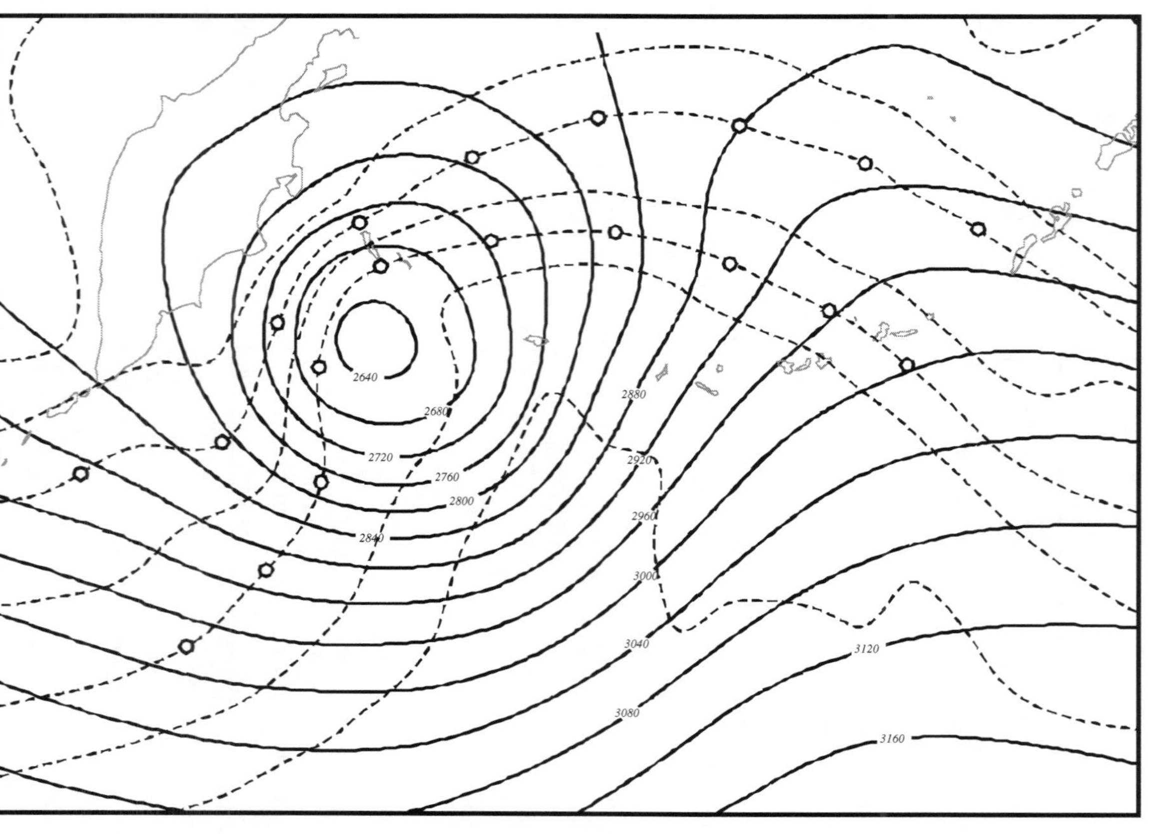

2640
2680
2720
2760
2800
2840
2880
2920
2960
3000
3040
3080
3120
3160

Figure 6.3A

6.9. Figure 6.4A shows the 850 hPa geopotential height (solid lines) and the 1000–850 hPa thickness (dashed lines) in a North Sea 'reverse shear' polar low.

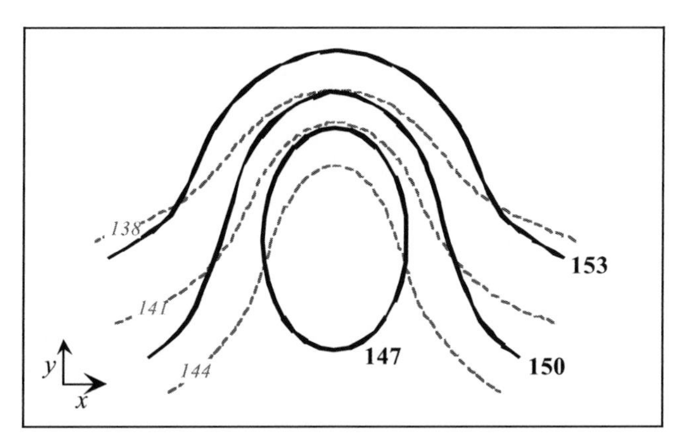

Figure 6.4A

(a) Indicate with a cross the location of the 850 hPa vorticity maximum.
(b) Indicate the direction of the 850–1000 hPa thermal wind.
(c) Use a + and a − to indicate the regions of synoptic-scale upward and downward vertical motions, respectively. Explain your reasoning in terms of the Sutcliffe development theorem.

6.10. Show that

$$\frac{2}{f}\left[J\left(\frac{\partial \phi}{\partial p}, \nabla^2 \phi\right) + J\left(\frac{\partial \phi}{\partial p}, ff_0\right)\right] = 2 f_0 \frac{\partial \vec{V}_g}{\partial p} \cdot \nabla\left(\zeta_g + f\right).$$

6.11. Demonstrate that the static stability parameter, σ, in the quasi-geostrophic omega equation

$$\sigma = -\frac{1}{\rho\theta}\frac{\partial \theta}{\partial p}$$

can be written in terms of the geopotential, ϕ, as

$$\sigma = \frac{\partial^2 \phi}{\partial p^2} + \frac{c_v}{p c_p}\frac{\partial \phi}{\partial p}.$$

6.12. Prove that, for adiabatic flow,

$$\vec{Q} = f\gamma \frac{d}{dt_g}\nabla\theta$$

where

$$\frac{d}{dt_g} = \frac{\partial}{\partial t_g} + \vec{V}_g \cdot \nabla.$$

6.13. Given that $Q = Q_n\hat{n} + Q_s\hat{s}$, do you expect that a component of quasi-geostrophic forcing for ascent in a region can result simply from curvature in the isotherms? Explain.

6.14. One of the many physical interpretations of the $\vec{Q}$-vector is that $\vec{Q}$ represents the degree of thermal wind imbalance. Why is this an acceptable statement?

Solutions

6.4. Advection is larger at 500 hPa.

7

The Vertical Circulation at Fronts

Objectives

One of the defining structural features of the mid-latitude cyclone is its asymmetric thermal structure manifest most clearly in the fronts that characterize the cyclone. Aside from their ubiquity, these fronts are vested with considerable sensible weather relevance as well since large variations of meteorological conditions exist across them and the precipitation distribution associated with a typical mid-latitude cyclone is often concentrated in their vicinity. Figure 7.1(a) shows analyses of the sea-level pressure and surface potential temperature for a typical mid-latitude cyclone. The characteristic comma-shaped cloud pattern from the same storm (Figure 7.1b) is anchored by the frontal structure identified in Figure 7.1(a). The vigilant reader will be able to establish, through daily inspection of surface, upper air, and satellite observations that the structural relationship demonstrated in Figure 7.1 is quite common in the middle latitudes.

Note that the across-front dimension of the cold front in Figure 7.1(a) (on the order of 100 km) is much smaller than its along-front dimension (on the order of 1000 km). Considering characteristic velocities given such length scales we can draw the preliminary conclusion that geostrophic balance exists in the along-front direction (where the Rossby number (R_o) is given by $R_o = 10\,\mathrm{m\,s^{-1}}/(10^{-4}\,\mathrm{s^{-1}})(10^6\,\mathrm{m}) = 0.1$. However, in the across-front direction $R_o = (10\,\mathrm{m\,s^{-1}})/(10^{-4}\,\mathrm{s^{-1}})(10^5\,\mathrm{m}) = 1.0$! Thus, mid-latitude fronts would appear to be hybrid phenomena characterized by along-front geostrophy but a fair degree of across-front ageostrophy. The mixture of scales that characterizes fronts makes them the focus of important scale interactions in the mid-latitude cyclone. For this reason, the purely quasi-geostrophic diagnostic perspective we have thus far developed will prove to be insufficient as a means to investigate fronts and it will have to be extended in order to incorporate additional, physically relevant processes that are fundamental to the frontal environment.

Mid-Latitude Atmospheric Dynamics Jonathan E. Martin

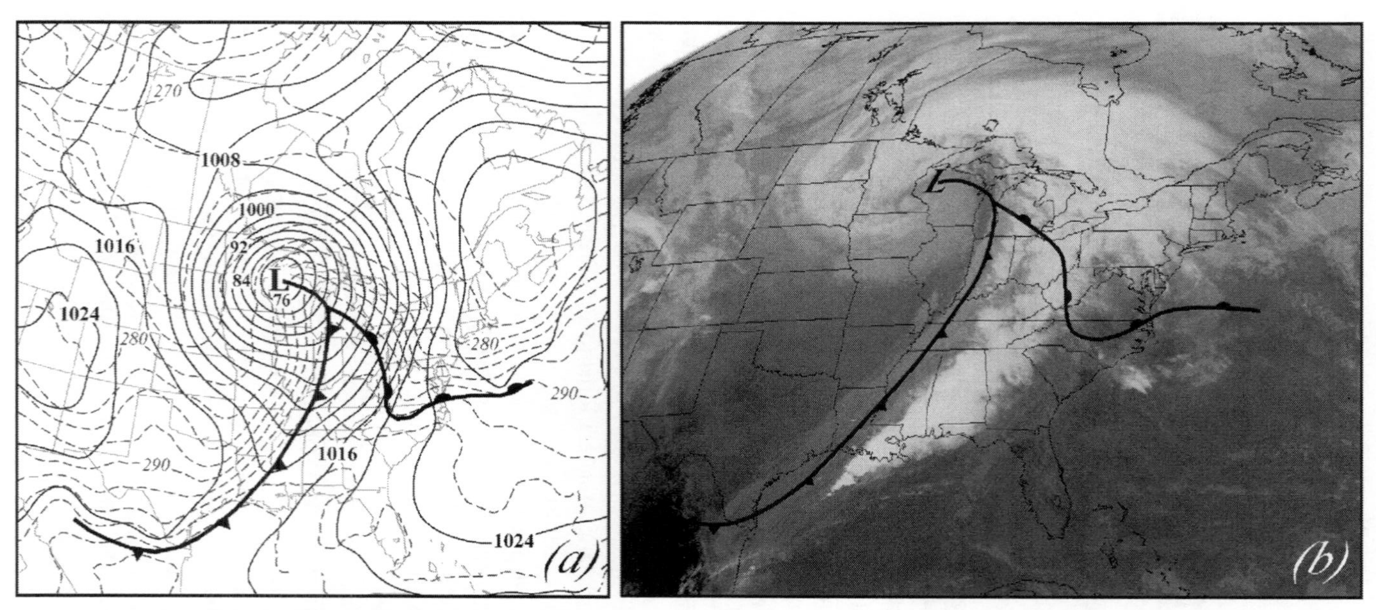

Figure 7.1 (a) Sea-level pressure and 950 hPa potential temperature analysis at 1800 UTC 10 November 1998. Solid lines are sea-level isobars, labeled in hPa and contoured every 4 hPa. Dashed lines are 950 hPa isentropes labeled in K and contoured every 2.5 K. Standard frontal symbols identify the cold and warm fronts while the occluded front is indicated as a thick black line. (b) Infrared satellite image (NOAA) of the same storm at 1815 UTC 10 November 1998. Frontal symbols as in (a)

Given the relationship between fronts and the cloud and precipitation distribution in mid-latitude cyclones that is suggested by Figure 7.1, a central question for our subsequent investigation is: *Why is such a relationship so prevalent?* In order to approach this question with precision, we first need to understand the essential elements of frontal structure. Next we consider how **frontogenesis**, the process of creating a front, leads to the vertical motions that characterize fronts. Adoption of a *semi-geostrophic* perspective in the Sawyer–Eliassen frontal circulation equation formally incorporates the interplay between the geostrophic and ageostrophic flows that characterizes the frontal environment. We then proceed to an investigation of fronts that form at the tropopause, known as upper-level fronts. Finally we consider some of the circumstances that conspire to produce the observed variation of precipitation intensity associated with fronts. We begin by establishing the essential characteristics of fronts in the next section.

7.1 The Structural and Dynamical Characteristics of Mid-Latitude Fronts

As demonstrated by Figure 7.1(a), a front is a boundary whose primary structural and dynamical characteristic is the larger-than-background temperature (or density) contrast associated with it. In order to determine some basic characteristics of fronts, from which we will create a working definition of a front, we will consider the somewhat unphysical case of the **zero-order front**. The zero-order front is characterized by *discontinuities* in the temperature and density across the frontal boundary. For this reason, it most closely approximates the notion of the knife-like polar front envisioned by the Bergen School in the Norwegian Cyclone Model. Real fronts, however, actually more closely resemble a first-order front, in which *gradients* of temperature and density, not the variables themselves, are discontinuous across the front. Since a front is a boundary between two different air masses and each air mass has a characteristic density, in a zero-order front density is discontinuous across the front (Figure 7.2). We will demand that pressure be continuous across

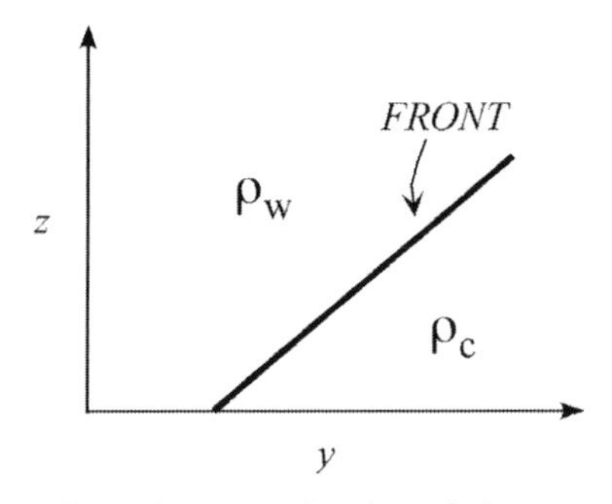

Figure 7.2 Vertical cross-section through the zero-order front

the zero-order front (so that the geostrophic winds are not infinite along the front!). Then, according to the gas law, temperature (T) must also be discontinuous across the front. Even though this will imply an infinite thermal wind, we will proceed anyway, the simplicity of the ensuing analysis being the motivation. If we take the x-axis as the along-front direction and further assume (1) that there is no along-front variation in any variable, and (2) that the pressure is steady state (i.e. $\partial p/\partial t = 0$), then the differential of pressure is given by

$$dp = \left(\frac{\partial p}{\partial y}\right) dy + \left(\frac{\partial p}{\partial z}\right) dz \tag{7.1}$$

on both sides of the front. This expression can be written for both the warm and the cold sides of the front as

$$dp_w = \left(\frac{\partial p}{\partial y}\right)_w dy + \left(\frac{\partial p}{\partial z}\right)_w dz \quad \text{and} \quad dp_c = \left(\frac{\partial p}{\partial y}\right)_c dy + \left(\frac{\partial p}{\partial z}\right)_c dz,$$

respectively. We can use the hydrostatic equation to substitute for $\partial p/\partial z$ in both expressions, set the expressions for dp equal to one another, and rearrange the result to get

$$0 = \left[\left(\frac{\partial p}{\partial y}\right)_c - \left(\frac{\partial p}{\partial y}\right)_w\right] dy - (\rho_c - \rho_w)\, g dz. \tag{7.2}$$

This can be solved for dz/dy, the slope of the zero-order front:

$$\frac{dz}{dy} = \frac{(\partial p/\partial y)_c - (\partial p/\partial y)_w}{g(\rho_c - \rho_w)}. \tag{7.3}$$

Since more dense fluid must lie beneath less dense fluid, as portrayed in Figure 7.2, in order that the frontal structure be statically stable and therefore sustainable, we note that $dz/dy > 0$. From (7.3), this implies that the across-front pressure gradient must be larger on the cold side of the front than on the warm side. Such a conclusion can be incorporated into constructing a physically accurate analysis of sea-level pressure in the vicinity of a front. Perhaps more enlightening for our investigation is to consider the along-front geostrophic winds which are related to the across-front pressure gradients. Recall that, in height coordinates,

$$u_g = -\frac{1}{\rho f}\frac{\partial p}{\partial y} \quad \text{or} \quad \frac{\partial p}{\partial y} = -f\rho u_g.$$

Using this expression we can recast (7.3) into

$$\frac{dz}{dy} = \frac{f(\rho_w u_{g_w} - \rho_c u_{g_c})}{g(\rho_c - \rho_w)}. \tag{7.4}$$

Now, in order for $dz/dy > 0$ we see that $u_{g_w} > u_{g_c}$; in other words, the front must be characterized by positive geostrophic relative vorticity ($\partial u_g/\partial y < 0$)! Thus, we

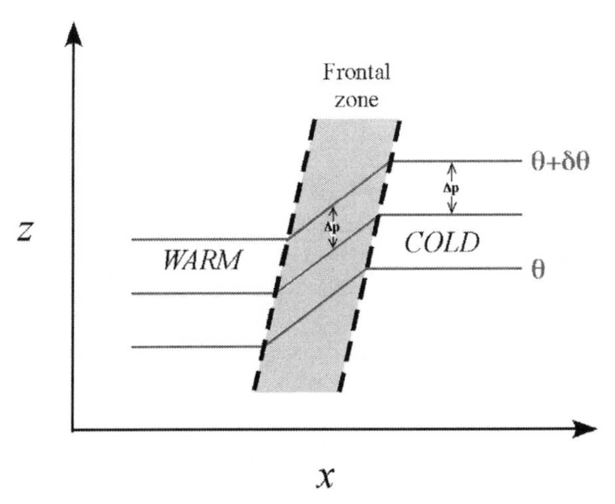

Figure 7.3 **Isentropes associated with a first-order front. Note that the static stability is largest in the frontal zone**

have discovered a fundamental dynamical characteristic of mid-latitude fronts – they are characterized by positive geostrophic relative vorticity. In fact, further inspection of (7.4) reveals that the stronger the density contrast across the front becomes, the more intense is the vorticity at the front.

In reality, the temperature cannot be discontinuous at a front, but the temperature *gradient* can be. In this more realistic case, we have a first-order discontinuity and the isentropes must appear as in Figure 7.3 in the first-order front. Careful examination of the isentropes in the frontal zone reveals that the frontal zone is also characterized by larger static stability ($-\partial\theta/\partial p$) than either the cold or warm side of the boundary. Thus, frontal zones are characterized by (1) larger-than-background horizontal temperature (density) contrasts, (2) larger-than-background relative vorticity, and (3) larger-than-background static stability. We will use these characteristics to define a front after we examine some observations of fronts.

A time series of rooftop observations at Madison, Wisconsin (known as a meteorogram) is shown in Figure 7.4(a). Note the $\sim 5°$C drop in temperature and corresponding 3°C drop in dewpoint temperature that occurred between 0510 and 0515 UTC. Simultaneously, the winds shifted from steady southwesterlies to steady northerlies. This time series clearly demonstrates the sharp temperature and moisture characteristics associated with a surface frontal passage (in this case, a cold frontal passage). It also reveals the strong cyclonic vorticity that must attend a mid-latitude frontal zone. Evidence for the enhanced static stability of a frontal zone is provided in Figure 7.4(b) which is a vertical cross-section through what is known as an upper-level front. Note that the static stability is elevated in the stratosphere, as expected, but also within the bundle of isentropes that extends beneath the jet maximum to nearly 700 hPa. This same bundle of isentropes constitutes the upper

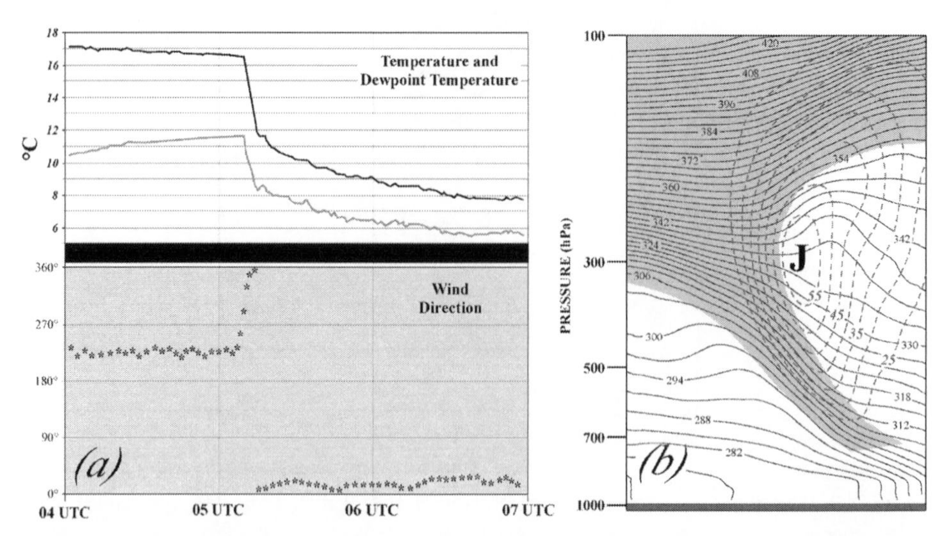

Figure 7.4 (a) Meteorogram of a surface cold frontal passage at Madison, WI between 0400 and 0700 UTC 30 April 2003. Black line is the temperature, gray line is the dewpoint, and asterisks are the wind direction time series, respectively. Note the coincidence of the temperature and dewpoint drops with the wind shift. (b) Vertical cross-section through an upper-level frontal zone at 1200 UTC 12 November 2003. Solid lines are isentropes labeled in K and contoured every 3 K. Dashed lines are isotachs labeled in m s^{-1} and contoured every 10 m s^{-1} starting at 25 m s^{-1}. Gray shading represents region of enhanced static stability which includes the upper-frontal zone itself

front itself and is clearly characterized by large horizontal temperature contrast as well as cyclonic vorticity (evidenced by the horizontal shear implied by the tight packing of the isotachs). Now we are prepared to establish a working definition of a front that is based upon the essential characteristics of mid-latitude frontal zones. When we use the term 'cold (warm) front' we will be referring to:

> *The leading edge of a transitional zone that separates advancing cold (warm) air from warm (cold) air, the length of which is significantly greater than its width. The zone is characterized by high static stability as well as larger-than-background gradients in temperature and relative vorticity.*

In nature, fronts defined in this way come in varying degrees of intensity but every front shares these fundamental physical and dynamical characteristics. Thus, the lack of a numerical designation here is not an oversight but rather an attempt to distinguish those features in the mid-latitude atmosphere that ought to be called fronts from those which should not. Of course, the intensity of a front *is* a meaningful distinction to make in terms of both scientific interest as well as sensible weather characteristics. One way of measuring the strength of one front against another is by considering the *magnitudes* of their respective horizontal temperature gradients. We will return to this important diagnostic in just a moment. First we will consider the intimate relationship between fronts and jets in the mid-latitude atmosphere.

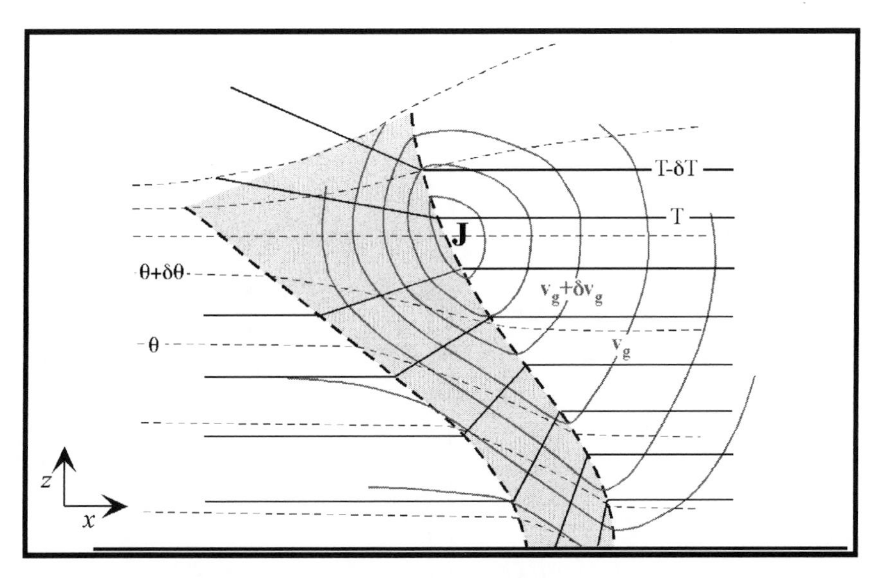

Figure 7.5 Idealized vertical cross-section through a frontal zone. Gray solid lines are isotachs of the geostrophic wind into the page with 'J' indicating the position of the wind maxima. Black solid lines are isotherms and thin dashed lines are isentropes. Gray shaded region with thick dashed border represents the idealized frontal zone

7.2 Frontogenesis and Vertical Motions

The thermal wind relation requires that fronts (regions of large ∇T) be associated with strong vertical shear of the geostrophic wind. Shown in Figure 7.5 is an idealized vertical cross-section through a frontal zone. Notice that the magnitude of ∇T is largest near the surface and that the frontal zone is characterized by the strongest vertical shear. Also notice that the leading edge of the zone (i.e. the front itself) is a maximum in geostrophic relative vorticity as we have previously suggested it should be. Recall from the frictionless vorticity equation that vorticity can change only as a result of divergence ($d\eta/dt = -f(\nabla \cdot \vec{V})$). By the continuity equation, divergence is accompanied by vertical motions ($\nabla \cdot \vec{V} = -\partial\omega/\partial p$). Using these two relationships we can establish the following logical argument. If, by some horizontal advective process, for instance, the magnitude of ∇T increases, then the wind shear and jet core wind speed necessarily increase as well. A more intense jet results in increased vorticity. Increased vorticity implies that some divergence is operating in the fluid. If divergence is operating, there must be some vertical motion as well. Therefore, an increase in the magnitude of ∇T requires the production of a vertical circulation in an atmosphere in approximate thermal wind balance. For the remainder of this chapter we will investigate various physical/mathematical formulations that seek to quantify this important physical relationship. The first step on this journey requires that we consider how an increase in the magnitude of ∇T can be accomplished.

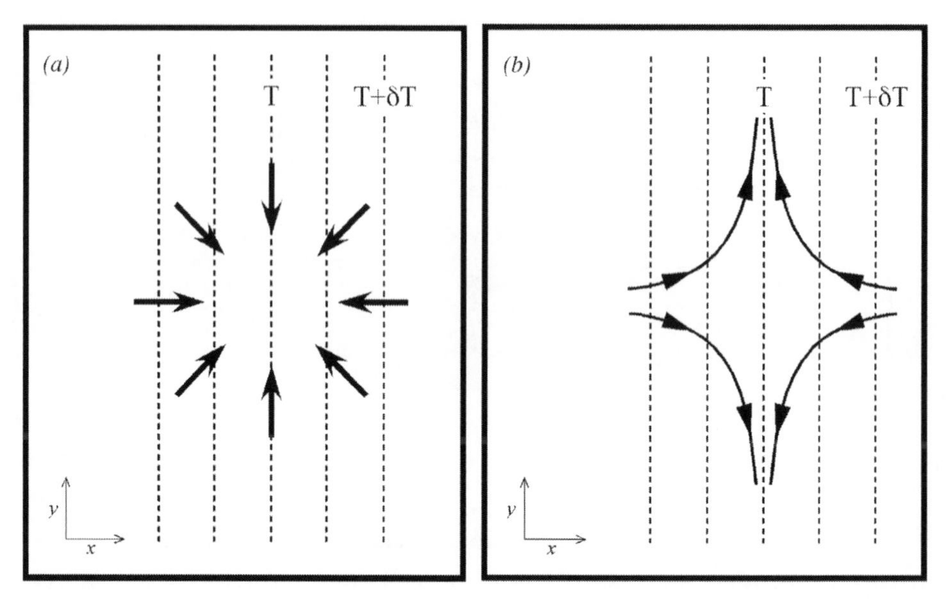

Figure 7.6 (a) Pure convergence superimposed upon a field of isotherms. (b) Horizontal deformation superimposed upon a field of isotherms. In both cases the horizontal wind will tend to intensify $|\nabla T|$

We shall broadly define as 'frontogenetic' any process that acts to increase the magnitude of ∇T. Such a process in action is known as **frontogenesis**. More specifically (for ease of physical interpretation later), we will refer to any horizontal advective process that acts to increase the magnitude of ∇T as **horizontal frontogenesis**. Some simple illustrations of horizontal frontogenetical processes are given in Figure 7.6. Given our verbal definition of frontogenesis, we can define a corresponding mathematical one (termed the **frontogenesis function**) as

$$\Im = \frac{d\left|\nabla_p \theta\right|}{dt}, \tag{7.5}$$

defining the Lagrangian rate of change of the magnitude of $\nabla_p \theta$ (the potential temperature gradient measured on an isobaric surface). Though it looks innocuous, (7.5) is a rather bulky expression (as we will see presently). Without loss of physical insight, we can consider the simpler 1-D version of (7.5) and gain some understanding of the nature of frontogenesis. Therefore, we will consider the processes that can change the magnitude of the x-direction temperature contrast using

$$\Im_x = \frac{d}{dt}\left(\frac{\partial \theta}{\partial x}\right).$$

The reader is asked to show that, given

$$\frac{d}{dt} = \frac{\partial}{\partial t} + u\frac{\partial}{\partial x} + v\frac{\partial}{\partial y} + \omega\frac{\partial}{\partial p},$$

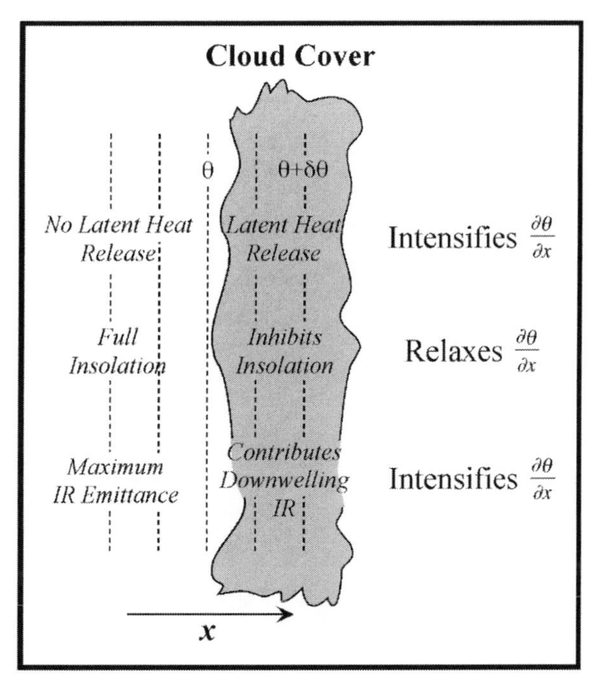

Figure 7.7 The diabatic effects of cloud cover on $\partial\theta/\partial x$. The effect of differential latent heat release can occur at any time of day. Differential insolation and infrared emittance are specific to day and night, respectively

then

$$\Im_x = \frac{d}{dt}\left(\frac{\partial\theta}{\partial x}\right) = \frac{\partial}{\partial x}\left(\frac{d\theta}{dt}\right) - \frac{\partial u}{\partial x}\frac{\partial\theta}{\partial x} - \frac{\partial v}{\partial x}\frac{\partial\theta}{\partial y} - \frac{\partial\omega}{\partial x}\frac{\partial\theta}{\partial p}. \tag{7.6}$$

Thus, there are four physical processes, represented by the four terms on the RHS of (7.6), that contribute to an increase in $\partial\theta/\partial x$. The first of these processes is the effect of across-front gradients in diabatic heating, represented by $\partial/\partial x(d\theta/dt)$. Consider the meridionally oriented isentropes illustrated in Figure 7.7. If there is latent heat release in ascending air on the warm side of this potential temperature gradient, then $\partial/\partial x(d\theta/dt) > 0$. Consequently, such a distribution of latent heat release is frontogenetical. Utilizing the same expression we can consider the effect of differential cloud cover on frontal strength. If the warm side of Figure 7.7 is cloudy and the cold side clear, then differential insolation during the day renders $\partial/\partial x(d\theta/dt) < 0$ and daytime heating is frontolytic under such circumstances. Under the same distribution of clouds during the night, the cold side cools more rapidly than the warm side so that $\partial/\partial x(d\theta/dt) > 0$ and so the cloud cover promotes frontogenesis.

The effect of confluence over the temperature gradient is represented by the second term on the RHS of (7.6),

$$-\frac{\partial u}{\partial x}\frac{\partial\theta}{\partial x}.$$

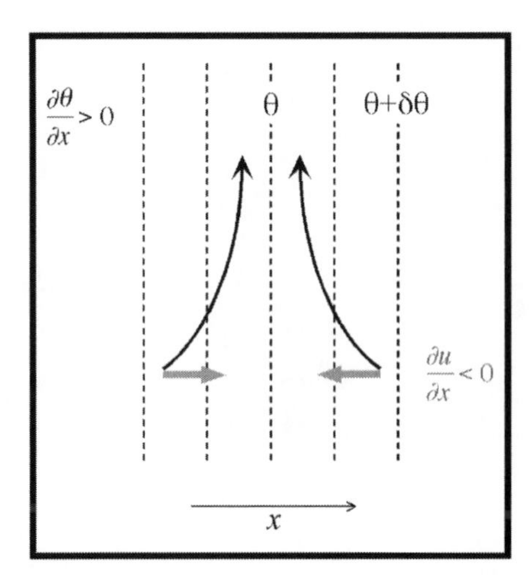

Figure 7.8 Confluent horizontal flow acting on meridionally oriented isentropes. The gray arrows represent the x-direction wind

Considering the confluent flow shown in Figure 7.8, we note immediately that $\partial\theta/\partial x > 0$. The winds are distributed such that $\partial u/\partial x < 0$. Overall, then, the effect of the confluent wind field depicted in Figure 7.8 is to promote frontogenesis. One can imagine the wind field acting to push the isentropes closer together in the horizontal, thereby increasing $|\partial\theta/\partial x|$.

The effect of horizontal shearing on $\partial\theta/\partial x$ is represented by the third term on the RHS of (7.6),

$$-\frac{\partial v}{\partial x}\frac{\partial \theta}{\partial y},$$

and is illustrated in Figure 7.9. In this instance, the isentropes are aligned at a slight angle to both the x- and y-axes in such a way that $\partial\theta/\partial y < 0$. Given the indicated

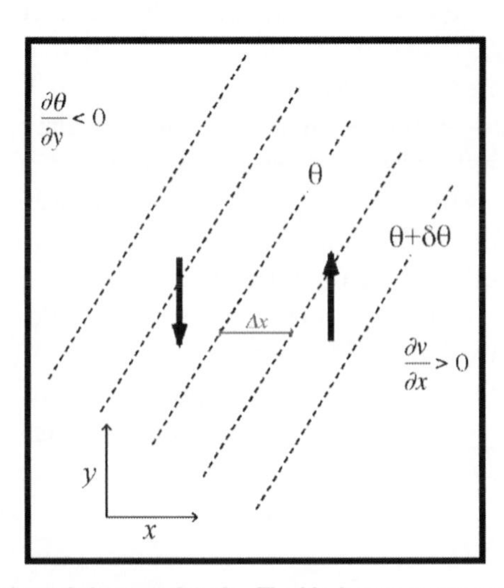

Figure 7.9 Effect of horizontal shear on $\partial\theta/\partial x$. The black arrows represent the y-direction wind

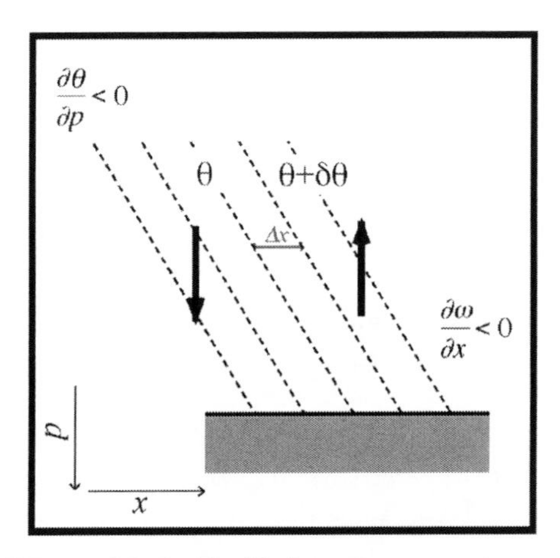

Figure 7.10 Effect of tilting on $\partial\theta/\partial x$. The black arrows represent upward and downward vertical motions

winds, it is clear that $\partial v/\partial x > 0$ as well, meaning that the entire shearing term is positive. Thus, such shearing will act to increase $\partial\theta/\partial x$ by rotating the isotherms into a more meridional orientation. This increase in $\partial\theta/\partial x$ does not, however, represent a decrease in the absolute distance between successive isentropes (as was the case for both of the prior physical mechanisms). We will show later, in consideration of the 2-D gradient of θ, that shear (more precisely, vorticity) does not modify the magnitude of $\nabla\theta$, but only changes its direction.

Finally, the effect of vertical tilting is represented by the fourth term on the RHS of (7.6),

$$-\frac{\partial\omega}{\partial x}\frac{\partial\theta}{\partial p}.$$

A thermally direct vertical circulation, along with a frontal bundle of isentropes, is illustrated in the vertical cross-section depicted in Figure 7.10. In a statically stable atmosphere, $\partial\theta/\partial p$ must be negative. Recalling that upward vertical motion is consistent with negative omega and vice versa, $\partial\omega/\partial x < 0$ for the situation depicted in Figure 7.10. Thus, the entire vertical tilting term is negative, suggesting that a thermally direct vertical circulation acts to decrease $\partial\theta/\partial x$ by rotating the isentropes into a more nearly horizontal orientation. This squares physically with the results of such a circulation considered in terms of temperature rather than potential temperature. From that perspective, the rising warm air cools by expansion while the sinking cold air warms by compression. Thus, the originally warm air is made colder while the originally cold air is made warmer under the influence of the thermally direct vertical motions.

The same physical reasoning can be applied to the more complicated, 3-D frontogenesis function given by (7.5). Using similar algebra as was used to derive (7.6), we find that

$$\begin{aligned}\Im_{3D} &= \frac{d}{dt}|\nabla\theta| = \frac{d}{dt}\left(\frac{\partial\theta}{\partial x}^2 + \frac{\partial\theta}{\partial y}^2\right)^{1/2} \\ &= \frac{1}{|\nabla\theta|}\left[\left(-\frac{\partial\theta}{\partial x}\right)\left(\frac{\partial u}{\partial x}\frac{\partial\theta}{\partial x} + \frac{\partial v}{\partial x}\frac{\partial\theta}{\partial y}\right) - \left(\frac{\partial\theta}{\partial y}\right)\left(\frac{\partial u}{\partial y}\frac{\partial\theta}{\partial x} + \frac{\partial v}{\partial y}\frac{\partial\theta}{\partial y}\right)\right. \\ &\quad \left. - \left(\frac{\partial\theta}{\partial p}\right)\left(\frac{\partial\omega}{\partial x}\frac{\partial\theta}{\partial x} + \frac{\partial\omega}{\partial y}\frac{\partial\theta}{\partial y}\right)\right]. \end{aligned} \tag{7.7}$$

In this more complete, 3-D expression, all terms with $\partial u/\partial x$ or $\partial v/\partial y$ are confluence terms, all terms with $\partial v/\partial x$ or $\partial u/\partial y$ are shearing terms, and all terms with derivatives of ω are tilting terms. The physical interpretation of each type of term is precisely the same as for our simpler expression (7.6). For many, but not all, types of frontal development it is sufficient to consider the 2-D version of (7.7) in which the tilting terms are neglected. The resulting expression,

$$\Im_{2D} = \frac{1}{|\nabla\theta|}\left[\left(-\frac{\partial\theta}{\partial x}\right)\left(\frac{\partial u}{\partial x}\frac{\partial\theta}{\partial x} + \frac{\partial v}{\partial x}\frac{\partial\theta}{\partial y}\right) - \left(\frac{\partial\theta}{\partial y}\right)\left(\frac{\partial u}{\partial y}\frac{\partial\theta}{\partial x} + \frac{\partial v}{\partial y}\frac{\partial\theta}{\partial y}\right)\right], \tag{7.8}$$

can be insightfully rewritten using the expression of the four kinematic components of the flow described in Chapter 1. Recalling that since the divergence, vorticity, stretching, and shearing deformations are defined as

$$D = \frac{\partial u}{\partial x} + \frac{\partial v}{\partial y}, \quad \zeta = \frac{\partial v}{\partial x} - \frac{\partial u}{\partial y}, \quad F_1 = \frac{\partial u}{\partial x} - \frac{\partial v}{\partial y}, \quad \text{and} \quad F_2 = \frac{\partial v}{\partial x} + \frac{\partial u}{\partial y},$$

respectively, the horizontal derivatives of the wind field appearing in (7.8) can be expressed as

$$\frac{\partial u}{\partial x} = \frac{D + F_1}{2}, \quad \frac{\partial v}{\partial y} = \frac{D - F_1}{2}, \quad \frac{\partial v}{\partial x} = \frac{\zeta + F_2}{2}, \quad \text{and} \quad \frac{\partial u}{\partial y} = \frac{F_2 - \zeta}{2}.$$

Substituting these expressions into (7.8) yields

$$\begin{aligned}\Im_{2D} = \frac{1}{|\nabla\theta|}&\left\{-\left(\frac{\partial\theta}{\partial x}\right)\left[\left(\frac{D + F_1}{2}\right)\frac{\partial\theta}{\partial x} + \left(\frac{\zeta + F_2}{2}\right)\frac{\partial\theta}{\partial y}\right]\right. \\ &\left. - \left(\frac{\partial\theta}{\partial y}\right)\left[\left(\frac{F_2 - \zeta}{2}\right)\frac{\partial\theta}{\partial x} + \left(\frac{D - F_1}{2}\right)\frac{\partial\theta}{\partial y}\right]\right\}. \end{aligned} \tag{7.9a}$$

Fully expanding the RHS of (7.9a) and grouping like terms results in

$$\Im_{2D} = \frac{-1}{2|\nabla\theta|}\left[D\left(\frac{\partial\theta}{\partial x}^2 + \frac{\partial\theta}{\partial y}^2\right) + F_1\left(\frac{\partial\theta}{\partial x}^2 - \frac{\partial\theta}{\partial y}^2\right) + 2F_2\left(\frac{\partial\theta}{\partial x}\frac{\partial\theta}{\partial y}\right)\right] \tag{7.9b}$$

demonstrating that only divergence and deformation can change $|\nabla\theta|$. Our suspicion that vorticity plays no role in frontogenesis (i.e. does not affect $|\nabla\theta|$), as suggested by our physical analysis of the simplified frontogenesis equation (7.5), is proven to be true.

Now, recall that deformation is not invariant (i.e. the stretching and shearing deformations 'look like' one another) so that the total deformation field can be represented by either one so long as the coordinate axes are rotated by an appropriate amount. By rotating the x- and y-axes counterclockwise by an angle ψ (where $\psi = \frac{1}{2}\tan^{-1}(F_2/F_1)$), we can rewrite (7.9b) as

$$\Im_{2D} = -\frac{1}{2\,|\nabla\theta|}\left\{D(|\nabla\theta|^2) + F_1'\left[\left(\frac{\partial\theta}{\partial x'}\right)^2 - \left(\frac{\partial\theta}{\partial y'}\right)^2\right]\right\} \tag{7.10a}$$

or

$$\Im_{2D} = -\frac{|\nabla\theta|}{2}\left\{D + \frac{F_1'[(\partial\theta/\partial x')^2 - (\partial\theta/\partial y')^2]}{|\nabla\theta|^2}\right\}. \tag{7.10b}$$

The geometry of the rotation of axes is illustrated in Figure 7.11. The angle β is the angle the isentropes make with the x'-axis (the axis of dilatation of the total

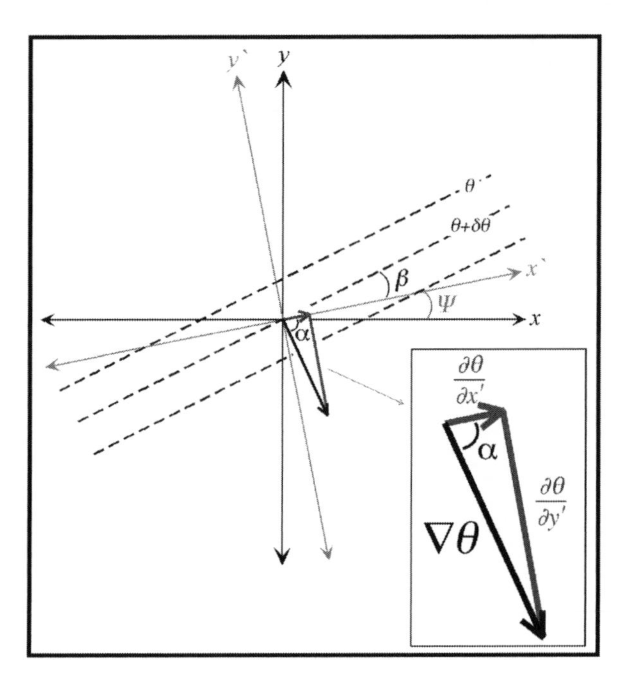

Figure 7.11 Geometry involved in formulating the kinematic form of the frontogenesis function (7.11). The gray axes are the rotated principal axes of the total deformation field with x' and y' representing the axes of dilatation and contraction, respectively. The angle Ψ is the rotation angle and β is the angle between the isentropes and the axis of dilatation of the total deformation field. The inset shows the angle α between the x' axis and the vector $\nabla\theta$. See text for additional explanation

deformation field). The angle α is the angle between the x'-axis and the vector $\nabla\theta$. Note also that $\partial\theta/\partial x'$ and $\partial\theta/\partial y'$ sum to $\nabla\theta$. From Figure 7.11 it is clear that

$$\cos\alpha = \frac{\partial\theta}{\partial x'}\Big/|\nabla\theta| \quad \text{and} \quad \sin\alpha = \frac{\partial\theta}{\partial y'}\Big/|\nabla\theta|.$$

Consequently,

$$\left(\frac{\partial\theta}{\partial x'}\right)^2 - \left(\frac{\partial\theta}{\partial y'}\right)^2 = |\nabla\theta|^2\left[\cos^2\alpha - \sin^2\alpha\right] = |\nabla\theta|^2\cos 2\alpha$$

so that (7.10b) can be rewritten as

$$\Im_{2D} = -\frac{|\nabla\theta|}{2}(D + F_1'\cos 2\alpha). \tag{7.10c}$$

Since $\alpha = 90^\circ - \beta$, and $\cos(\delta - \varepsilon) = \cos\delta\cos\varepsilon + \sin\delta\sin\varepsilon$ by a trigonometric identity, then $\cos 2\alpha = -\cos 2\beta$. Thus, (7.10c) can finally be expressed as

$$\Im_{2D} = \frac{|\nabla\theta|}{2}(F\cos 2\beta - D) \tag{7.11}$$

where F is the total deformation of the flow ($F = (F_1^2 + F_2^2)^{1/2}$). Based upon (7.11) it is clear that two kinematic environments will promote frontogenesis. Frontogenesis will occur whenever non-zero $|\nabla\theta|$ is coincident with convergence ($D < 0$). Frontogenesis will also occur if the total deformation field (F) acts upon isentropes that are between 0° and 45° (β) of the axis of dilatation of the total deformation field. Anytime β is between 46° and 90°, the deformation promotes frontolysis. A couple of hypothetical flow fields superposed with isentropes are illustrated in Figure 7.12.

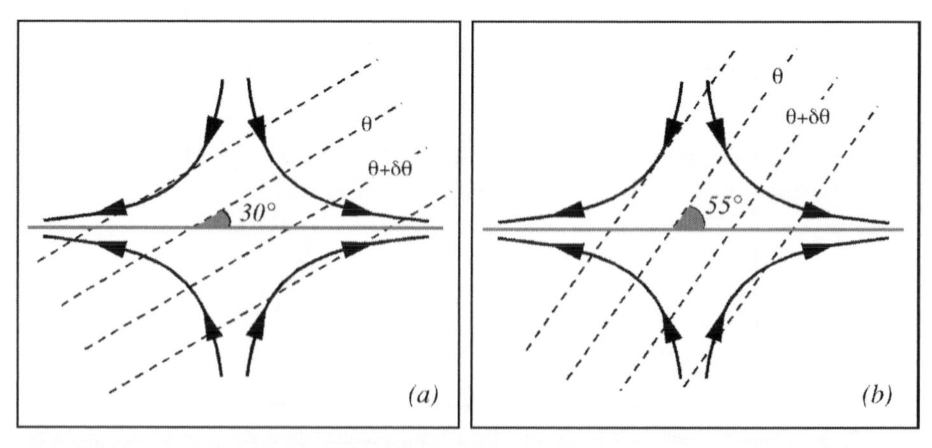

Figure 7.12 (a) Bundle of isentropes in a field of deformation. The thick gray line is the axis of dilatation of the deformation field with which the isentropes make an angle of 30°. The result is frontogenesis. (b) Same bubdle of isentropes at an angle of 55° with the axis of dilatation of the deformation field. The result is frontolysis

This geometric form of the 2-D frontogenesis function is very useful in that it distills the essential physics of frontogenesis into an expression that is rather easy to apply to real weather maps. Such application involves quick identification of the angle between the isentropes and the axis of dilatation of the total deformation field and can be used to identify regions of frontogenesis. It is not, however, amenable to quick calculation of the magnitude of the frontogenesis. In the modern computer era, with the availability of gridded data sets of observations and forecasts, the numerical calculation of frontogenesis using (7.7) or (7.8) is equally simple and much more precise, though perhaps not as physically insightful to the young scientist.

As we have already discussed, there is a physical relationship between changes in $|\nabla\theta|$ and the production of vertical circulations in the middle latitudes. We can use R. C. Sutcliffe's ideas to put some mathematical rigor to that argument. Recall that in our discussion of the diagnosis of vertical motions in mid-latitudes we began with

$$\frac{d\vec{V}}{dt} - \left(\frac{d\vec{V}}{dt}\right)_0 = \vec{V}_s \cdot \nabla\vec{V}_0 + \frac{d\vec{V}_s}{dt}. \tag{7.12}$$

Let us now concentrate on the physical meaning of the second term on the RHS of (7.12) which describes the rate of change of the vertical shear vector, $\vec{V}_s$. If we make the assumption that the vertical shear is geostrophically balanced, then $\vec{V}_s$ is directly related to $\nabla\theta$ by the thermal wind relationship. In such a case, an increase in $\vec{V}_s$ (i.e. $d\vec{V}_s/dt > 0$) is associated with an increase in $|\nabla\theta|$ and is therefore a result of positive horizontal frontogenesis. As discussed in Chapter 6, when $d\vec{V}_s/dt$ is positive, a thermally direct vertical circulation results. Thus, we can conclude that a thermally direct (indirect) vertical circulation will attend positive (negative) horizontal frontogenesis. This relationship underlies the ubiquity of clouds and precipitation in the vicinity of mid-latitude frontal zones! Of course, in making this connection, we are implicitly asserting that such fronts are characterized by positive horizontal frontogenesis, an assertion that is readily verified by observations.

Finally, we might consider an alternative version of (7.8) in which all winds are geostrophic,

$$\Im_{2D_g} = \frac{1}{|\nabla\theta|}\left[\frac{\partial\theta}{\partial x}\left(-\frac{\partial u_g}{\partial x}\frac{\partial\theta}{\partial x} - \frac{\partial v_g}{\partial x}\frac{\partial\theta}{\partial y}\right) + \frac{\partial\theta}{\partial y}\left(-\frac{\partial u_g}{\partial y}\frac{\partial\theta}{\partial x} - \frac{\partial v_g}{\partial y}\frac{\partial\theta}{\partial y}\right)\right]. \tag{7.13a}$$

The terms inside the parentheses on the RHS of (7.13a) are equal to

$$\frac{1}{f\gamma}Q_1 \quad \text{and} \quad \frac{1}{f\gamma}Q_2,$$

the components of the $\vec{Q}$-vector, respectively. Thus, (7.13a) can be expressed as

$$\Im_{2D_g} = \left(\frac{1}{f\gamma}\right)\frac{\vec{Q}\cdot\nabla\theta}{|\nabla\theta|}, \tag{7.13b}$$

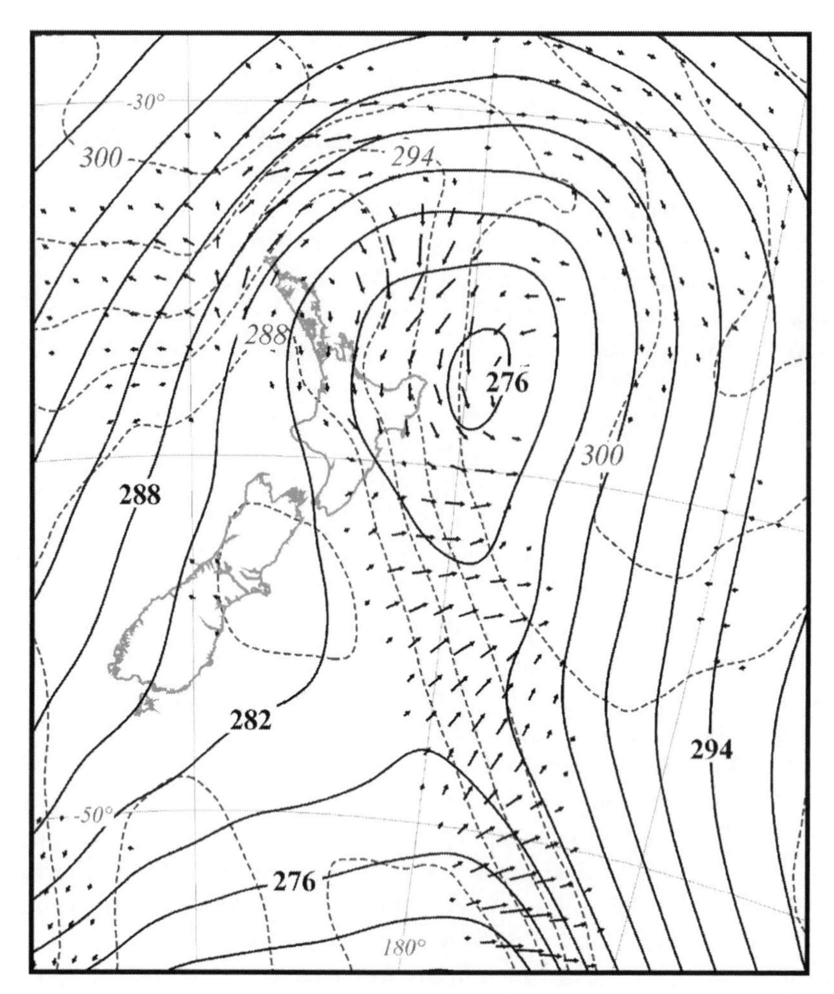

Figure 7.13 The 700 hPa geopotential heights (solid lines), isentropes (dashed lines), and $\vec{Q}$-vectors near New Zealand at 0600 UTC 16 August 2004. Geopotential heights are labeled in dam and contoured every 3 dam. Isentropes are labeled in K and contoured every 3 K. For clarity, only $\vec{Q}$-vectors larger than 2×10^{-10} m^2 kg^{-1} s^{-1} are plotted

a scalar multiple of the magnitude of the across-isentrope component of $\vec{Q}$, as shown in Chapter 6. Shown in Figure 7.13 is a set of $\vec{Q}$-vectors and isentropes at 700 hPa. From (7.13b), any place where $\vec{Q}$-vectors point across the isentropes from cold to warm air will be associated with horizontal frontogenesis (i.e. $\Im_{2D_g} > 0$). In such locations, the geostrophic winds are advecting θ in such a way as to increase $|\nabla\theta|$ and we should expect a thermally direct vertical circulation to respond. Figure 7.13 illustrates that in such a setting the $\vec{Q}$-vectors will be convergent somewhere, and to some degree, on the warm side of the baroclinic zone. This implies that the warm air will rise and the cold air, in which the $\vec{Q}$-vectors are divergent, will be sinking – precisely what we expected.

As physically compelling as this is, the geostrophic frontogenesis function only references the influence of geostrophic advection on forcing the secondary circulation. We might reasonably ask if this is enough to describe nature accurately. Recall that the geostrophic balance is fairly well obeyed in the along-front direction but not so well obeyed in the mesoscale, across-front direction. In nature, it is entirely possible that across-front advections of temperature and geostrophic momentum might accomplish a considerable amount of frontal intensification and that a large fraction of the across-front winds will not be in geostrophic balance. Indeed, (7.11) makes clear that a sizeable portion of the frontogenetical forcing resides in the divergence (D) of the ageostrophic wind. We now examine whether or not the geostrophic frontogenesis function is a reasonable diagnostic of what actually occurs at fronts.

In order to make this assessment, let us consider the effect of geostrophic confluence on the evolution of the temperature contrast illustrated in Figure 7.14. Let us assume that

$$\frac{d}{dt}\left(\frac{\partial \theta}{\partial x}\right) = -\frac{\partial u_g}{\partial x}\frac{\partial \theta}{\partial x} = k\frac{\partial \theta}{\partial x} \tag{7.14a}$$

where k is a constant, characteristic value of geostrophic confluence ($k = -\partial u_g/\partial x = 10^{-5}\,\mathrm{s}^{-1}$). Given these assumptions we proceed by first noting that (7.14a) can be solved explicitly: $d\ln(\partial\theta/\partial x)/dt = k$ can be rewritten as $d\ln(\partial\theta/\partial x) = k\,dt$. This can be integrated to yield

$$\left(\frac{\partial \theta}{\partial x}\right)_t = \left(\frac{\partial \theta}{\partial x}\right)_0 e^{kt} \tag{7.14b}$$

thus suggesting that, for typical conditions at middle-latitudes, it takes 10^5 seconds ($\sim$1 day) for pure geostrophic confluence to increase the intensity of a frontal temperature contrast by a factor of e. Such an intensification rate is much slower than what is actually observed in nature as illustrated by the observations in Figure 7.14. Why should nature be able to accomplish frontogenesis so much faster than our geostrophic confluence model? In making our case for geostrophic confluence, we have adopted a view of frontal intensification in which the secondary circulation forced by the geostrophic frontogenesis does not feed back upon the across-front advection of temperature (or momentum). We have, thus, not considered a truly *dynamical* approach to the problem as the neglected across-front ageostrophic temperature advection will produce some ageostrophic frontogenesis which, when added to the geostrophic frontogenesis, will accomplish greater total frontogenesis. Nature, of course, includes such ageostrophic feedbacks on the frontal intensification rate. Thus, in order to describe nature more accurately, we need to include these across-front, ageostrophic advections of temperature and momentum in our diagnostic equations for frontogenesis. The so-called **semi-geostrophic equations**, which we will now develop, will include these important missing processes in a more comprehensive and physically accurate picture of frontogenesis.

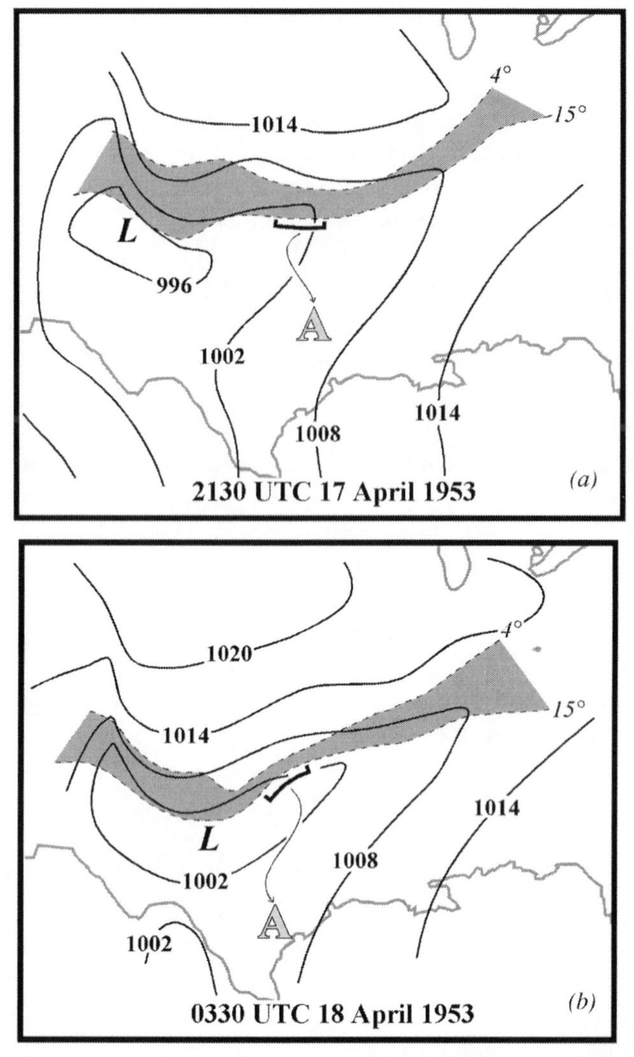

Figure 7.14 (a) Sea-level pressure (solid lines) and temperature (dashed lines) analyses at 2130 UTC 17 April 1953. Isobars are labeled in hPa and contoured every 6 hPa. The isothermal band between 4°C and 15°C is shaded. (b) As for (a) but for 0330 UTC 18 April 1953. The region of temperature gradient labeled 'A' has intensified by more than a factor of 2 in 6 hours. Adapted from Sanders (1955)

7.3 The Semi-Geostrophic Equations

J. S. Sawyer[1] investigated a large number of frontal passages in the United Kingdom in the early 1950s and came to the conclusion that active fronts (those fronts associated

[1] John S. Sawyer was born in Wembley, England on 19 June, 1916. He joined the Meteorological Office in 1938 and was a forecaster during World War II in two distinct theaters: Western Europe from 1942 to 1943 and in the Middle East from 1943 to 1945. After the war he worked under R. C. Sutcliffe in the new Forecasting

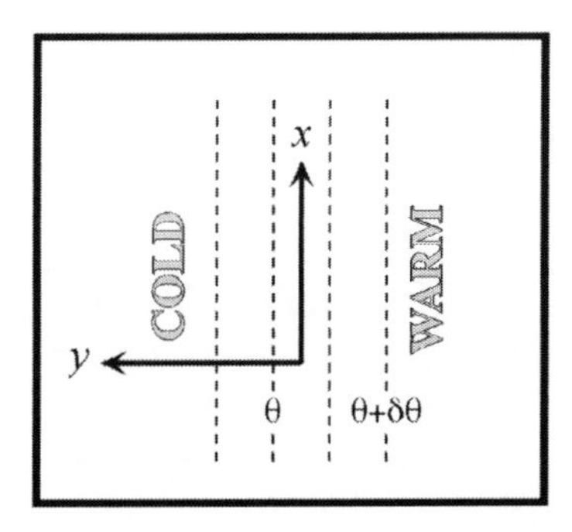

Figure 7.15 Coordinate system orientation to isentropes adopted for development of the Sawyer–Eliassen equation

with clouds and precipitation) were invariably associated with frontogenesis. We have already seen that bands of considerable baroclinicity can be produced by differential horizontal advection in non-divergent (i.e. geostrophic) deformation fields. But such non-divergent flows cannot account for the production of the characteristic frontal horizontal wind shear or the jet stream because these features are characterized by vorticity which can only be produced by divergent motions. Let us consider a front aligned such that the x-axis is along the front (i.e. along the isentropes) and the y-axis points directly into the cold air as depicted in Figure 7.15. Recall that the geostrophic wind relations are given by

$$U_g = -\frac{1}{f}\frac{\partial \phi}{\partial y} \quad \text{and} \quad V_g = \frac{1}{f}\frac{\partial \phi}{\partial x}.$$

The hydrostatic equation, as we saw at the end of Chapter 6, can be written as

$$\frac{1}{f}\frac{\partial \phi}{\partial p} = -\gamma\theta$$

where

$$\gamma = \frac{R}{f p_0}\left(\frac{p_0}{p}\right)^{c_v/c_p},$$

with $p_0 = 1000\,\text{hPa}$, is a function of pressure only. This expression for hydrostatic balance results in simplified expressions for the thermal wind components,

$$\frac{\partial U_g}{\partial p} = \gamma\frac{\partial \theta}{\partial y} \quad \text{and} \quad \frac{\partial V_g}{\partial p} = -\gamma\frac{\partial \theta}{\partial x}. \tag{7.15}$$

Research Division of the Met Office and did extensive work on the calculation of vertical motions and numerical weather prediction. His famous contribution to the theory of frontal circulations, in which a 1-D version of the so-called Sawyer–Eliassen equation was first derived, was published in 1956. For his lifelong contributions to dynamic meteorology he was honored by both the Royal Meteorological Society and the World Meteorological Organization. He died on 19 September 2000.

Since the front is assumed to be a 2-D one, there is no along-front geopotential height gradient so the equation of motion in the along-front (x) direction is given by

$$\frac{dU_g}{dt} + \frac{du}{dt} = fv, \tag{7.16a}$$

where u and v (U_g and V_g) are the x- and y-direction ageostrophic (geostrophic) winds, respectively. If we assume that the along-front flow is nearly geostrophic (i.e. u is small with respect to U_g) then we can make the **geostrophic momentum approximation**. This approximation simply implies that there is no systematic increase in the magnitude of the along-front ageostrophic wind (i.e. $\left|dU_g/dt\right| \gg |du/dt|$). Using the geostrophic momentum approximation, (7.16a) is simplified to

$$\frac{dU_g}{dt} = fv, \tag{7.16b}$$

or, fully expanded,

$$\frac{dU_g}{dt} = \frac{\partial U_g}{\partial t} + U_g\frac{\partial U_g}{\partial x} + u\frac{\partial U_g}{\partial x} + V_g\frac{\partial U_g}{\partial y} + v\frac{\partial U_g}{\partial y} + \omega\frac{\partial U_g}{\partial p} = fv. \tag{7.17}$$

Similarly, the thermodynamic energy equation becomes

$$\frac{d\theta}{dt} = \frac{\partial \theta}{\partial t} + U_g\frac{\partial \theta}{\partial x} + u\frac{\partial \theta}{\partial x} + V_g\frac{\partial \theta}{\partial y} + v\frac{\partial \theta}{\partial y} + \omega\frac{\partial \theta}{\partial p}. \tag{7.18}$$

Now, since the along-front flow is presumed to be largely in geostrophic balance, we will hereafter ignore the along-front ageostrophic advection terms (i.e. $u\,\partial/\partial x$ terms). We will also introduce a new variable, the absolute geostrophic momentum (M), defined as

$$M = U_g - fy \tag{7.19}$$

noting that M is conserved, under the given assumptions, according to (7.16b). Using (7.19), (7.17) can be rewritten as

$$\frac{\partial U_g}{\partial t} + U_g\frac{\partial U_g}{\partial x} + V_g\frac{\partial U_g}{\partial y} + v\frac{\partial M}{\partial y} + \omega\frac{\partial M}{\partial p} = 0. \tag{7.20}$$

Now, taking $\partial/\partial p$ of (7.20) and adding it to $-\gamma\partial/\partial y$ of (7.18), and using the thermal wind relationships and the non-divergence of the geostrophic wind, we get

$$-\frac{\partial}{\partial y}\left(\gamma v\frac{\partial \theta}{\partial y} + \gamma\omega\frac{\partial \theta}{\partial p}\right) + \frac{\partial}{\partial p}\left(v\frac{\partial M}{\partial y} + \omega\frac{\partial M}{\partial p}\right)$$
$$= -2\left(\frac{\partial U_g}{\partial p}\frac{\partial U_g}{\partial x} + \frac{\partial V_g}{\partial p}\frac{\partial U_g}{\partial y}\right) - \gamma\frac{\partial}{\partial y}\left(\frac{d\theta}{dt}\right). \tag{7.21a}$$

The continuity equation in isobaric coordinates ($\nabla \cdot \vec{V} = 0$) can be simplified to

$$\frac{\partial v}{\partial y} + \frac{\partial \omega}{\partial p} \approx 0$$

by assuming that the along-front derivative of the along-front ageostrophic flow $(\partial u/\partial x)$ is negligible. If we then set $v = -\partial\psi/\partial p$ and $\omega = \partial\psi/\partial y$ (7.21a) can be rewritten in terms of a streamfunction, ψ, for the ageostrophic flow in the y–p plane as

$$\left(-\gamma\frac{\partial\theta}{\partial p}\right)\frac{\partial^2\psi}{\partial y^2} + \left(2\frac{\partial M}{\partial p}\right)\frac{\partial^2\psi}{\partial p\partial y} + \left(-\frac{\partial M}{\partial y}\right)\frac{\partial^2\psi}{\partial p^2} = Q_g - \gamma\frac{\partial}{\partial y}\left(\frac{d\theta}{dt}\right) \quad (7.21b)$$

where

$$Q_g = -2\left(\frac{\partial U_g}{\partial y}\frac{\partial V_g}{\partial p} - \frac{\partial V_g}{\partial y}\frac{\partial U_g}{\partial p}\right) \quad (7.22)$$

is the geostrophic forcing function. Equation (7.21b) is known as the **Sawyer–Eliassen circulation equation** as it is based upon pioneering work by J. S. Sawyer and A. Eliassen.[2] The Sawyer–Eliassen equation is a linear, second-order partial differential equation for the 2-D, across-front (transverse) ageostrophic streamfunction, ψ. The general form of such an equation is

$$A\frac{\partial^2 u}{\partial x^2} + B\frac{\partial^2 u}{\partial x\partial y} + C\frac{\partial^2 u}{\partial y^2} + D\frac{\partial u}{\partial x} + E\frac{\partial u}{\partial y} + Fu = G$$

and its solution characteristics can be assessed by considering the discriminant, $B^2 - 4AC$. The following conditions can be determined from the discriminate: if

$$B^2 - 4AC \quad \begin{array}{ll} < 0 & \textit{Elliptic} \\ = 0 & \textit{Parabolic} \\ > 0 & \textit{Hyperbolic.} \end{array}$$

In general form, elliptic solutions are those in which u is uniquely determined from the forcing function, G. More specifically, for the Sawyer–Eliassen equation, solutions for ψ (the transverse ageostrophic streamfunction) will arise entirely as a consequence of the frontogenetic forcing provided that

$$\gamma\left(\frac{\partial\theta}{\partial p}\frac{\partial M}{\partial y} - \frac{\partial\theta}{\partial y}\frac{\partial M}{\partial p}\right) > 0.$$

Physically, this is the condition that the quasi-geostrophic potential vorticity is greater than zero in the solution domain. If this condition is not met, then there is either

[2] Arnt Eliassen was born in Oslo, Norway on 9 September 1915. He was introduced to meteorology, rather by accident, in a course taught by Sverre Pettersen in autumn 1938. He later worked as an assistant under Tor Solberg and, later still, under C. G. Rossby in Chicago in the 1950s. He was one of the giants of modern dynamical meteorology involved in pioneering work on such diverse topics as the development of the QG system of equations, potential vorticity applications to atmospheric flow, numerical weather prediction, tropical cyclone development, and mid-latitude frontal circulations. In 1962 he published a generalization of Sawyer's earlier work on fronts, presenting the 2-D version of the Sawyer–Eliassen equation in a paper that I consider to be the most clearly written scientific paper I have ever read. I had the opportunity to meet Prof. Eliassen in Bergen in 1994 at which time I relayed these thoughts to him. With characteristic humility he replied, 'I had to write very clearly and carefully as my English was not so good.' He died on 22 April 2000.

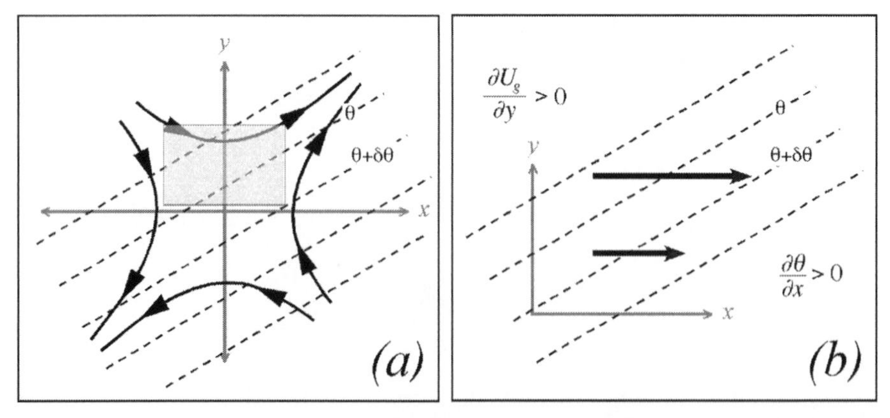

Figure 7.16 (a) Isentropes in a frontogenetic field of deformation. (b) Close-up of the lightly shaded, boxed area in (a) illustrating the effect of geostrophic shearing deformation

inertial instability, or static instability somewhere in the domain. The presence of either instability in the solution domain will allow growth of non-unique solutions, arising from the release of the instability, thus prohibiting clear attribution of the resulting ageostrophic motions to the process of frontogenesis.

The Sawyer–Eliassen equation is a bulky expression and yet rather simple conceptual interpretations are possible using it. We will concentrate on the geostrophic forcing term, Q_g, in our exploration of the physical interpretation of the Sawyer–Eliassen equation, as we can determine the sense of the circulation by considering Q_g alone. Using the thermal wind relationships we can rewrite (7.22) as

$$Q_g = 2\gamma \left(\frac{\partial U_g}{\partial y} \frac{\partial \theta}{\partial x} + \frac{\partial V_g}{\partial y} \frac{\partial \theta}{\partial y} \right) \tag{7.23}$$

where the first term on the RHS of (7.23) is known as the geostrophic shearing deformation while the second term is called the geostrophic stretching deformation. We will investigate each of these terms in isolation beginning with the geostrophic shearing deformation. Figure 7.16 illustrates an example of geostrophic shearing deformation. Before considering the mathematical underpinnings of this problem, let us consider the physics of the situation depicted in Figure 7.16. It is clear that the apparent shear zone in Figure 7.16(b) is a small portion of the larger-scale deformation field shown in Figure 7.16(a). That deformation field will rotate the isentropes into an alignment that is parallel to the axis of dilatation (x-axis) over time. Simultaneously, the isentropes will be pushed closer together so positive horizontal frontogenesis is implied. As we have seen earlier in this chapter, positive horizontal frontogenesis is associated with a thermally direct vertical circulation. The geostrophic shearing deformation term itself has the form

$$Q_{g_{SH}} = 2\gamma \frac{\partial U_g}{\partial y} \frac{\partial \theta}{\partial x}.$$

For the situation depicted in Figure 7.16(b), both $\partial U_g/\partial y$ and $\partial \theta/\partial x$ are positive; thus, $Q_{g_{SH}} > 0$.

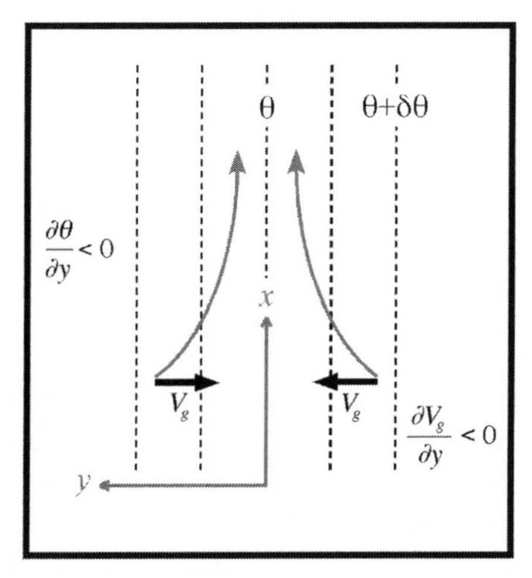

Figure 7.17 Illustration of the stretching deformation term of the Sawyer–Eliassen equation

The geostrophic stretching deformation can be investigated with the use of Figure 7.17 in which the confluent entrance region of a jet streak is depicted. The confluent geostrophic flow is clearly tending to increase the baroclinicity in the region. As a consequence, there is positive horizontal frontogenesis occurring there, and a thermally direct vertical circulation results. The geostrophic stretching deformation term itself has the form

$$Q_{g_{ST}} = 2\gamma \frac{\partial V_g}{\partial y} \frac{\partial \theta}{\partial y}.$$

For the confluent jet entrance, both $\partial V_g / \partial y$ and $\partial \theta / \partial y$ are negative; thus, $Q_{g_{ST}} > 0$. As it turns out, all we need to know in order to discern the sense of the 2-D ageostrophic circulation using the Sawyer–Eliassen equation is the sign of the RHS of (7.21b) – in this particular example, the geostrophic forcing function, Q_g. Anytime Q_g is positive (negative), a thermally direct (indirect) circulation is diagnosed.

Notice that Q_g can be written to look like a part of the $\vec{Q}$-vector since

$$Q_g = 2\gamma \left(\frac{\partial U_g}{\partial y} \frac{\partial \theta}{\partial x} + \frac{\partial V_g}{\partial y} \frac{\partial \theta}{\partial y} \right) = 2\gamma \left(\frac{\partial \vec{V}_g}{\partial y} \cdot \nabla \theta \right)$$

and the $\hat{j}$ component of $\vec{Q}$ is equal to

$$Q_2 = -f\gamma \left(\frac{\partial \vec{V}_g}{\partial y} \cdot \nabla \theta \right).$$

Thus, the geostrophic forcing function, Q_g, of the Sawyer–Eliassen equation is a scalar multiple of the $\hat{j}$ component of $\vec{Q}$; $Q_2 = -(f/2) Q_g$.

The LHS of the Sawyer–Eliassen equation (7.21b) is a rather complex-looking expression but considerable physical insight into the process of frontogenesis can be garnered by considering each term in some detail. Since (7.21b) is an equation for ψ, we will consider the physical interpretations of both the derivatives of ψ as well as their coefficients. The first term on the LHS of (7.21b) is $(-\gamma \partial\theta/\partial p)\partial^2\psi/\partial y^2$. This term represents the product of the static stability $(-\gamma \partial\theta/\partial p)$ and across-front gradients in ω ($\partial^2\psi/\partial y^2$, since $\omega = \partial\psi/\partial y$). The only way that across-front gradients in vertical motion can have any effect on $|\nabla\theta|$ is if they act upon the static stability via the tilting term. The second term on the LHS of (7.21b) represents the product of the across-front baroclinicity $(2\partial M/\partial p)$ and the across-front, ageostrophic divergence ($\partial^2\psi/\partial p\partial y$, or $-\partial v/\partial y$ since $v = -\partial\psi/\partial p$). Clearly, if there is ageostrophic convergence in the presence of baroclinicity, the frontal intensity is increased. Finally, the third term on the LHS of (7.21b) represents the product of the vorticity $(-\partial M/\partial y)$ and the across-front vertical shear of the ageostrophic wind ($\partial^2\psi/\partial p^2$, or $-\partial v/\partial p$ since $v = -\partial\psi/\partial p$). The tilting of vortex tubes by the across-front vertical shear of the ageostrophic wind will modulate the tilt of the frontal zone, an observable characteristic of fronts in nature. Notice that each of the coefficients of these three terms represents one of the three essential dynamical characteristics of a front and that each of the three terms themselves represents an aspect of the secondary ageostrophic circulation that responds to the frontogenetical forcing. This suggests that in solving (7.21b) for ψ, a successive overrelaxation procedure (SOR) would execute the following solution steps: (1) assess the RHS forcing in (7.21b), (2) make a first guess for ψ in the solution domain, (3) use ψ to compute the first-guess ageostrophic circulation, (4) allow the ageostrophic circulation to advect temperature and momentum (as in the terms just discussed on the LHS of (7.21b)), (5) iterate to a state of balance between the RHS and LHS. In this way, the ageostrophic secondary circulation feeds back into the final frontogenesis process as originally intended with the introduction of the increased complexity of the Sawyer–Eliassen equation. Thus, solution of the Sawyer–Eliassen equation mimics nature and suggests that frontogenesis is a two-step process. First, the non-divergent, geostrophic deformation tightens the temperature gradient resulting in the production of secondary, ageostrophic transverse circulation. Second, the ageostrophic circulation itself advects temperature and momentum in the frontal zone, produces the characteristic vorticity, and further intensifies the temperature contrast leading to the sometimes rapid production of the sharp frontal boundaries; observed in the mid-latitude atmosphere.

Up to this point in our discussion of frontogenesis we have been solely focused on the development of fronts at or near the surface of the Earth. The surface of the Earth represents a physical boundary. Fronts, however, are not confined to form only at physical boundaries; they may also form at thermodynamic boundaries across which there is very little mixing. One such boundary in the Earth's atmosphere is the tropopause boundary. Next we will investigate the development of fronts at the tropopause boundary, examining both the processes by which these fronts form and the consequences their development has on the array of weather systems that parade across the middle latitudes.

7.4 Upper-Level Frontogenesis

A vertical cross-section through a modest local wind speed maximum (labeled with a 'J') in the upper troposphere and lower stratosphere is shown in Figure 7.18. The tropopause is easily identified by the sudden increase in the vertical gradient in potential temperature corresponding to increased static stability. In accordance with the thermal wind relationship, the local wind speed maximum sits atop a column of air in which a modest horizontal temperature contrast exists, particularly in the upper troposphere. As a consequence of the vertical wind shear resulting from the presence of the local wind speed maximum, horizontal vortex tubes with the indicated spin are present beneath the maximum. With this physical background, let us now consider what might develop *if* a thermally indirect circulation straddling the wind speed maximum can be generated.

First, since the lower stratosphere is characterized by high static stability, the hypothetical thermally indirect circulation will tilt a bundle of closely packed isentropes from their original horizontal orientation into a more vertical orientation. Since this change is associated with vertical tilting, as we have already seen, the absolute distance between isentropes in the tilted bundle will not change. As the isentropes acquire a more vertical orientation, the magnitude of the horizontal θ gradient ($|\nabla\theta|_H$) increases. The hypothetical thermally indirect circulation also acts upon the horizontal vortex tube, gradually tilting a component of that vorticity into the vertical direction. Thus, a local increase in cyclonic vorticity results in precisely the same region in which a local increase in $|\nabla\theta|$ is accomplished. These ingredients, along with the high static stability present in the developing baroclinic zone owing to its lower stratospheric origin, constitute the essential dynamical characteristics of a front. Thus, imposition of a thermally indirect circulation onto an environment such as that illustrated in Figure 7.18 leads to the development of a frontal zone along the tropopause boundary. Such frontal zones are known as **upper-level fronts**. As a consequence of their characteristic association with intensifying wind speed maxima, they are also often referred to as **upper-level jet/front systems**.

It is important to note that such upper-level fronts do not separate air masses with different origins in the horizontal as is the case with surface-based frontal zones. Instead, upper-level fronts separate tropospheric air (beneath them) from stratospheric air (above them). In fact, the tropopause boundary often becomes 'folded' over itself in the vicinity of the upper-level front. A number of observable quantities are present in the atmosphere that allow for a clear distinction between tropospheric and stratospheric air to be made. In the late 1940s and 1950s it was common practice to test new weapons systems by exploding nuclear devices at high altitude in the stratosphere, the idea being that the dangerous radioactive by-products of the devices would rain out in the highly stratified stratosphere and would not quickly or easily mix into the troposphere. Some of the pioneering work done in identifying and diagnosing upper-level fronts used analysis of radioactivity, an indisputable criterion

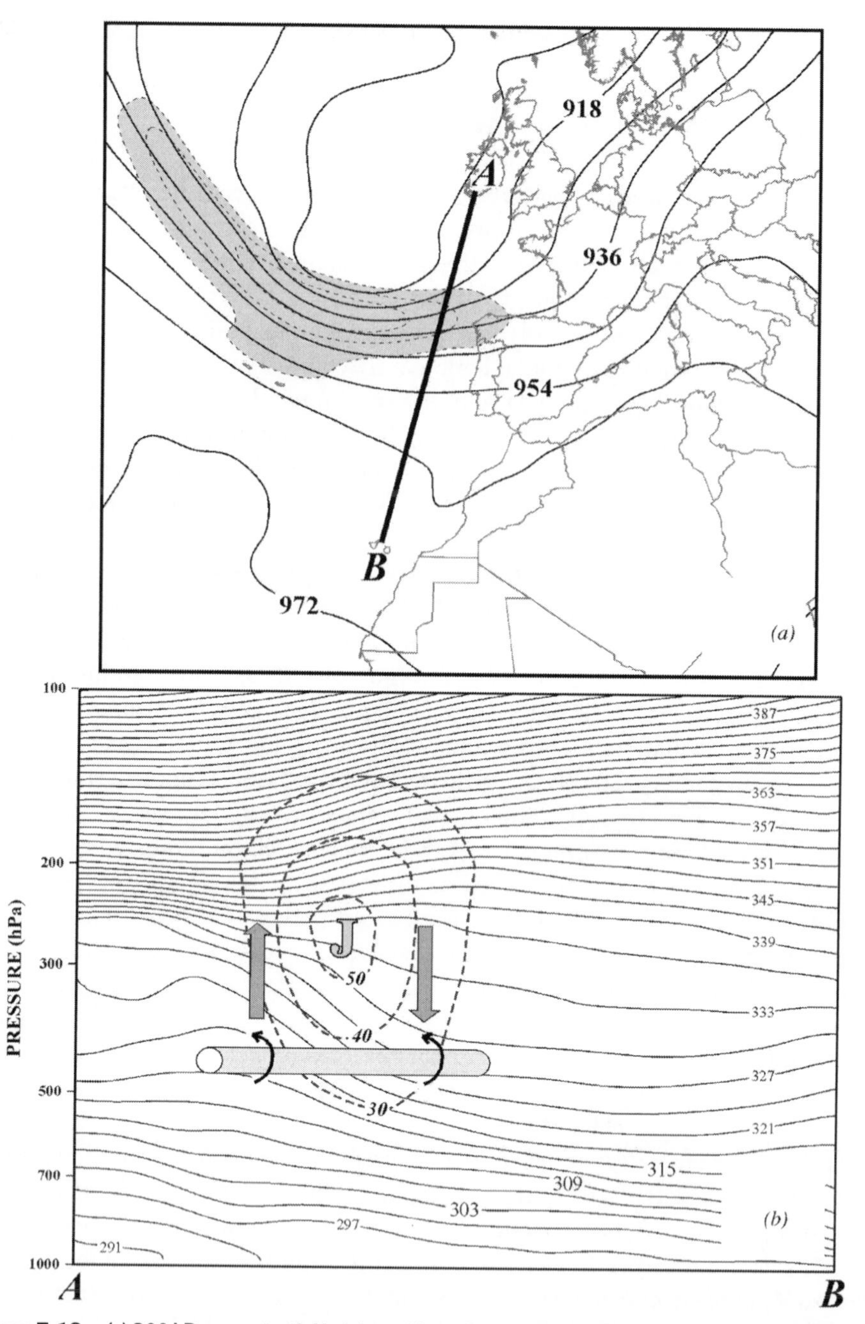

Figure 7.18 (a) 300 hPa geopotential height and isotachs from the NCEP AVN model valid at 1800 UTC 17 August 2004. Geopotential heights labeled in dam and contoured every 9 dam. Isotachs contoured every 10 m s^{-1} starting at 30 m s^{-1}. Cross-section along line A–B shown in (b). (b) Vertical cross-section along line A–B in (a). Solid lines are isentropes labeled in K and contoured every 3 K. Dashed lines are isotachs labeled in m s^{-1} and contoured every 10 m s^{-1} starting at 30 m s^{-1}. 'J' indicates the position of the jet core. Light shaded tube with arrows illustrates the horizontal vorticity associated with the vertical wind shear of the jet. Dark shaded arrows are the upward and downward branches of a hypothetical thermally indirect circulation. The consequences of such a circulation in this environment are discussed in the text

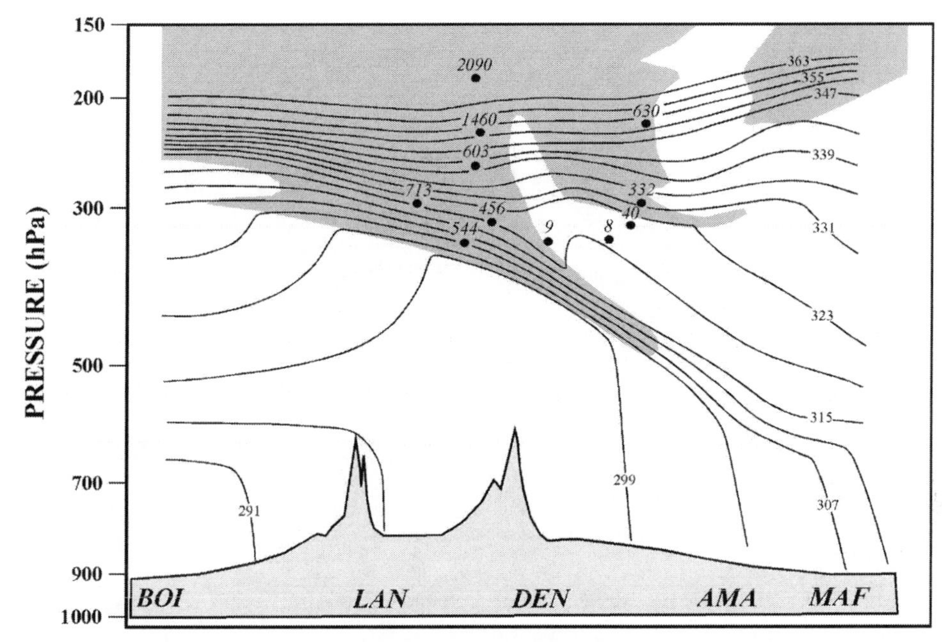

Figure 7.19 Vertical cross-section taken at 0000 UTC 22 April 1963 from Boise, ID (BOI), to Lander, WY (LAN), to Denver, CO (DEN), to Amarillo, TX (AMA), to Midland, TX (MAF). Solid lines are isentropes labeled in K and contoured every 4 K. Shading highlights regions in which potential vorticity (PV) is greater than or equal to 2.5 PVU (1PVU $= 10^{-6}$ K m^2 kg^{-1} s^{-1}) safely indicating stratospheric air. Dots are measurements of radioactive decay of strontium (^{90}Sr) in units of disintegrations per minute per 1000 cubic feet of air. Note the high radioactivity that exists within the high-PV stratospheric air as well as within the upper frontal zone. Adapted from Danielsen (1964)

for establishing the presence of stratospheric air in upper-level fronts. Upper-level fronts are also characterized by high ozone mixing ratios – again, indicative of the fact that air of stratospheric origin is present in upper-level frontal zones. Finally, since the lower stratosphere is also a region of high potential vorticity (PV) as compared to the upper troposphere, the fact that upper-level frontal zones are characterized by high PV is yet another clear indication that upper-level fronts separate stratospheric from tropospheric air. Some of these analysis elements are illustrated in Figure 7.19, a vertical cross-section through an upper-level front. It is clear from Figure 7.18 that a thermally indirect vertical circulation in the vicinity of an upper tropospheric wind speed maximum is of vital importance to the development of such upper-level frontal zones. Naturally, then, we must determine under what synoptic-scale conditions such a thermally indirect vertical circulation can be produced. Understanding this question will unlock more secrets concerning upper-level fronts.

We will consider this question from the perspective of the Sawyer–Eliassen equation. Let us first consider the vertical circulations that accompany the entrance and exit regions of an upper tropospheric jet streak such as the one illustrated in

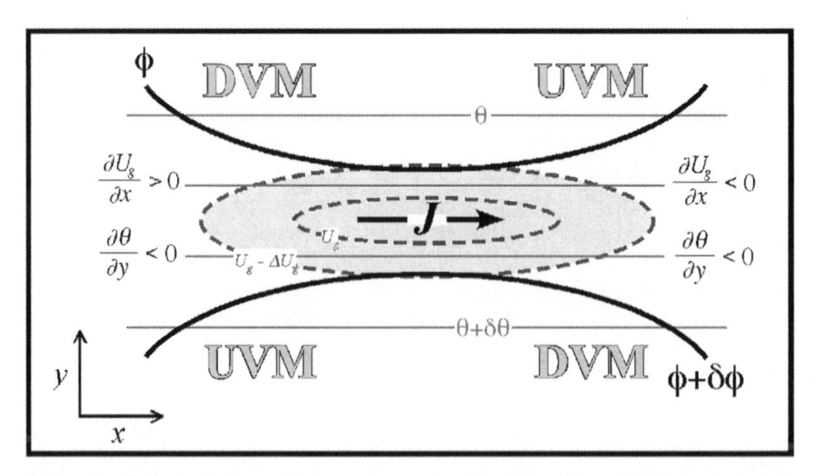

Figure 7.20 Vertical circulations at the entrance and exit regions of a straight jet streak as diagnosed by the stretching deformation term of the Sawyer–Eliassen equation. 'DVM' and 'UVM' correspond to downward and upward vertical motions, respectively

Figure 7.20. The non-divergence of the geostrophic wind allows the geostrophic stretching deformation term to be written as

$$Q_{g_{ST}} = 2\gamma \frac{\partial V_g}{\partial y}\frac{\partial \theta}{\partial y} = -2\gamma \frac{\partial U_g}{\partial x}\frac{\partial \theta}{\partial y}.$$

As we have already seen, a thermally direct vertical circulation arises in the jet entrance region where $\partial U_g/\partial x > 0$ and, consequently, $Q_{g_{ST}} > 0$. The jet exit region, however, is characterized by $\partial U_g/\partial x < 0$ and $Q_{g_{ST}} < 0$, consistent with a thermally indirect vertical circulation. Thus, the exit region of a jet streak is a preferred location for the development of an upper-level front. Note that the diffluent horizontal wind field superimposed upon the isentropes in the exit region clearly promotes **horizontal frontolysis** (i.e. $d|\nabla\theta|_H/dt_g < 0$) which is associated with a thermally indirect circulation. When that circulation is able to draw upon the high static stability of the lower stratosphere, as it can if the thermally indirect circulation occurs in the upper troposphere, then the frontolytic effect of the horizontal winds is superseded by the tendency of the tilting to increase $|\nabla\theta|_H$.

Let us now consider the effect of geostrophic shearing deformation on upper frontogenesis. Recall that the geostrophic shearing deformation term from the Sawyer–Eliassen equation is given by

$$Q_{g_{SH}} = 2\gamma \frac{\partial U_g}{\partial y}\frac{\partial \theta}{\partial x}.$$

Figure 7.21 illustrates an example of horizontal cold air advection in the presence of cyclonic shear. In such a case it is clear that the horizontal winds are tending to decrease $|\nabla\theta|_H$ by forcing a horizontal separation of the isentropes. Consistent with this fact, $Q_{g_{SH}} < 0$ and a thermally indirect vertical circulation is the result. Consequently, the development of an upper-level front is promoted in such an environment. The circumstance of *warm air advection in anticyclonic shear* will also render $Q_{g_{SH}} < 0$

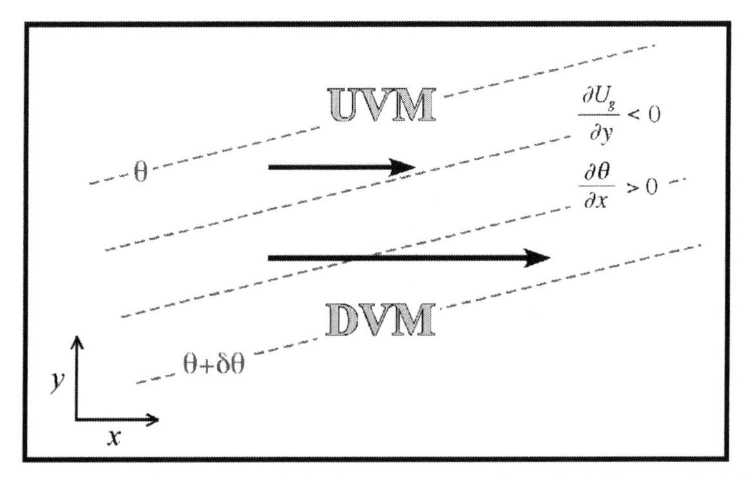

Figure 7.21 Cold air advection in cyclonic shear producing a thermally indirect vertical circulation. Arrows are the geostrophic winds. 'DVM' and 'UVM' correspond to downward and upward vertical motions, respectively

and thus promote upper frontogenesis. Conversely, cold (warm) air advection in the presence of anticyclonic (cyclonic) shear will be associated with $Q_{g_{SH}} > 0$, thermally direct circulations, and an upper-frontolytic tendency.

We will now develop a natural coordinate version of the geostrophic shearing deformation forcing term to make diagnosis of these four combinations of horizontal temperature advection and shear even more transparent. Figure 7.22 shows simple schematics of all four of the relevant combinations along with the corresponding Cartesian expressions for the temperature gradient and horizontal shear as well as the resulting sign of $Q_{g_{SH}}$. We will define the $\hat{s}$ direction as the direction along the isotherms (with cold air to the left) and U_g as the flow in the $\hat{s}$ direction. Further, we will define $\hat{n}$ to point into the cold air, perpendicular to the isentropes. If we then assess the magnitude of the horizontal temperature contrast as $|\partial\theta/\partial n|$, we can rewrite the geostrophic shearing deformation term as

$$Q_{g_{SH}} = 2\gamma \left|\frac{\partial \theta}{\partial n}\right| \frac{\partial U_g}{\partial s}. \tag{7.24}$$

Thus, if U_g increases (decreases) along an isentrope, there must be a thermally direct (indirect) vertical circulation. We now exploit the simplicity of this expression to examine the influence of along-flow temperature advection in straight jet streaks.

First, let us consider the situation in which there is no along-flow temperature advection in the vicinity of a straight jet as illustrated in Figure 7.23(a). Concentrating on the isentrope in the middle of the bundle that lies parallel to the jet axis, we see that $\partial U_g/\partial s > 0$ from the entrance region to the jet core itself. As a consequence of (7.24), $Q_{g_{SH}} > 0$ and the circulation in that portion of the jet/front system must be thermally direct. To the east of the jet core, $\partial U_g/\partial s < 0$ along the central isentrope so that, by (7.24), $Q_{g_{SH}} < 0$ and the circulation must be thermally indirect in the exit portion of the jet/front system. The combination of these diagnoses leads to the

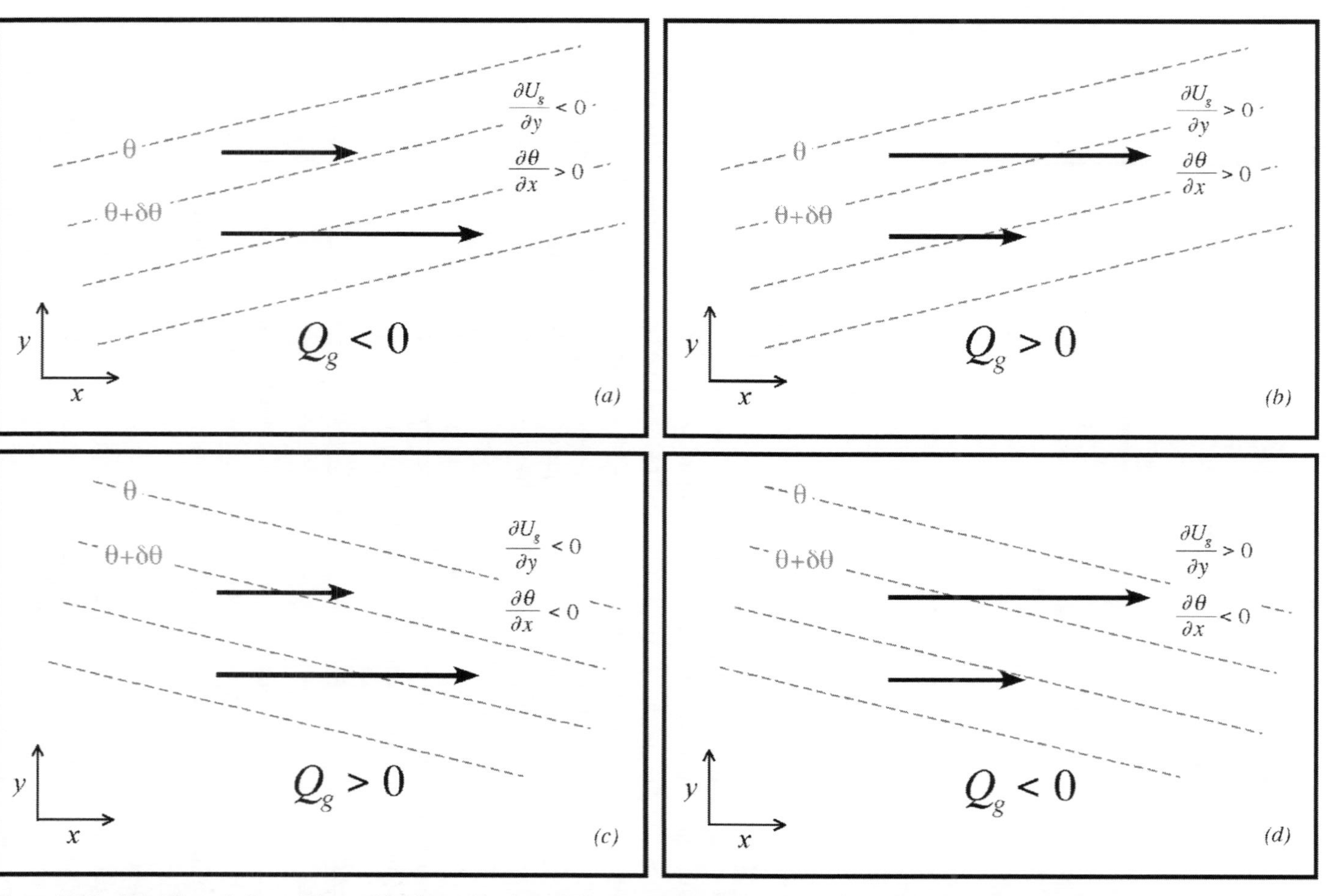

Figure 7.22 The effect of all possible combinations of horizontal shear and temperature advection. (a) Cold air advection in cyclonic shear, leading to a thermally indirect circulation. (b) Cold air advection in anticyclonic shear, leading to a thermally direct circulation. (c) Warm air advection in cyclonic shear, leading to a thermally direct circulation. (d) Warm air advection in anticyclonic shear, leading to a thermally indirect circulation

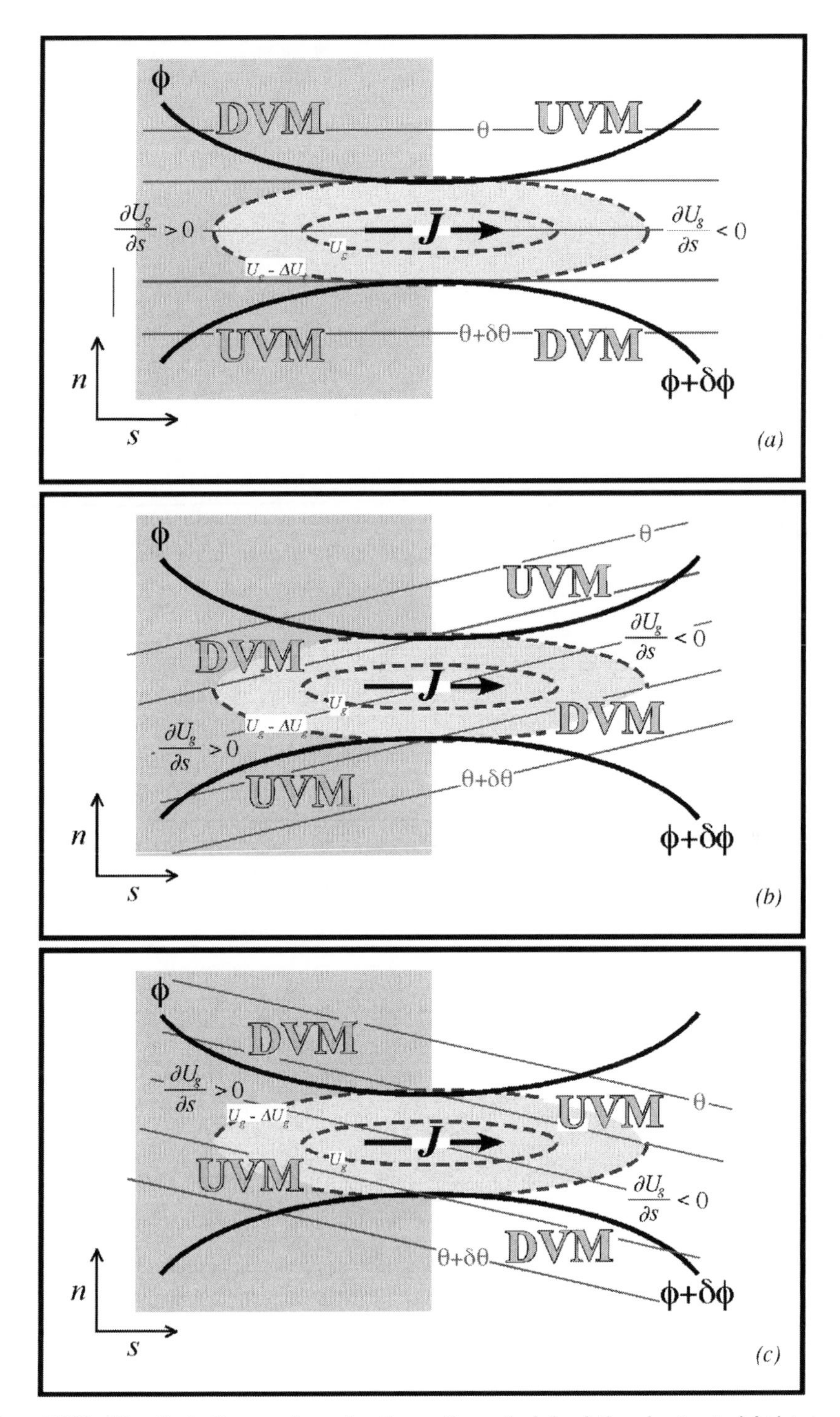

Figure 7.23 The effect of temperature advection on the vertical circulation about a straight jet streak. (a) Straight jet with no temperature advection along its axis. (b) Straight jet with cold air advection along its axis. (c) Straight jet with warm air advection along its axis. Gray shading on left isolates the jet entrance region. Right side is the jet exit region

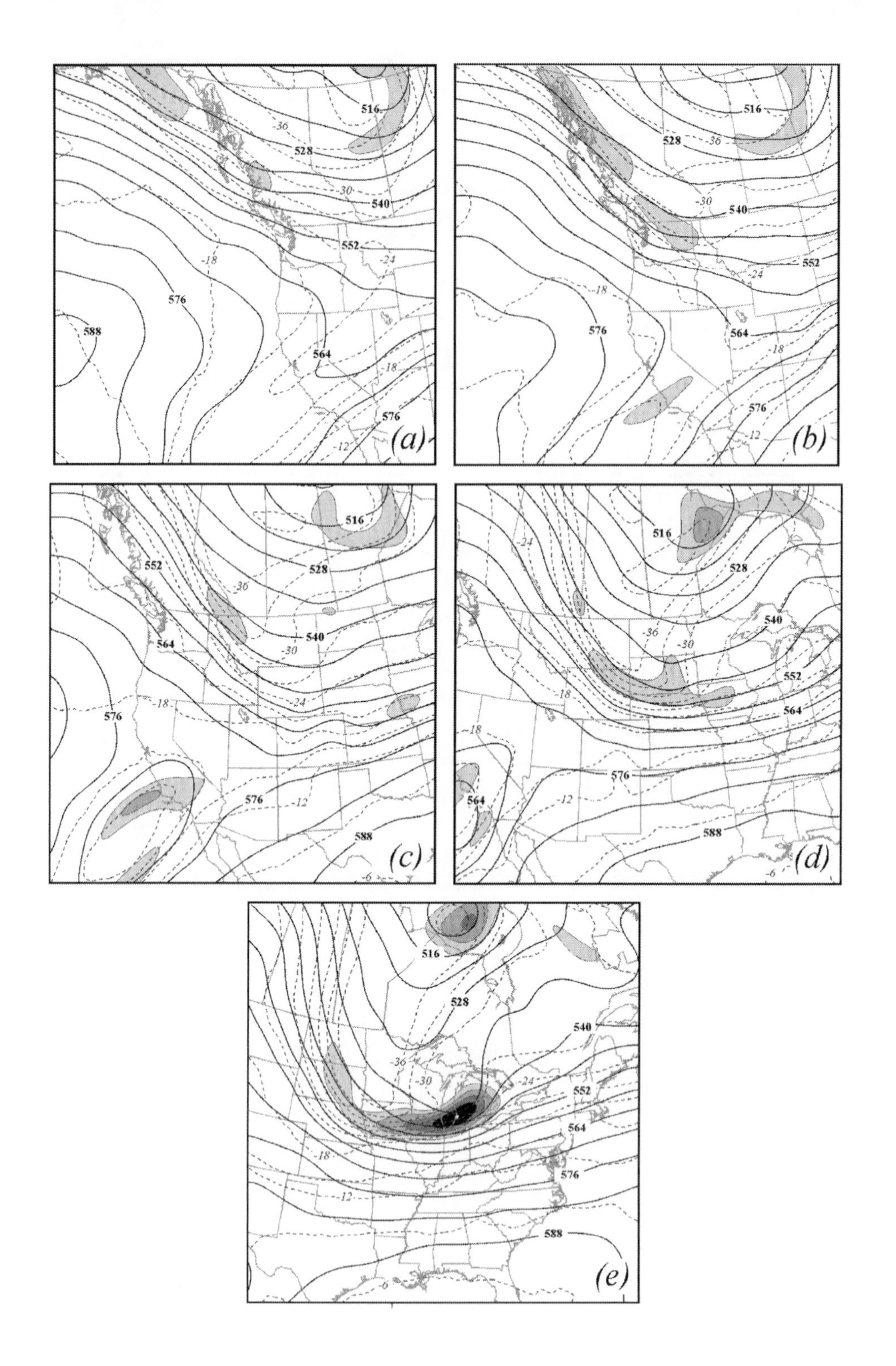

516
528
540
552
564
576
588
-36
-30
-24
-18
-12
-6
(a)
(b)
(c)
(d)
(e)

familiar four-quadrant model of vertical motions in the vicinity of a jet streak. Now, if the jet crosses the same bundle of isentropes at a slight angle, then significant along-flow temperature advection will result. In Figure 7.23(b) we see a jet characterized by cold air advection along its axis. Once again, we can use (7.24) to diagnose the distribution of vertical motions through consideration of the middle isentrope in the bundle. In the entrance region of the jet portrayed in Figure 7.23(b), $\partial U_g/\partial s > 0$ and so $Q_{g_{SH}} > 0$ there. The associated thermally direct circulation, however, must straddle the middle isentrope and is thus shifted toward the anticyclonic side of the jet in the entrance region. In the exit region, a similar analysis yields $\partial U_g/\partial s < 0$ and $Q_{g_{SH}} < 0$. The associated thermally indirect circulation must straddle the middle isentrope, however, and is consequently shifted toward the cyclonic shear side of the jet axis. The combination of these two diagnoses is that subsidence is maximized directly along the jet axis when the jet is characterized by cold advection along its axis. Another way to diagnose this vertical motion distribution is to consider the Sutcliffe/Trenberth form of the omega equation in which thermal wind advection of absolute geostrophic vorticity determines ω. Since the isentropes in Figure 7.23(b) are a surrogate for the direction of the thermal wind, it is clear that anticyclonic vorticity advection by the thermal wind will be maximized right at the position of the jet core with corresponding regions of cyclonic vorticity advection on the flanking cyclonic and anticyclonic sides. This distribution of vertical motion will tend to increase both the horizontal temperature contrast as well as the vertical component of vorticity through tilting, thus intensifying the upper-level jet/front system. The exact opposite situation occurs when there is warm air advection along the axis of the jet, as illustrated in Figure 7.23(c). The Sutcliffe/Trenberth diagnostic also suggests upward vertical motion through the jet core for the case of warm air advection along the jet, a circumstance that promotes the weakening of an upper-level front.

For the case of cold air advection along the jet axis, a particularly important result of the subsidence maxima near the jet core is the resulting downward advection of high PV into the upper troposphere. In the previous chapter we considered the important effect that vorticity advection in the middle and upper troposphere has on the development of mid-latitude weather systems. In that discussion there was no reference made to the origin of the important middle tropospheric vorticity features themselves. It may now seem viable that some of these features arise from downward advection of high PV associated with the development of upper front/jet systems. Observations suggesting such origins are plentiful. Consider, for instance, the case illustrated in Figure 7.24. At 0000 UTC 11 November 2003, modest confluence between a high-latitude trough and a ridge at upper tropospheric levels

Figure 7.24 (a) 500 hPa geopotential height (solid lines), temperature (dashed lines), and absolute vorticity (shading) at 0000 UTC 11 November 2003. Geopotential height is labeled in dam and contoured every 6 dam. Temperature is labeled in °C and contoured every 3°C. Absolute vorticity is labeled in 10^{-5} s^{-1} and contoured every 5×10^{-5} s^{-1} starting at 20×10^{-5} s^{-1}. (b) As for (a) but for 1200 UTC 11 November 2003. (c) As for (a) but for 0000 UTC 12 November 2003. (d) As for (a) but for 1200 UTC 12 November 2003. (e) As for (a) but for 0000 UTC 13 November 2003 was evident just off the Alaskan panhandle (Figure 7.24a). This confluence began to concentrate the horizontal temperature contrast and, as a consequence, to intensify the jet. Only a modest 500 hPa absolute vorticity maximum was associated with this developing feature at this time. Twelve hours later, the modest vorticity maximum had migrated southeastward to coastal British Columbia along with a significant horizontal temperature contrast (Figure 7.24b). By 0000 UTC 12 November, an emerging jet/front system was embedded in rather straight northwesterly flow characterized by geostrophic cold air advection along the leading portion of the axis of the jet (Figure 7.24c), a configuration that leads to upper frontogenesis. The 500 hPa absolute vorticity had intensified only slightly to this point in the development of this upper-level front. A more intense baroclinic zone had developed by 1200 UTC 12 November with an attendant increase in the absolute vorticity (Figure 7.24d). Finally, by 0000 UTC 13 November, an intense upper-level front, characterized by a large horizontal temperature contrast as well as a significant 500 hPa vorticity maximum, had developed over the lower Great Lakes States (Figure 7.24e). As the upper-level front moved toward the downstream side of its associated upper-level short-wave disturbance by this time, a powerful surface cyclone had begun to develop downshear of the upper-level front. Though not apparent in this case, in some instances, by the time the upper front/jet system is downstream of the upper short-wave axis there can be warm air advection along the jet axis leading to strong ascent beneath the jet core and commencement of the decay of the upper front.

7.5 Precipitation Processes at Fronts

Though we have spent a great deal of effort describing the relationship between the vertical circulations at fronts and their relationship to the process of frontogenesis, the vertical motions themselves are correctly viewed as *necessary but insufficient* to produce the precipitation often found in the vicinity of fronts. The simplest way to verify this statement is to consider the case of a strong frontal circulation occurring in the complete absence of water vapor. Clearly in such a case no precipitation could possibly occur. Thus, it would appear that the correct *thermodynamic* conditions are as vital an ingredient in the production of precipitation in mid-latitude cyclones as the dynamical ones.

The canonical Norwegian Cyclone Model (to be described in the next chapter) correctly portrayed fronts as the seat of much of the precipitation in mid-latitude cyclones. But, as a result of observational constraints, the model also suggested that the frontal precipitation intensity was uniform in the frontal regions and exhibited no cellular substructure. With the advent of weather radar in the late 1950s, this presumption was exposed as a misconception: considerable cellular substructure exists in the distribution of frontal precipitation. Naturally, the question of what controls the mesoscale distribution of precipitation in the vicinity of frontal zones became a major research area. In this section we will broadly consider this question, concentrating on the simple idea that the juxtaposition of dynamical forcing

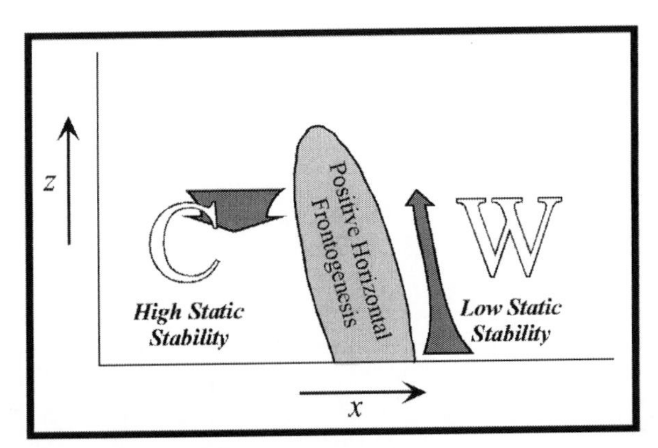

Figure 7.25 Schematic vertical cross-section through a region of positive horizontal frontogenesis. The warm side of the region is characterized by lower static stability than the cold side. The dark arrows represent the resulting thermally direct vertical circulation: a narrow, intense updraft and a widespread, benign downdraft

with appropriate thermodynamic conditions goes a long way to explaining the gross characteristics of the mesoscale distribution of precipitation at fronts.

We begin this investigation by considering the effect of variations of static stability across a schematic frontal zone as depicted in Figure 7.25. Recall that both the QG omega and Sawyer–Eliassen circulation equations related the geostrophic forcing ($-2\nabla \cdot \vec{Q}$ or Q_g) to ω via some measure of the static stability: σ in the QG omega equation and QG PV in the Sawyer–Eliassen equation. In essence, the stability functions as an amplitude modulator for ω in both expressions, for when the stability is weak (strong) there is little (considerable) resistance to vertical displacement and for a given amount of forcing, a large (only a small) vertical displacement may arise. For the schematic frontal cross-section in Figure 7.25, let us assume that the forcing for ascent results from horizontal frontogenesis ($F_{2D} > 0$). The first-order response to this forcing is, as we have seen, the production of a thermally direct vertical circulation with ascent on the warm side and descent on the cold side of the frontal zone. If we assume that the resistance to vertical displacement is smaller (larger) on the warm (cold) side of the frontal zone, then the response to the frontogenetic forcing will be modulated so as to produce intense ascent and much more benign descent. Since continuity of mass requires that the amount of mass that ascends be equal to the amount that descends, the updraft on the warm side, being more intense, must also be horizontally restricted as compared to the downdraft on the cold side. The consequence of these influences is that the updraft on the warm side of such a front will be narrow and intense, producing a narrow linear band of precipitation, while the downdraft will be wider and more benign. Thus, it is generally the case that the narrow precipitation bands commonly observed in association with frontogenetically active frontal zones are a consequence of the modulation of the response to frontogenetic forcing for vertical motions by cross-frontal differences in static stability. Next we investigate the synoptic-scale conditions favorable for the production of a particular type of static instability known as **convective** or **potential instability**.

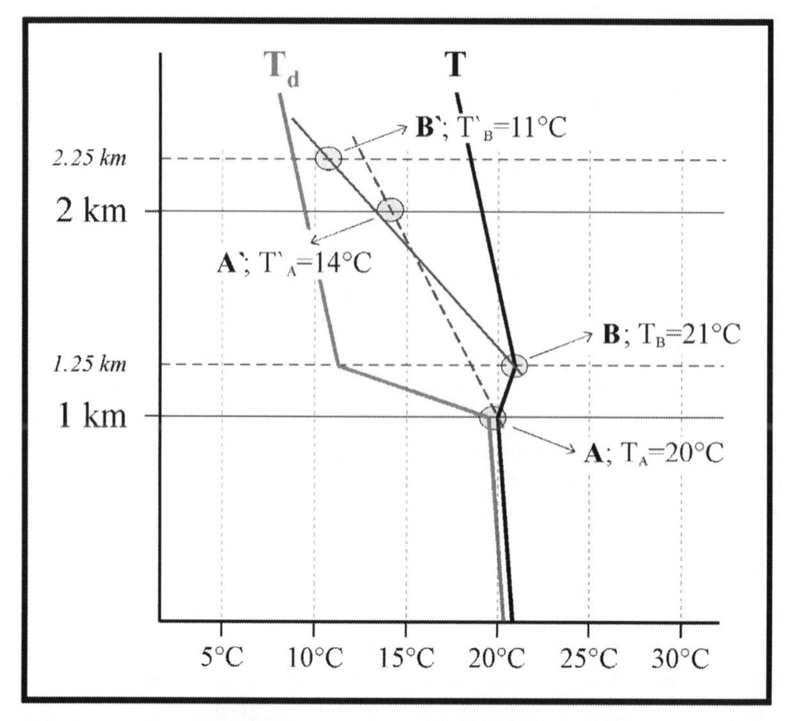

Figure 7.26 Pseudo-adiabatic diagram portraying a hypothetical convectively unstable sounding. The dashed line represents the moist adiabatic lapse rate (6°C km^{-1}) while the solid line represents the dry adiabatic lapse rate (9.8°C km^{-1}). Layer A–B (located between 1 and 1.25 km) is characterized by a temperature inversion. Parcel A is saturated while parcel B is very dry. Upon being lifted 1 km, this layer develops an absolutely unstable lapse rate

Convective instability is a particularly potent form of gravitational instability. Consider the hypothetical sounding shown in Figure 7.26. Given the temperature and humidity characteristics of layer A–B, parcels A and B will be subjected to differential adiabatic cooling rates upon being lifted with A cooling at ∼ 6°C km^{-1} (the moist adiabatic rate) and B cooling at 9.8°C km^{-1} (the dry adiabatic rate). Upon being lifted 1 km, layer A–B will become absolutely unstable since its lapse rate will then exceed the dry rate and free convection will ensue. Thus, rapid and intense convection, leading to a narrow updraft, is likely to result from such a configuration. It can be shown that temperature and dewpoint profiles such as those in Figure 7.26 conspire to render $\partial\theta_e/\partial z < 0$ in layer A–B. The necessary condition for convective instability is, in fact, that $\partial\theta_e/\partial z$ be less than zero. Such a stratification can develop as a result of differential moisture advection during cyclogenesis. In fact, it is very common in cyclogenesis over the southern plains of the United States in which southeasterly low-level, high-θ_e flow from the Gulf of Mexico occurs beneath southwesterly low-θ_e flow at upper levels off the Mexican Plateau as illustrated schematically in Figure 7.27(a).

Another preferred region for the development of convective instability is the poleward edge of the so-called 'dry slot' in cyclones. The dry slot is a region of desiccated

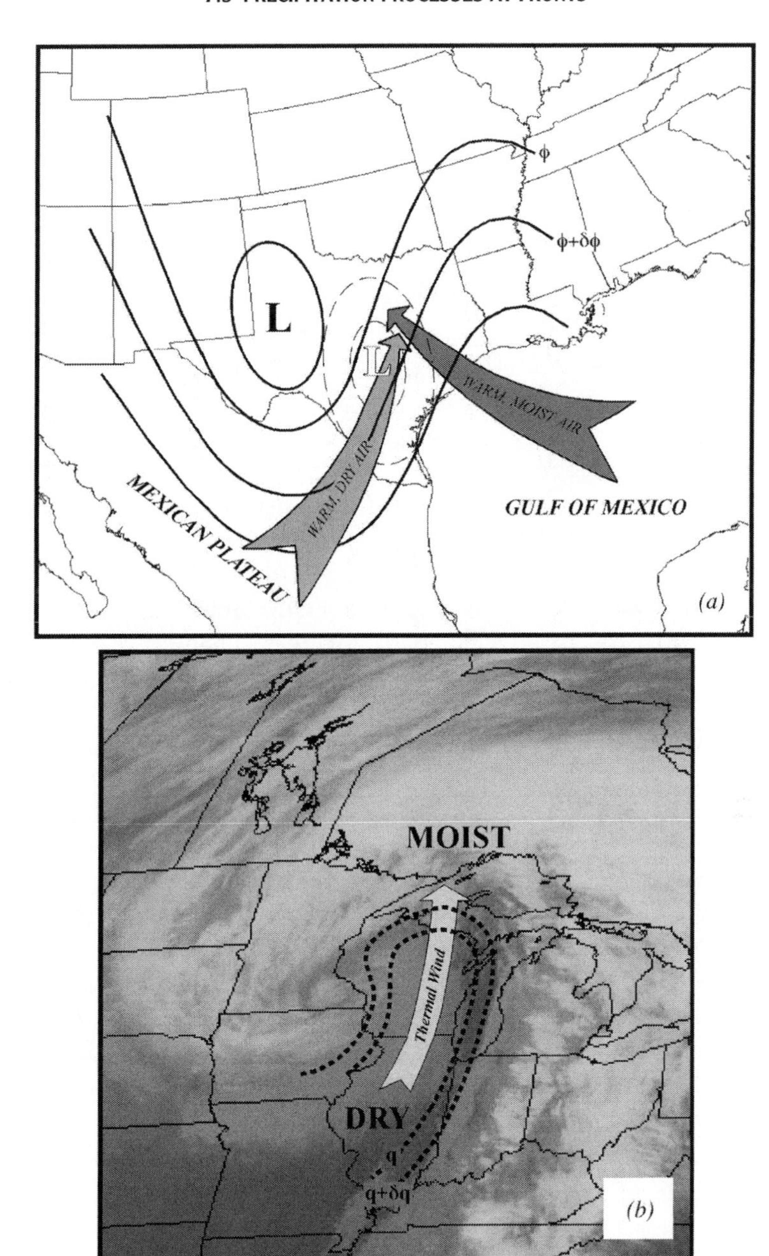

Figure 7.27 (a) Schematic of the synoptic conditions leading to the development of convective instability in the Central Plains of the United States. Dashed lines are sea-level isobars, solid lines are geopotential height lines at 500 hPa. (b) Illustration of the differential vertical advection of dry air characteristic of the poleward edge of the dry slot. Infrared satellite image (NOAA) is from 1815 UTC 10 November 1998. Thick dashed lines are schematic isopleths of mixing ratio (q) and the white arrow is the tropospheric thermal wind vector

air, often with a history of subsidence, that develops in the upper troposphere to the west of the cold frontal cloud shield. At the poleward edge of this feature, a significant moisture contrast exists through a deep layer of the middle and upper troposphere. The proximity of the cold frontal baroclinicity ensures that a significant vertical wind shear is present in the vicinity of the poleward edge of the dry slot. Thus, as illustrated in Figure 7.27(b), there is a region of strong dry air advection increasing with height that can lead to the development of narrow regions of mid-level convective instability and the consequent narrow bands of convective precipitation.

Though upright gravitational instabilities such as convective instability play a role in the development of frontal precipitation, another type of instability called **symmetric instability**, in which narrow, front parallel circulations can develop in response to *slantwise* motions, may also be responsible for the production of the linear bands of precipitation often observed in association with fronts. In order to develop a theory regarding the stability of such slantwise motions (i.e. motions with a component in both the horizontal and vertical), we must consider a rather specific environment in which (1) the vertical shear is geostrophically balanced, and (2) the flow is 2-D (i.e. not curved). Such conditions are not unlike those we assumed in developing the Sawyer–Eliassen equation so we are, to a fair degree, describing the flow in the vicinity of a frontal zone. Consider the depiction in Figure 7.28(a) of 2-D flow in the $y-z$ plane. In such a case, there is no x-direction pressure gradient so the frictionless equation of motion, under the geostrophic momentum approximation, reduces to

$$\frac{dU_g}{dt} - fv = 0 \quad \text{or} \quad \frac{d}{dt}(U_g - fy) = 0. \tag{7.25}$$

If we now define M_g, the absolute geostrophic momentum from the Sawyer–Eliassen discussion in Section 7.3, as $M_g = U_g - fy$, then (7.25) becomes a statement of the conservation of M_g. Next we imagine that some perturbation in the atmosphere accomplishes an exchange of the tubes of air labeled 1 and 2 in Figure 7.28(a) along a θ_e surface and consider the consequences. If the original velocity of tube 1 is U_{g_1}, then $M_{g_1} = U_{g_1} - fy$. Upon moving tube 1 to the original position $(y + \Delta y)$ of tube 2, conservation of M_g implies that

$$\begin{aligned} M_{g_1} = U_{g_1} - fy &= U'_{g_1} - f(y + \Delta y) \\ &= U'_{g_1} - fy - f\Delta y \end{aligned}$$

so that

$$U'_{g_1} = U_{g_1} + f\Delta y. \tag{7.26a}$$

Similarly, if the original velocity of tube 2 is U_{g_2}, then $M_{g_2} = U_{g_2} - f(y + \Delta y)$. Upon moving tube 2 to the original position (y) of tube 1, conservation of M_g yields,

$$\begin{aligned} M_{g_2} = U_{g_2} - f(y + \Delta y) &= U'_{g_2} - f(y) \\ &= U'_{g_2} - fy \end{aligned}$$

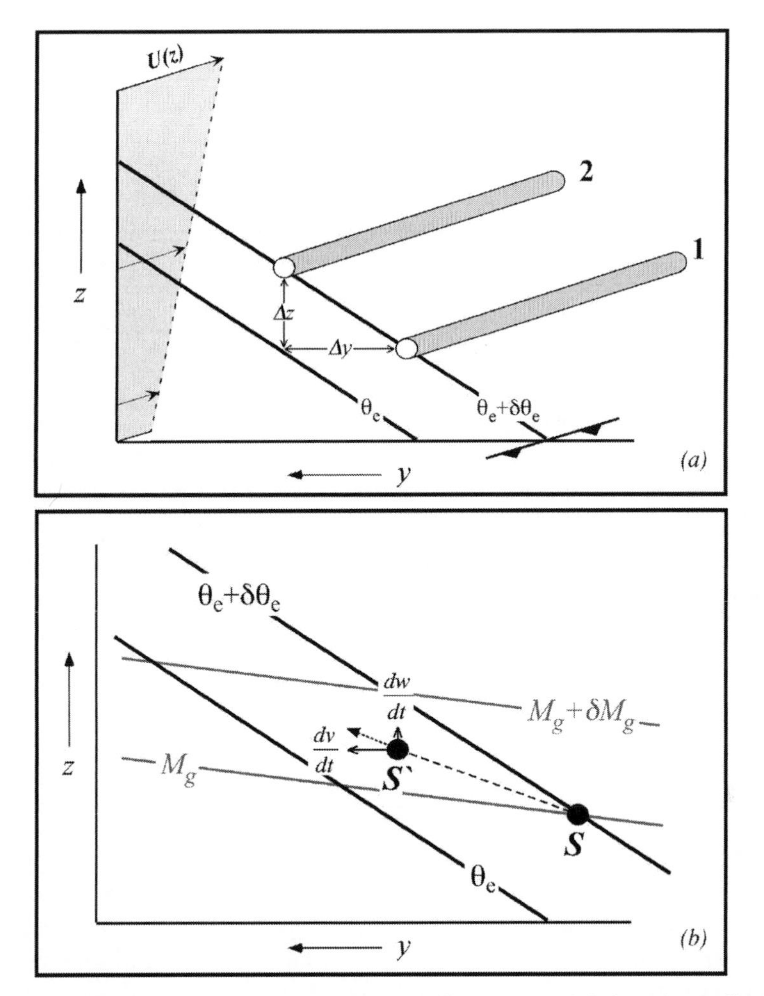

Figure 7.28 (a) Schematic vertical cross-section perpendicular to a straight frontal zone aligned along the *y*-axis. Thick solid lines are isopleths of θ_e. Tubes 1 and 2, infinite in the *x* direction and separated by distances Δy and Δz in the horizontal and vertical directions, are described in the text. (b) Distribution of M_g and θ_e isopleths in a cross-section like that shown in (a). A parcel, S, displaced along a slanted path (dashed line) will experience vertical and horizontal accelerations (solid arrows) as indicated. These accelerations arise from the dual constraints of conserving θ_e and M_g. The dashed black arrow represents the resultant acceleration

so that

$$U'_{g_2} = U_{g_2} - f\Delta y. \tag{7.26b}$$

Now, after this exchange, we measure the change in kinetic energy of the environment from the original state to the perturbed state. This change in kinetic energy (KE) is given by

$$\Delta KE = \frac{1}{2}m\left(U'^2_{g_1} + U'^2_{g_2}\right) - \frac{1}{2}m\left(U^2_{g_1} + U^2_{g_2}\right). \tag{7.27}$$

Substituting (7.26a) and (7.26b), the reader can show that (7.27) reduces to

$$\Delta KE = mf\Delta y(M_{g_1} - M_{g_2}). \tag{7.28}$$

If $\Delta KE < 0$, this implies that the environment has given up energy to the disturbance that initiated the tube exchange, thus the environment was unstable to that disturbance. Therefore, we conclude that the environment is unstable to such slantwise displacements when $M_{g_2} > M_{g_1}$, or, with reference to Figure 7.28(a), when $(\partial M_g/\partial y)_{\theta_e} > 0$ which translates to

$$(\zeta_g + f)_{\theta_e} < 0. \tag{7.29a}$$

If the slope of the M_g surfaces is shallower than the slope of the θ_e surfaces (in a saturated atmosphere) then the atmosphere is unstable to such disturbances and the instability will result in convection of heat and momentum approximately along sloping θ_e surfaces. If the 2-D flow in Figure 7.28 is saturated then a parcel conserves both M_g and θ_e so that for displacement paths that lie between the sloping M_g and θ_e surfaces, horizontal and vertical forces compel the parcel back toward its original M_g and θ_e surfaces. Thus, as indicated in Figure 7.28(b), the displaced parcel accelerates away from its origin along that slanted path. Notice also that the stability criterion (7.29a) can be expressed equivalently as

$$\left(\frac{\partial \theta_e}{\partial z}\right)_{M_g} < 0. \tag{7.29b}$$

Thus, along the slanted path indicated in Figure 7.28(b), a parcel is equivalently (1) inertially unstable on a θ_e surface (7.29a), or (2) convectively unstable on an M_g surface (7.29b) even though the environment depicted in Figure 7.28 is inertially and convectively stable! Such an instability is known as **conditional symmetric instability (CSI).**[3]

Another means of determining the necessary conditions for this CSI arises from consideration of a quantity called the geostrophic moist potential vorticity (PV_{e_g}) defined as

$$PV_{e_g} = -(f\hat{k} + \nabla \times \vec{V}_g) \cdot \nabla \theta_e. \tag{7.30a}$$

Upon expansion, (7.30a) takes the form

$$PV_{e_g} = \frac{\partial v_g}{\partial p}\frac{\partial \theta_e}{\partial x} - \frac{\partial u_g}{\partial p}\frac{\partial \theta_e}{\partial y} - \left(\frac{\partial v_g}{\partial x} - \frac{\partial u_g}{\partial y} + f\right)\frac{\partial \theta_e}{\partial p}. \tag{7.30b}$$

[3] The adjective 'symmetric' is used here since the theory is the isentropic equivalent of a theory derived by Rayleigh concerning convection of angular momentum in circularly symmetric, barotropic fluid flows. The adjective 'conditional' refers to the condition that the air must be saturated in order that the instability be realized under the specified necessary conditions.

To be consistent with the 2-D assumption we adopted in formulating the necessary condition for CSI, we neglect x variations and (7.30b) becomes

$$PV_{e_g} = -\frac{\partial u_g}{\partial p}\frac{\partial \theta_e}{\partial y} - \left(f - \frac{\partial u_g}{\partial y}\right)\frac{\partial \theta_e}{\partial p}$$

or, since $M_g = U_g - fy$,

$$PV_{e_g} = \frac{\partial M_g}{\partial y}\frac{\partial \theta_e}{\partial p} - \frac{\partial M_g}{\partial p}\frac{\partial \theta_e}{\partial y}. \tag{7.31}$$

Returning now to Figure 7.28(b) in which the necessary condition for CSI is portrayed, we see that anytime isopleths of M_g are less steeply sloped in the y–p plane than isopleths of θ_e, then (7.31) will be negative. Thus, whenever the 2-D PV_{e_g} is negative the necessary condition for CSI is met. It is believed that when PV_{e_g} becomes negative and such air is lifted to saturation in a frontal environment, CSI is released. This slantwise-directed, free convective overturning produces roll circulations parallel to the thermal wind. As the rolls grow, they create regions of convective instability with the resulting convection leading to the banded nature of the precipitation.

Recall that the ellipticity condition for the Sawyer–Eliassen equation, namely

$$\gamma\left(\frac{\partial \theta}{\partial p}\frac{\partial M}{\partial y} - \frac{\partial \theta}{\partial y}\frac{\partial M}{\partial p}\right) > 0,$$

is the dry adiabatic version of (7.31). Thus, strictly speaking, the Sawyer–Eliassen equation is hyperbolic when $PV_{e_g} < 0$ and the air is saturated. If, however, cross-front variations of PV_{e_g} exist in which PV_{e_g} is positive everywhere, the resulting vertical circulation will still be affected. With smaller PV_{e_g} on the warm side of a thermally direct frontal circulation, the vertical motions will be intense since the resistance to slantwise displacement (proportional to the magnitude of PV_{e_g}) is lower than on the cold side. The updraft will also be horizontally restricted since mass continuity must be obeyed. Thus, even in the absence of any actual instability, sound arguments can be made that account for the banded nature of precipitation in the vicinity of active frontal zones.

Having established the value of using PV_{e_g} as a measure of the response to frontogenetic forcing, let us investigate the means by which this variable may change with time. The Lagrangian rate of change of the full PV_{e_g} (7.30b) is given by

$$\frac{d}{dt}(PV_{e_g}) = -\vec{\eta}\cdot\nabla\dot{\theta}_e + f\frac{\partial \vec{V}_g}{\partial p}\cdot\nabla\theta_e \tag{7.32}$$

which describes two processes that can contribute to a change in PV_{e_g}. The first term on the RHS of (7.32) relates the influence of diabatic effects on PV_{e_g}. Of particular interest is the vertical component of that diabatic term given by

$$-\left(\frac{\partial v_g}{\partial x} - \frac{\partial u_g}{\partial y} + f\right)\frac{\partial \dot{\theta}_e}{\partial p}$$

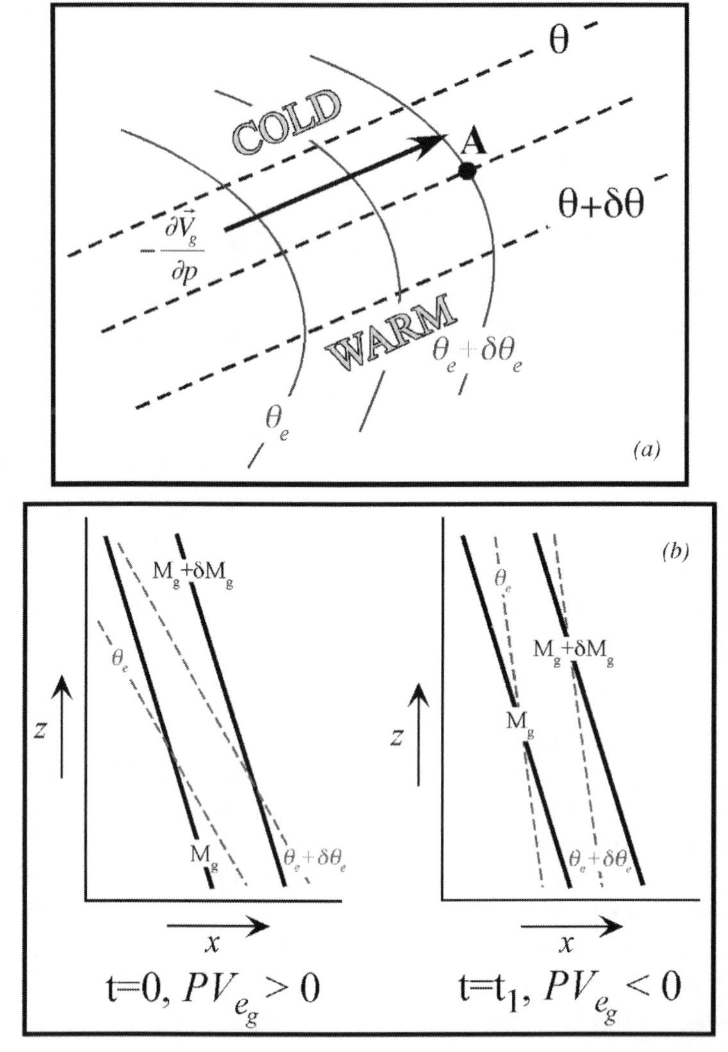

Figure 7.29 (a) Distribution of θ and θ_e isopleths associated with differential dry air advection and reduction of PV_{e_g}. (b) Vertical cross-section illustrating the effect of such differential dry air advection on the slope of M_g and θ_e isopleths

where $-\partial\dot{\theta}_e/\partial p$ is the vertical gradient of diabatic heating. Diabatic processes redistribute PV_{e_g} in the vertical, effectively 'destroying' PV_{e_g} above the level of maximum heating and 'creating' it below that level. The second term on the RHS of (7.32) describes the effect of vertically differential horizontal θ_e advection on changing the PV_{e_g}, a purely adiabatic mechanism. Consider the distribution of θ and θ_e isopleths illustrated in Figure 7.29(a). It is clear that in that scenario, there will be negative θ_e advection by the thermal wind. Physically, this means that at station A, θ_e will be decreased more rapidly aloft than near the surface. As a consequence, any given θ_e isopleth will be tilted into a more vertical orientation over time (Figure 7.29b).

Since the actual temperature field is represented in the thermal wind $(-\partial \vec{V}_g/\partial p)$, then the θ_e advection described by this term is actually just moisture advection. Such moisture advection, though it does affect the orientation of the moist isentropes, hardly affects the mass field at all. Consequently, the slopes of M_g isopleths are essentially unaffected by this process. Thus, a gradual increase in the slope of θ_e surfaces relative to M_g surfaces – that is, a reduction in PV_{e_g} – will be produced by the process described by this term. It is worth noting that the distribution of dry and moist isentropes shown in Figure 7.29(a) is characteristic of the poleward edge of the 'dry slot' discussed previously.

The frontal processes we have studied in this chapter are only a single component of the multi-scaled circulation system known as the mid-latitude cyclone. These fascinating disturbances, ubiquitous features of the mid-latitude circulation, can simultaneously affect millions of square kilometers over a life cycle that typically lasts from 3 to 7 days. In the next chapter we will investigate the nature of the mid-latitude cyclone by employing the many diagnostic tools we have thus far developed in our study.

Selected References

Margules (1906) provides the derivation of the zero-order frontal slope formulas.
Bergeron (1928) is among the first to discuss a frontogenesis function.
Bluestein, *Synoptic-Dynamic Meteorology in Midlatitudes, Volume II*, provides a clear derivation of the trigonometric form of the frontogenesis function starting with the algebraic form.
Hoskins and Pedder (1980) describe the relationship between the $\vec{Q}$-vector and quasi-geostrophic frontogenesis.
Hoskins and Bretherton (1972) describe the role of ageostrophic motions in the collapse of a frontal zone to very small scale.
Sanders (1955) is the seminal observational paper concerning surface frontogenesis.
Eliassen (1962) introduces the 2-D form of the so-called Sawyer–Eliassen equation – the best scientific paper ever read by the author.
Keyser and Shapiro (1986) provide a comprehensive review of observational and dynamical research on upper-level fronts and frontogenesis.
Sanders and Bosart (1985) offer a clear physical description of the influence of cross-front gradients of stability on determining some characteristics of frontal precipitation bands.
Emanuel (1979) and Bennetts and Hoskins (1979) are seminal papers on the theory of conditional symmetric instability (CSI).
Schultz and Schumacher (1999) offer a careful review of CSI theory and application.

Problems

7.1. (a) Prove that

$$\frac{d}{dt}\left(\frac{\partial\theta}{\partial x}\right) = \frac{\partial}{\partial x}\left(\frac{d\theta}{dt}\right) - \frac{\partial u}{\partial x}\frac{\partial\theta}{\partial x} - \frac{\partial v}{\partial x}\frac{\partial\theta}{\partial y} - \frac{\partial\omega}{\partial x}\frac{\partial\theta}{\partial p}$$

given that

$$\frac{d}{dt} = \frac{\partial}{\partial t} + u\frac{\partial}{\partial x} + v\frac{\partial}{\partial y} + \omega\frac{\partial}{\partial p}.$$

(b) For adiabatic, frictionless flow prove that

$$\Im_{3D} = \frac{d}{dt}|\nabla\theta|$$
$$= \frac{1}{|\nabla\theta|}\left[-\left(\frac{\partial\theta}{\partial x}\right)\left(\frac{\partial\vec{V}}{\partial x}\cdot\nabla\theta\right)-\left(\frac{\partial\theta}{\partial y}\right)\left(\frac{\partial\vec{V}}{\partial y}\cdot\nabla\theta\right)-\left(\frac{\partial\theta}{\partial p}\right)(\nabla\omega\cdot\nabla\theta)\right].$$

7.2. Figure 7.1A shows zonally oriented isentropes superposed with some geostrophic winds on an isobaric surface.

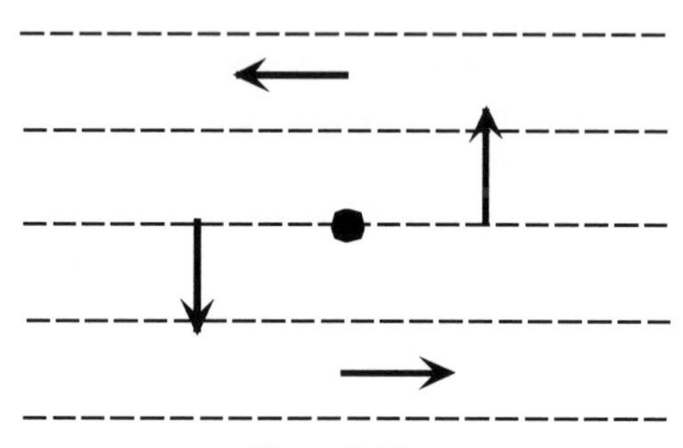

Figure 7.1A

(a) Use the natural coordinate form of $\vec{Q}$,

$$\vec{Q} = -\frac{R}{f}\left|\frac{\partial T}{\partial n}\right|\left(\hat{k}\times\frac{\partial\vec{V}_g}{\partial s}\right)$$

to draw the $\vec{Q}$-vector at the indicated point.

(b) Is there any geostrophic frontogenesis implied by this setting? Explain your answer with reference to your result in (a) and the $\vec{Q}$-vector form of the QG frontogenesis function ($F_{geo} = \vec{Q}\cdot\nabla\theta/|\nabla\theta|$).

(c) Given that the geostrophic frontogenesis function is also equal to

$$\Im_{geo} = \frac{|\nabla\theta|}{2}(F\cos 2\beta)$$

(where F is the total deformation field) does the result you obtained in (b) surprise you or not? Explain your answer.

7.3. Figure 7.2A is a schematic of 500 mb Φ and θ.

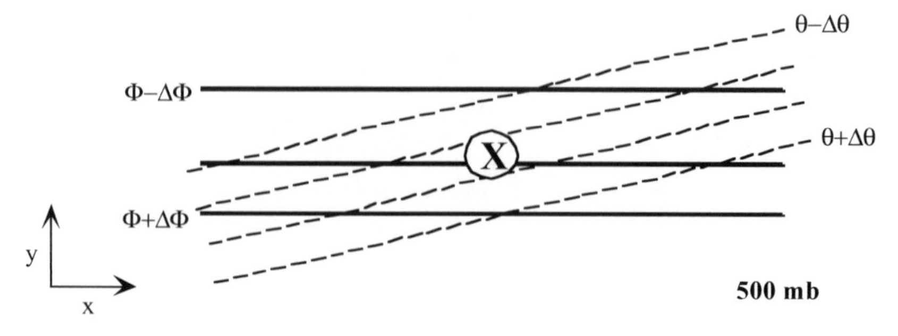

Figure 7.2A

Recall that the geostrophic shearing deformation forcing term of the Sawyer–Eliassen equation is given by

$$Q_{shear} = 2\gamma \frac{\partial U_g}{\partial y}\frac{\partial \theta}{\partial x}.$$

By applying this forcing term in Figure 7.2A, answer the following questions.

(a) What is the instantaneous effect of the geostrophic, horizontal flow on $|\nabla\theta|$?
(b) Given your answer in (a), indicate the orientation of the $\vec{Q}$-vectors with respect to the isotherms in that region. Explain your answer. (You do *not* have to *carefully* draw $\vec{Q}$-vectors for this answer!)
(c) What type of vertical circulation is forced by this circumstance?
(d) What will the effect of the forced vertical circulation be on $|\nabla\theta|$ in the same region? Explain your answer.
(e) The 'X' in Figure 7.2A marks a 500 mb absolute vorticity maximum. How will its magnitude change with time? Defend your answer.

7.4. Given the 700 hPa isentropes and $\vec{Q}$-vectors in Figure 7.3A, what is the sign of Q_{SE} (the geostrophic forcing function in the Sawyer–Eliassen equation) in that region? Defend your answer.

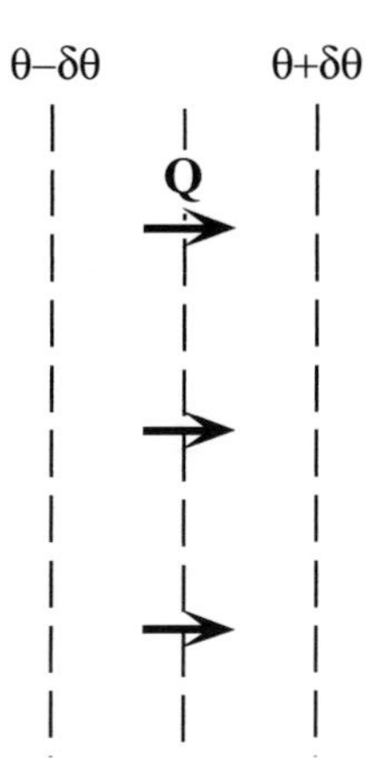

Figure 7.3A

7.5. (a) In an atmosphere with only geostrophic horizontal flow, would fronts be stronger or weaker than those observed in the real atmosphere? Give a physical explanation of your answer.
(b) Describe a mathematical rationale for the answer in (a) by considering the kinematic form of the frontogenesis function

$$\Im = \frac{|\nabla\theta|}{2}(F\cos 2\beta - D)$$

where F is the total deformation and D is the divergence.
(c) What is the major difference between the semi-geostrophic (SG) and quasi-geostrophic (QG) approximations? Why is SG more appropriate for investigating real fronts?

7.6. Recall that in the derivation of the Sawyer–Eliassen equation, we considered the quantity $M = U_g - fy$ (the absolute geostrophic momentum).

(a) What do vertical gradients of M represent physically?

(b) What do across-front $(\partial/\partial y)$ gradients of M represent physically?

(c) Describe why the product $(-\partial M/\partial y)(\partial M/\partial p)$ might be a useful parameter to consider in identifying frontal zones.

(d) Can the differences in static stability across a front be quantified in terms of M? Explain your answer.

7.7. In saturated, 2-D geostrophic flow parallel to the x-axis, the y and z equations of motion reduce to

$$\frac{dv}{dt} = -f(M_{parcel} - M_{env})$$

$$\frac{dw}{dt} = \kappa(\theta_{e\,parcel} - \theta_{e\,env})$$

where κ is a positive constant and $M = U_g - fy$.

(a) For a parcel originally at point P in the cross-section in Figure 7.4A, which of the hypothetical parcel trajectories will result in parcel acceleration away from P? Explain your answer.

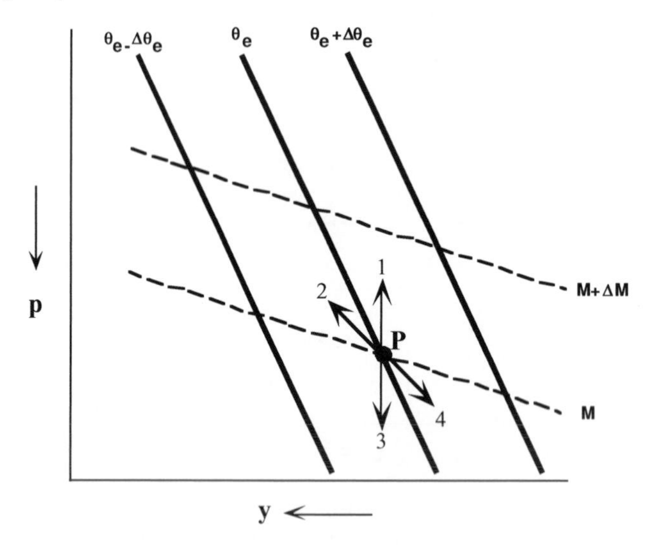

Figure 7.4A

(b) What is the sign of the geostrophic moist potential vorticity (PV_{e_g}) in this cross-section? Explain your choice. (This could be a one-sentence answer.)

7.8. Imagine that the surface horizontal velocities and temperature on a certain day can be described by the following functional expressions:

$$U = -2 - (4 \times 10^{-5})x - (2 \times 10^{-5})y$$

$$V = 4 + (2 \times 10^{-5})x + (1 \times 10^{-5})y$$

$$T = 270 - (\sqrt{3} \times 10^{-5})x - (1 \times 10^{-5})y.$$

(a) What are the values of the vorticity, divergence, and shearing and stretching deformations?
(b) What is the direction of the axis of dilatation of the total deformation field?
(c) What is value of the frontogenesis function in units of K $(100\,\text{km})^{-1}$ $(10^5\,\text{s})^{-1}$?

7.9. In the decade after World War II, it became common practice to test nuclear weapons by exploding them in the stratosphere. The belief at the time was that the stable stratosphere provided a safe reservoir for the radioactive residue of such explosions. Given the ubiquity of upper-level jet/front systems in the middle latitudes, comment on the prudence of this practice. Do you suspect the discovery of upper-level jet/front systems had a significant role in the adoption of the Nuclear Test Ban Treaty of 1960? Defend your answer.

7.10. (a) Prove that the change in kinetic energy (ΔKE) of the environment resulting from the exchange of tubes 1 and 2 in Figure 7.28(a) is given by

$$\Delta KE = mf\Delta y(M_{g_1} - M_{g_2}).$$

(b) Explain why the instability condition ($\Delta KE < 0$) *must* imply that the M_g lines are less steeply sloped than the θ_e lines.

7.11. Figure 7.5A depicts a typical synoptic setting over the south central United States with a 500 hPa trough axis centered over west Texas and a surface low-pressure center over south central Texas.

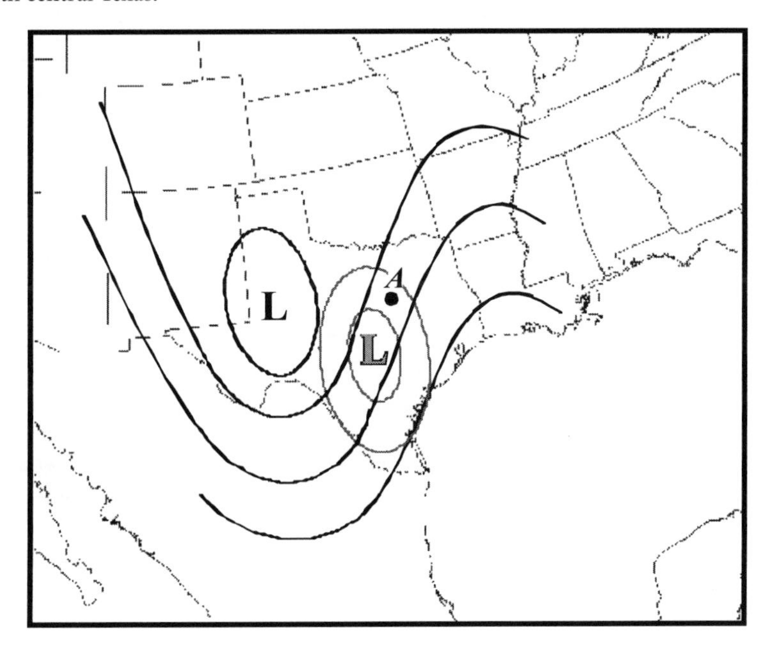

Figure 7.5A

(a) Sketch the 500 hPa *and* surface geostrophic winds at A.
(b) What are the source regions of the air that resides above A? (That is, what are the sources of the 500 hPa and surface air?)

(c) What type of instability will result over A from this synoptic setting? Be specific and draw a sketch of the likely temperature and dewpoint sounding over A.
(d) Would you expect general rising or sinking of air in the vicinity of A? Explain your answer.
(e) Given all of the above, what forecast would you make for northern Texas/southern Oklahoma on this day? Explain your answer.

7.12. Given that the Lagrangian rate of change of the geostrophic moist potential vorticity (PV_{e_g}) is given by

$$\frac{d}{dt}(PV_{e_g}) = -\vec{\eta}_g \cdot \nabla\dot{\theta}_e + f\frac{\partial \vec{V}_g}{\partial p} \cdot \nabla\theta_e.$$

(a) Show that the adiabatic production term ($f(\partial \vec{V}_g/\partial p) \cdot \nabla\theta_e$) can be written as

$$f\frac{\partial \vec{V}_g}{\partial p} \cdot \nabla\theta_e = \gamma\hat{k} \cdot (\nabla\theta_e \times \nabla\theta).$$

(b) Use the right hand rule to determine the sign of this term at the schematic 'dry slot' of a mid-latitude cyclone depicted in Figure 7.29(a).

7.13. Figure 7.6A illustrates the positively sloped interface, *F*, between two regions (labeled 1 and 2) of a fluid characterized by a *first-order discontinuity* in potential temperature, θ. Show that the slope of this interface in the *x*–*z* plane is given by

$$\left(\frac{dz}{dx}\right)_F = [(\partial\theta/\partial x)_1 - (\partial\theta/\partial x)_2]/[(\partial\theta/\partial z)_2 - (\partial\theta/\partial z)_1].$$

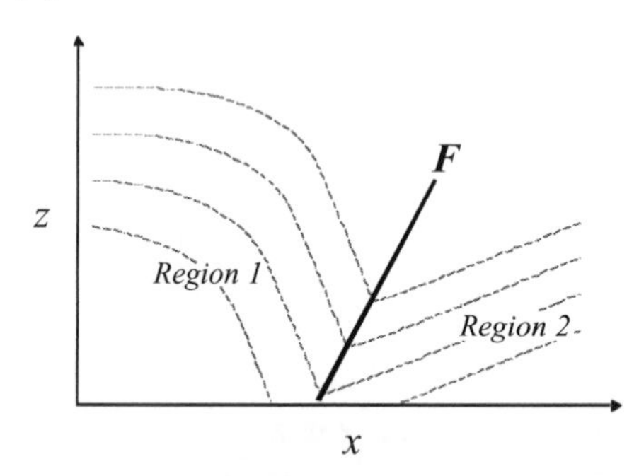

Figure 7.6A

7.14. Show that the vertical motions associated with the across-isentrope component of $\vec{Q}$ ($\vec{Q}_n$) are dependent on both the geostrophic frontogenesis as well as *across-front gradients* of the geostrophic frontogenesis.

Solutions

7.2. (b) There is no geostrophic frontogenesis implied.

7.3. (a) Forces a decrease in $|\nabla\theta|$.
(c) Thermally indirect

7.4. Positive

7.6. (a) Horizontal temperature contrasts (b) $-\zeta_g$

7.7. (a) Parcels 2 and 4 (b) Negative

7.8. (a) $\zeta = 4 \times 10^{-5}\,\mathrm{s}^{-1}$, $D = -3 \times 10^{-5}\,\mathrm{s}^{-1}$, $F_2 = 0$, and $F_1 = -5 \times 10^{-5}\,\mathrm{s}^{-1}$(b) $0°$ (c) $5.5\,\mathrm{K}\,(100\,\mathrm{km})^{-1}\,(10^5\,\mathrm{s})^{-1}$

8

Dynamical Aspects of the Life Cycle of the Mid-Latitude Cyclone

Objectives

The single most common weather element in the middle-latitudes is the frontal cyclone. As a consequence of this fact, the mid-latitude cyclone has been the subject of scientific scrutiny for well over 200 years. In this chapter we will employ the diagnostic tools and dynamical insights thus far developed and apply them to gain an understanding of the structure, evolution, and underlying dynamics of the mid-latitude cyclone life cycle.

This life cycle consists of various stages. We will pursue our investigation of several of these stages by adopting the perspective that the cyclone is the product of development initiated by finite, identifiable disturbances in the flow, not a manifestation of unstable growth of an infinitesimal perturbation. Consistent with this choice of perspective, we will make use of the quasi-geostrophic diagnostics developed in previous chapters to consider the dynamics of the cyclogenesis, post-mature, and decay stages of the cyclone life cycle. Examination of the post-mature stage will involve consideration of the structural and dynamical nature of the occlusion. Though research regarding the mid-latitude cyclone stretches back into the eighteenth century, we will begin our investigation by considering the broad structural characteristics of these storms starting with the synthesis of prior observations made manifest in the so-called polar front theory of cyclones.

8.1 Introduction: The Polar Front Theory of Cyclones

Much of the understanding of mid-latitude cyclones that existed before the turn of the twentieth century was fragmentary and lacked an organizing conceptual framework. Just after the end of World War I, meteorologists at the University of Bergen in

Norway, under the leadership of Vilhelm Bjerknes,[1] developed the polar front theory of the structure and life cycle of mid-latitude cyclones, now known colloquially as the Norwegian Cyclone Model (NCM). The essential genius of this conceptual model, which represented a grand synthesis of prior insights concerning the cyclone, was that it described the instantaneous structure of the cyclone while placing that structure into an identifiable life cycle. At the conceptual heart of the NCM was the existence of a globe-girdling, tropospheric deep, knife-like boundary known as the polar front which separated cold polar air from warm tropical air (Figure 8.1a). For reasons that were not discussed in the seminal paper by Bjerknes and Solberg (1922) that introduced the NCM, perturbation vortices occasionally developed along this polar front (Figure 8.1b). The existence of such vortices would then serve to deform the polar front, locally ushering tropical air poleward and polar air equatorward (Figure 8.1b). The precise mechanism by which the perturbation vortex would grow in intensity is not well explained in the NCM, but the continued growth of the perturbation was thought to lead to further deformation of the polar front (Figure 8.1c) and a lower sea-level pressure at the center of the perturbation. By this so-called mature stage of the life cycle, the deformation of the polar front had become so extreme as to lend the cyclone its now familiar characteristic frontal structure: a cold front extending equatorward and a warm front extending eastward from the sea-level pressure minimum. The region of homogeneous temperature between the two fronts was deemed the warm sector. Continued intensification of the cyclone compelled the cold front to encroach upon, and subsequently overtake, the warm front. Two important results of this process were that (1) the sea-level pressure minimum was removed from the peak of the warm sector and (2) an occluded front developed to connect the cyclone center to the peak of the warm sector (Figure 8.1d). It was thought that this process could result in the development of two varieties of occluded fronts in cyclones. One of these was the so-called warm occlusion in which the cold front would ascend the warm front upon overtaking it, leading to a vertical structure similar to that portrayed in Figure 8.2(a). Conversely, a so-called cold occlusion would result if the encroaching cold front was able to undercut the warm front and a vertical structure similar to that portrayed in Figure 8.2(b) would result. The warm (cold) occlusion was thought to occur when the air poleward of the warm front was more (less) dense than the air west of the cold front. Note that in either case, the development of the occluded front was associated with the denser air lifting the less

[1] Vilhelm Bjerknes was born on 14 March 1862 in Christiania, Norway (now Oslo). He was the son of a professor of mathematics and the father of Jacob Bjerknes whose seminal paper written with Halvor Solberg in 1922 established the Norwegian Cyclone Model. He earned an MS in 1888 and then moved to Bonn, Germany where he collaborated with Heinrich Hertz and received his Ph.D. in 1892. In 1897 he discovered the circulation theorem that bears his name and thereafter pursued research aimed at employing the circulation theorem to scientific weather forecasting. He was the driving force behind the establishment of the Bergen Geophysical Institute, colloquially known as the Bergen School, which eventually attracted such giants as Solberg, Bergeron, Petterssen, and Rossby. He died in 1951.

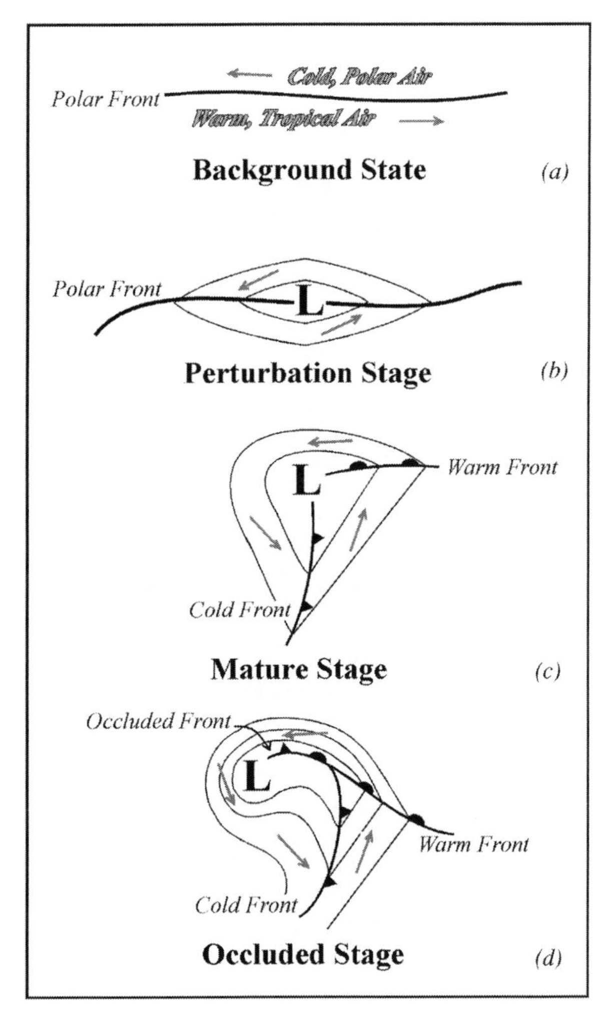

Figure 8.1 Evolution of a mid-latitude cyclone according to the Norwegian Cyclone Model. (a) The polar front as a background state. (b) The initial cyclonic perturbation. (c) The mature stage. (d) The occluded stage. The thin solid lines are isobars of sea-level pressure and the arrows are surface wind vectors

dense air aloft. In so doing, the horizontal density contrast originally characterizing the cyclone (manifest in the horizontal temperature gradient associated with the polar front) was reduced and a stable vertical stratification near the cyclone center was gradually put in place. As illustrated in Figure 8.3, transformation of an originally horizontal density contrast into a purely vertical one reduces the center of gravity of a fluid system gradually driving the system to its lowest potential energy state. Based upon this type of energetics argument, the NCM proposed that the development of the occluded front heralded the post-mature phase for a mid-latitude cyclone, a

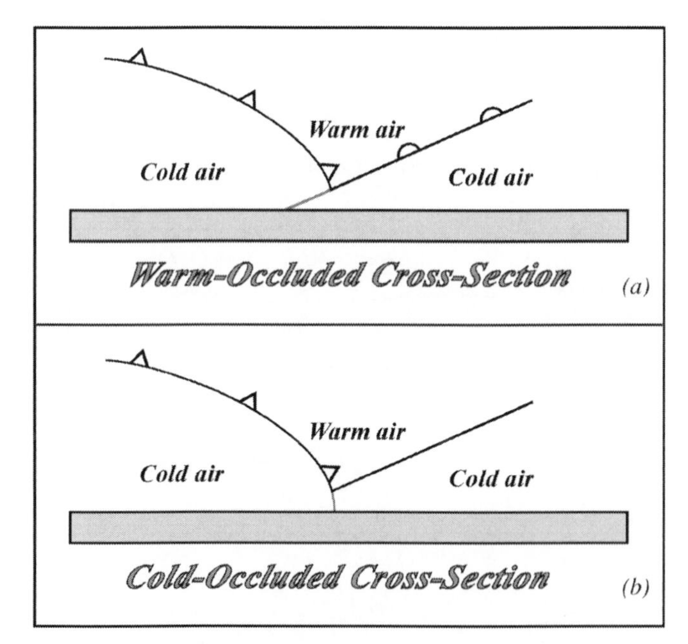

Figure 8.2 (a) Vertical cross-section through a warm occlusion in which the cold front ascends the warm front leaving a warm occluded front near the surface (gray line). (b) Vertical cross-section through a cold occlusion in which the warm front ascends the cold front leaving a cold occluded front near the surface (gray line)

cessation of intensification, and the commencement of cyclone decay. The nature of the cyclone decay was not described in the NCM beyond mention of the fact that the post-mature phase cyclone would eventually succumb to frictional dissipation associated with the surface of the Earth.

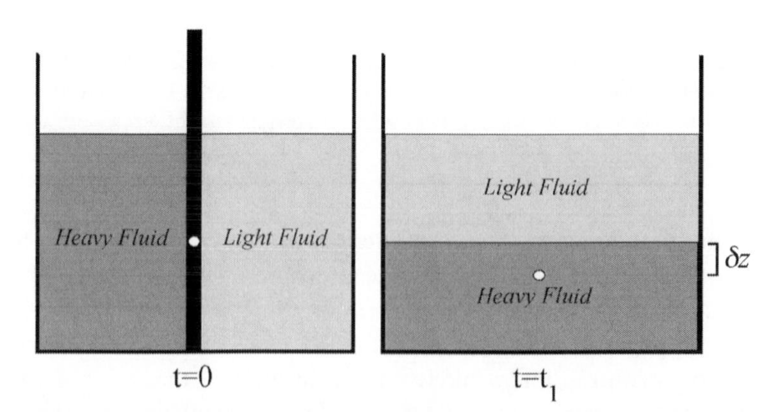

Figure 8.3 Fluids of different densities separated horizontally in a container by a dividing wall (thick black line) at $t = 0$. The white dot represents the height of the center of gravity of the two-fluid system. At $t = t_1$, after the divider has been removed, the height of the center of gravity of the fluid system has been lowered by an amount δz

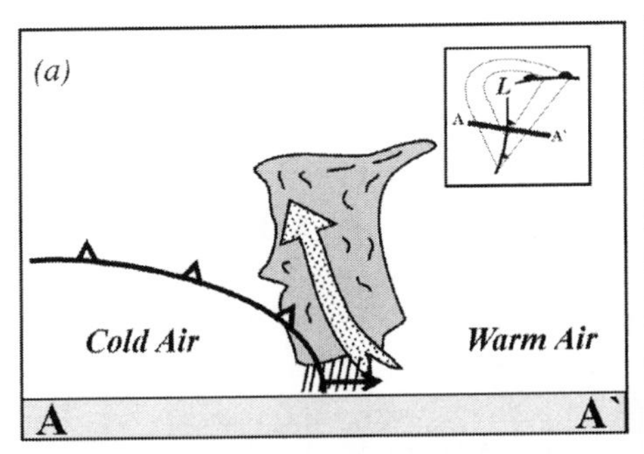

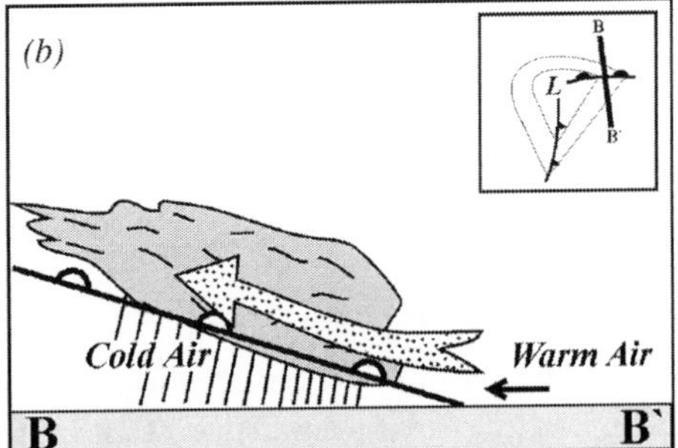

Figure 8.4 (a) Vertical cross-section through a cold front according to the NCM. Dotted arrow represents the updraft of air at the frontal boundary. Inset indicates the location of A–A′, referring to Figure 8.1(c). (b) Vertical cross-section through a warm front according to the NCM. Inset indicates the location of B–B′, referring to Figure 8.1 (c). Dotted arrow represents the updraft of air at the frontal boundary

The NCM accounted for the typical cloud and precipitation distribution associated with a mid-latitude cyclone with reference to the vertical structure of the fronts themselves. The cold front was described as a steeply sloped boundary between polar and tropical air masses that steadily advanced into the tropical air. The advance produced upgliding motions along the boundary itself and, as a consequence of its steep slope, the updrafts were vigorous and horizontally restricted leading to a narrow, sometimes squally precipitation distribution (Figure 8.4a). The warm front, on the other hand, was a less steeply sloped boundary between advancing tropical air and gradually retreating polar air (Figure 8.4b). The upgliding motions along the warm frontal surface were considered to be less intense as a consequence of the shallower slope. As a result, the cloudiness associated with the warm front was more horizontally widespread and the precipitation more benign.

Despite its great insights, the NCM, like all great conceptual leaps, certainly has its limitations. For instance, the nature of, and relationship between, the perturbations which grow into cyclones and the large-scale environment that promotes such growth are not addressed in the NCM and yet are clearly at the heart of understanding the mid-latitude cyclone life cycle. In addition, as discussed in Chapter 7, much more dynamically compelling arguments exist for explaining the production of vertical circulations at fronts. The fronts themselves, in fact, are not knife-like discontinuities as suggested by the NCM but zones of contrast across which temperature, density, and pressure are continuous. Neither are the frontal zones continuous through the depth of the troposphere. Note also that the NCM does not describe the physical mechanisms by which the cyclone intensifies from its incipient stage (Figure 8.1b) through its post-mature stage (Figure 8.1d). Also the development of the surface occluded front, though partly described through reference to the vertical frontal structure, is more correctly viewed as part of a process of occlusion that needs to be more comprehensively considered. Additionally, the decay of the cyclone is barely discussed in the NCM and yet clearly represents a major component of the cyclone life cycle. Finally, given the lack of available upper air observations at the time of its development, the NCM does not describe the vertical structure of cyclones and the manner in which that vertical structure supports the cyclone and evolves throughout its life cycle. In the remainder of this chapter we will investigate (1) the nature of cyclogenesis (the intensification of a cyclone), (2) the process of occlusion, and (3) the nature of cyclolysis (the decay of a cyclone). In order to provide a broad background to these discussions, and to illuminate some characteristics of a basic hydrodynamic instability that underlies the existence of mid-latitude cyclones, we first explore the basic environmental conditions that prevail at middle latitudes and then explore the characteristic vertical structure of a developing mid-latitude cyclone building our model literally from the ground up.

8.2 Basic Structural and Energetic Characteristics of the Cyclone

The uneven heating of the spherical Earth results in a pole-to-equator temperature gradient on the planet. As a consequence of the dominance of the thermal wind balance outside of the tropics, such a temperature gradient is manifest as a baroclinic westerly vertical shear at middle latitudes. If we consider the rather hypothetical situation in which the mid-latitude flow is purely zonal and in thermal wind balance, then at some middle or upper tropospheric level the geopotential height lines and isotherms would be everywhere parallel. Imagine that a wave-like perturbation were introduced into this flow and that the speed of the wave exactly equaled the speed of the background zonal flow. In such a case, only the meridional motions associated with the perturbation would be discernible. Those meridional motions would promote warm air advection downstream of the trough axis and cold air advection upstream of the trough axis as shown in Figure 8.5, eventually producing a wave in the thermal field that would lag the wave in the momentum field by one-quarter

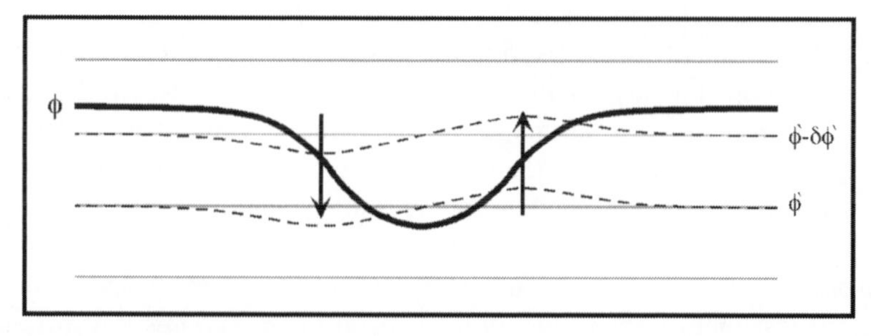

Figure 8.5 Effect of introducing a wave in the momentum field into a zonally oriented bundle of column-averaged isotherms. Light gray lines are undisturbed thickness isopleths of the mean state. Dashed lines are the disturbed thickness isopleths after the meridional motions of the wave (arrows) have distorted them. The thick black line shows a schematic geopotential height line. Note that the resulting thickness wave is a quarter wavelength out of phase with the wave in the geopotential height

wavelength. In order for this wave-like perturbation to grow, two conditions must be met: (1) the positive and negative zonal temperature anomalies must become larger, and (2) the kinetic energy associated with the wave motions must increase.

The pole-to-equator temperature gradient represents a horizontal density contrast conceptually analogous to that shown in the left panel of Figure 8.3. If, by some mechanism, the dense fluid ends up beneath the less dense fluid (as shown in the right panel of Figure 8.3) then the center of mass of the fluid system has been reduced and there has been a conversion of *some* of the initial potential energy into the kinetic energy of the fluid motions involved in the rearrangement. That fraction of the total potential energy that can be converted into kinetic energy is known as the **available potential energy** (APE). Were our hypothetical wave-like disturbance able to convert the APE of the background zonal baroclinic shear into the kinetic energy of its own motions then the wave-like perturbation would grow at the expense of the basic flow. In such a case, we would designate the background flow as unstable to the introduction of such a disturbance.

Mid-latitude cyclones and anticyclones are wave phenomena. As a result, any regional sea-level pressure analysis, such as the example shown in Figure 8.6, will display an alternating sequence of surface high- and low-pressure disturbances. In order that a surface low- (high-)pressure system remain a region of relative low (high) pressure, air must be extracted from (stuffed into) the atmospheric column above the surface. Thus, an alternating sequence of highs and lows, each associated with sinking or rising air in their respective columns, characterizes a mid-latitude wave train as shown in Figure 8.7(a). Recall that based on simple curvature arguments alone, we know that upward (downward) vertical motions occur downstream of trough (ridge) axes at upper tropospheric levels. Consequently, regions of low (high) geopotential height must be located to the west of the rising (sinking) air columns as shown in Figure 8.7(b). Thus, we know that for developing mid-latitude disturbances, the geopotential height axes tilt westward, into the vertical shear, with increasing height.

Recall that at the mature stage of the mid-latitude cyclone, the low-pressure center is located at the peak of the warm sector. The surface anticyclone lies to the west of the surface cyclone with its center close to the center of minimum temperature at sea level. Now, since the hypsometric equation relates thickness to column-averaged temperature, upper tropospheric geopotential minima (maxima) must lie atop relatively cold (warm) columns. Thus, as shown in Figure 8.7(c), the thermal axes of developing mid-latitude waves tilt eastward with increasing height. Finally, note that since the air is rising through the warm column and sinking through the cold column, developing mid-latitude disturbances are characterized by thermally direct vertical circulations which convert the APE of the background baroclinicity, which is itself manifest in the westerly vertical shear of the large-scale flow, into the kinetic energy of the disturbances.

The fact that the structure of the mid-latitude cyclone results in spontaneous conversion of APE to kinetic energy implies that the background zonal baroclinic

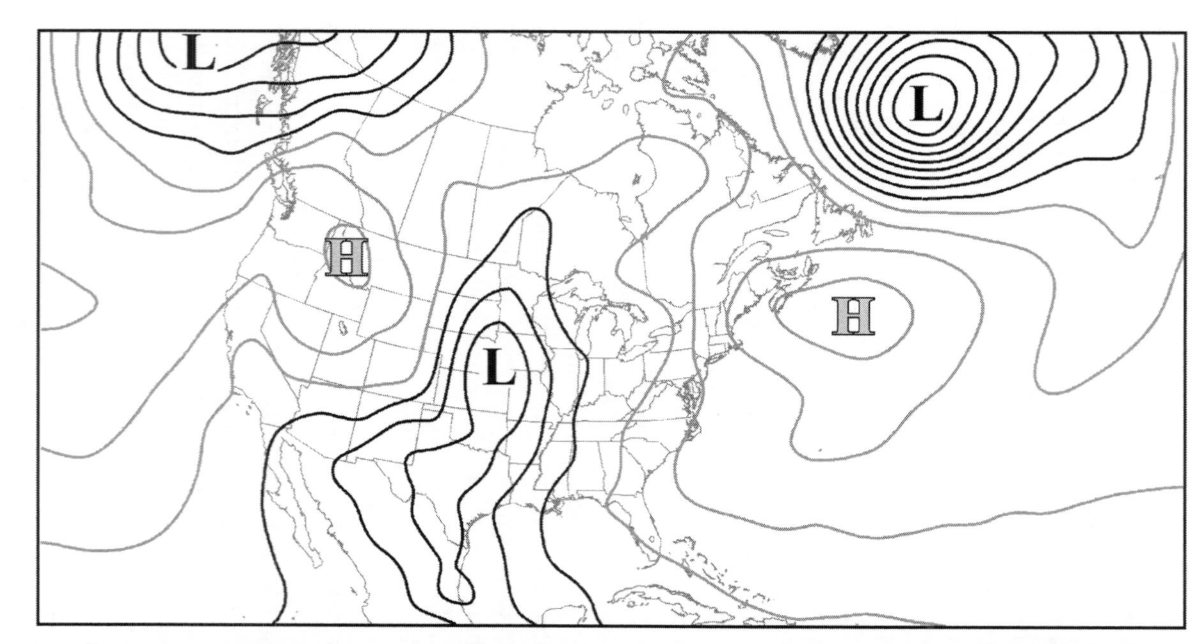

Figure 8.6 Sea-level pressure analysis over North America at 1200 UTC 16 April 2004. Solid lines are isobars contoured every 4 hPa with black (gray) lines corresponding to pressures less than (greater than) or equal to 1012 (1016) hPa. The L and H identify centers of surface low- and high-pressure systems, respectively

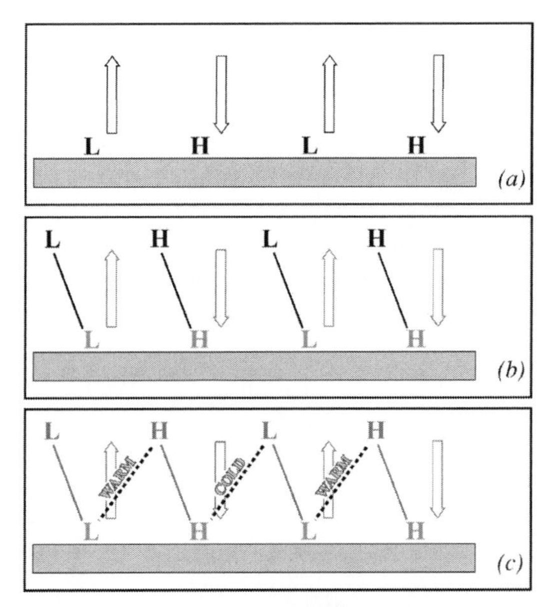

Figure 8.7 Vertical structure of a developing mid-latitude cyclone. (a) Alternating sequence of surface high- and low-pressure systems with ascent (descent) slightly downstream from the lows (highs). (b) Upper tropospheric lows and highs are displaced to the west with height (see text for explanation). Thick solid lines represent geopotential axes connecting surface and upper tropospheric features. (c) Thick dashed lines are thermal axes which tilt slightly to the east with height (see text for explanation). Note that warm air is ascending and cold air is descending in this wave train

shear is, indeed, unstable to certain wave-like perturbations and that mid-latitude cyclones are a primary manifestation of this instability. A more fully developed version of this baroclinic instability[2] theory suggests that disturbances of the scale of mid-latitude short waves (3000 to 4500 km in wavelength), in environments characterized by observed values of vertical shear, are those that exhibit the most efficient growth by this mechanism.

Though elements of the foregoing characteristic vertical structure of cyclones were known in the late nineteenth century, almost no mention was made of the vertical wave structure of cyclones in the NCM. The goal in the subsequent sections will not be to provide a comprehensive review of the theory and supporting observations regarding the various stages of the mid-latitude cyclone life cycle,[3] but instead to demonstrate that the diagnostic tools we have developed thus far can be gainfully employed in developing an understanding of the basic elements of that life-cycle evolution.

[2] Baroclinic instability theory was discovered independently by Jule Charney (1947) and Eric Eady (1949) using approaches to the problem that were significantly different from one another. Charney was also the first to derive rigorously the quasi-geostrophic system of equations upon which much of this book is based.

[3] A vast literature exists on the related subjects and to review it is an enormous job. A number of seminal references from that literature are given at the end of the chapter.

8.3 The Cyclogenesis Stage: The QG Tendency Equation Perspective

Cyclogenesis is the process by which a surface cyclone initially develops and subsequently intensifies. Intensification is often measured in terms of sea-level pressure decreases following the cyclone center. A consequence of this semi-Lagrangian negative pressure tendency is an increase in the low-level geostrophic vorticity. Thus, cyclogenesis can be viewed as a process of low-level vorticity production. Vorticity production necessitates the presence of divergence and vertical motions, as we have already seen. Consideration of the isobaric form of the continuity equation leads to

$$\int_0^{p_s} \partial\omega = -\int_0^{p_s} (\nabla \cdot \vec{V})\partial p \quad \text{or} \quad \omega_{p_s} = -\int_0^{p_s} (\nabla \cdot \vec{V})\partial p. \tag{8.1}$$

Now, since

$$\omega = \frac{dp}{dt} = \frac{\partial p}{\partial t} + \vec{V}_a \cdot \nabla p + w\frac{\partial p}{\partial z}$$

and both w and $\vec{V}_a \cdot \nabla p$ are nearly zero *at the surface of the Earth*,[4] then (8.1) can be rewritten as

$$\frac{\partial p_s}{\partial t} \approx -\int_0^{p_s} (\nabla \cdot \vec{V})\partial p. \tag{8.2}$$

This expression, known as the pressure tendency equation, dictates that the surface pressure tendency at a point is a consequence of the total convergence of mass into the vertical column of atmosphere above that point. Thus, net mass divergence (convergence) in the column is responsible for sea-level pressure falls (rises) at a given location. As we have already seen, however, measuring the divergence cannot be done with a great degree of accuracy. Thus, approximations to (8.2) must be made in order to render useful results. The simplest set of approximate equations are the quasi-geostrophic set that we derived in Chapter 5. Recall that the quasi-geostrophic vorticity and thermodynamic energy equations were given by

$$\frac{\partial \zeta_g}{\partial t} = -\vec{V}_g \cdot \nabla(\zeta_g + f) + f_0\frac{\partial \omega}{\partial p}$$

$$\frac{\partial}{\partial t}\left(-\frac{\partial \phi}{\partial p}\right) = -\vec{V}_g \cdot \nabla\left(-\frac{\partial \phi}{\partial p}\right) + \sigma\omega,$$

---

[4] Ageostrophic pressure advection ($\vec{V}_a \cdot \nabla p$) can be different from zero just above the surface, however.

respectively. If we represent the geopotential tendency as $\chi = \partial\phi/\partial t$, then the geostrophic vorticity tendency can be expressed as

$$\frac{\partial \zeta_g}{\partial t} = \frac{1}{f_0}\nabla^2 \chi$$

and the above two expressions can be rewritten as

$$\nabla^2 \chi = -f_0 \vec{V}_g \cdot \nabla \left(\frac{1}{f_0}\nabla^2 \phi + f \right) + f_0^2 \frac{\partial \omega}{\partial p} \tag{8.3a}$$

$$\frac{\partial \chi}{\partial p} = -\vec{V}_g \cdot \nabla \left(\frac{\partial \phi}{\partial p} \right) - \sigma\omega. \tag{8.3b}$$

Now, in order to eliminate the ω terms in (8.3) we take $(f_0^2/\sigma)\partial/\partial p$ of (8.3b) and add it to (8.3a) to get

$$\left(\nabla^2 + \frac{f_0^2}{\sigma}\frac{\partial^2}{\partial p^2} \right) \chi = -f_0 \vec{V}_g \cdot \nabla \left(\frac{1}{f_0}\nabla^2 \phi + f \right) - \frac{f_0^2}{\sigma}\frac{\partial}{\partial p}\left(\vec{V}_g \cdot \nabla \left(\frac{\partial \phi}{\partial p} \right) \right) \tag{8.4}$$

which is known as the **quasi-geostrophic height tendency equation.** The operator on the LHS of (8.4) is exactly the same as the operator on the LHS of the QG omega equation and can be interpreted similarly. When

$$\left(\nabla^2 + \frac{f_0^2}{\sigma}\frac{\partial^2}{\partial p^2} \right) \chi$$

is less than (greater than) zero, then χ itself is greater than (less than) zero. The RHS of (8.4) suggests that there are two processes that can contribute to local geopotential height changes. The first of these is represented by

$$-f_0 \vec{V}_g \cdot \nabla \left(\frac{1}{f_0}\nabla^2 \phi + f \right)$$

which describes the effect of geostrophic vorticity advection on height falls. Figure 8.8 shows a schematic upper tropospheric trough with a cyclonic vorticity maxima at its base. Immediately to the east (west) of the trough axis, there is positive (negative) geostrophic vorticity advection. In the absence of other processes, PVA is associated with height falls ($\chi < 0$) while NVA is associated with height rises ($\chi > 0$). Interestingly, however, since the geostrophic vorticity advection is precisely zero at the axis of the trough (since the gradient of geostrophic vorticity is zero there), there is no height tendency at that point. Since that point represents the location of lowest geopotential height in the first place, we see that the geostrophic vorticity advection term can only propagate an already existing disturbance – it cannot intensify it! It is interesting to note that since the geostrophic absolute vorticity is

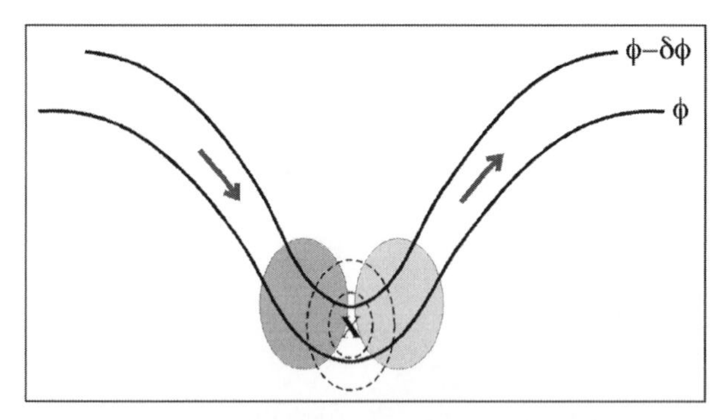

Figure 8.8 Upper tropospheric trough in the northern hemisphere. Thick black lines are isopleths of geopotential. Gray arrows are geostrophic wind vectors and the dashed lines are contours of cyclonic vorticity with the 'X' indicating the vorticity maximum. Light (dark) shading represents region of positive (negative) vorticity advection

given by $\eta_g = (1/f_0)\nabla^2\phi + f$, the horizontal geostrophic vorticity advection can be written as

$$-f_0\vec{V}_g \cdot \nabla\left(\frac{1}{f_0}\nabla^2\phi + f\right) = -f_0\nabla \cdot (\vec{V}_g\eta_g). \tag{8.5}$$

Thus, the geostrophic absolute vorticity flux convergence (divergence) is associated with height falls (rises).

The second term on the RHS of (8.4) can be rewritten as

$$-\frac{f_0^2}{\sigma}\frac{\partial}{\partial p}\left[-\vec{V}_g \cdot \nabla\left(-\frac{\partial\phi}{\partial p}\right)\right]. \tag{8.6a}$$

Since $-\partial\phi/\partial p = RT/p$ and $-\partial/\partial p$ represents the vertical derivative, (8.6a) can be expressed as

$$\frac{Rf_0^2}{p\sigma}\left[-\frac{\partial}{\partial p}(-\vec{V}_g \cdot \nabla T)\right] \tag{8.6b}$$

which is easily recognized as the vertical derivative of geostrophic temperature advection. It is this term that controls the *development* of disturbances from the perspective of the QG height tendency equation. We see that geostrophic temperature advection increasing (decreasing) upward is associated with height falls (rises). In a developing mid-latitude cyclone, middle tropospheric height rises (manifest as ridge building) typically occur to the east of the sea-level pressure (SLP) minimum, in the vicinity of the cyclone's warm front. From the tendency equation perspective, this is a result of the fact that warm air advection is strong in the lower troposphere and weaker in the middle and upper troposphere in that region of the storm. The result is warm air advection decreasing with height (phenomenologically equivalent to cold

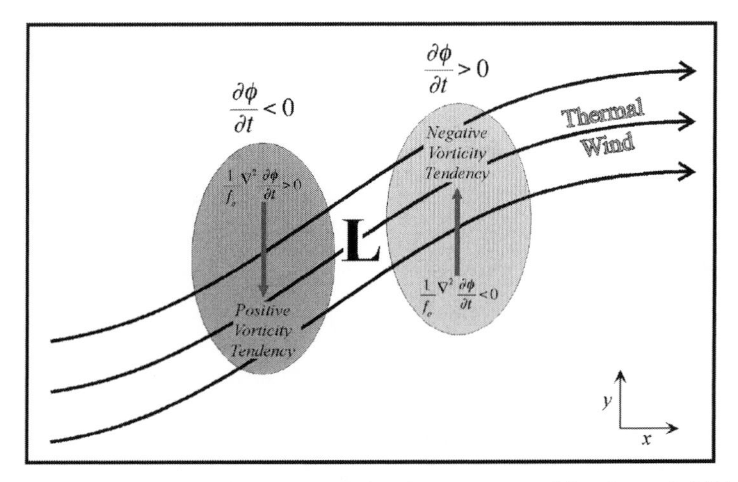

Figure 8.9 The effect of horizontal temperature advection on geopotential tendency. Solid black arrows are streamlines of the lower tropospheric thermal wind. Surface low-pressure center is indicated with 'L' and the gray arrows represent the lower tropospheric winds associated with the storm. Light (dark) shaded region identifies an area where warm (cold) air advection decreases with height leading to height rises (falls) and a negative (positive) vorticity tendency

air advection increasing with height) and height rises. To the west of the SLP minimum, in the vicinity of the surface cold front, middle tropospheric height falls occur consistent with the fact that the lower tropospheric cold air advection associated with the surface cold front is stronger than the middle and upper tropospheric cold air advection in the same location. Thus, the geostrophic temperature advection increases with height in that vicinity and, consequently, mid-tropospheric height falls occur there. The height falls to the west of the developing SLP minimum lead to a positive, mid-tropospheric geostrophic vorticity tendency there, while the height rises to the east of the SLP minimum are associated with a negative, mid-tropospheric geostrophic vorticity tendency there, as illustrated in Figure 8.9. The juxtaposition of positive and negative geostrophic vorticity tendencies, in turn, promotes more intense positive vorticity advection by the thermal wind (and consequent upward vertical motion) in the vicinity of the SLP minimum leading to continued lower tropospheric cyclogenesis. In this way, the asymmetric temperature advection field associated with a developing mid-latitude cyclone makes a significant contribution to the dynamics of cyclogenesis. Of course, nature is nearly always more complicated than the simplest example and careful analysis of the geostrophic temperature advection profile is necessary to employ the QG tendency diagnostic usefully in any given real case.

If we employ the chain rule on (8.6b) we notice that

$$\frac{Rf_0^2}{p\sigma}\left[-\frac{\partial}{\partial p}(-\vec{V}_g \cdot \nabla T)\right] = \frac{Rf_0^2}{p\sigma}\left[\frac{\partial \vec{V}_g}{\partial p} \cdot \nabla T - \vec{V}_g \cdot \nabla\left(-\frac{\partial T}{\partial p}\right)\right]. \tag{8.7}$$

The first term on the RHS of (8.7) describes the thermal wind advection of temperature which is identically zero. Hence, the QG tendency equation becomes

$$\left(\nabla^2 + \frac{f_0^2}{\sigma}\frac{\partial^2}{\partial p^2}\right)\chi = -f_0\vec{V}_g \cdot \nabla\left(\frac{1}{f_0}\nabla^2\phi + f\right) - \frac{Rf_0^2}{p\sigma}\left(-\vec{V}_g \cdot \nabla\frac{\partial T}{\partial p}\right)$$

or, since

$$T = -\frac{p}{R}\frac{\partial \phi}{\partial p}$$

from the hydrostatic equation,

$$\left(\nabla^2 + \frac{f_0^2}{\sigma}\frac{\partial^2}{\partial p^2}\right)\chi = -f_0\vec{V}_g \cdot \nabla\left(\frac{1}{f_0}\nabla^2\phi + f\right) - \frac{f_0^2}{\sigma}\left(\vec{V}_g \cdot \nabla\frac{\partial^2 \phi}{\partial p^2}\right). \tag{8.8}$$

We will return to this form of the tendency equation in Chapter 9.

8.4 The Cyclogenesis Stage: The QG Omega Equation Perspective

Now that we have seen both the QG omega and height tendency equations we are ready to consider the physical sequence of events that characterize the adjustment of the mass and temperature fields to a canonical cyclogenesis event. Nearly all cyclogenesis events proceed from a precursor upper-level disturbance in the flow. This disturbance manifests itself as a relative vorticity maxima as illustrated in Figure 8.10(a). Given the influence of vorticity advection in the tendency equation, the disturbance will propagate in the direction of the flow. Since the disturbance is often initially largest at middle and upper tropospheric levels, where the geostrophic winds are often largest as well, there will be upward-increasing positive (negative) vorticity advection (PVA (NVA)) downstream (upstream) of the disturbance. From the QG omega equation, this circumstance is associated with upward (downward) vertical motion downstream (upstream) of the trough axis as shown in Figure 8.10(b). An alternative way to view this forcing for vertical motion is through the approximate Trenberth form of the QG omega equation in which PVA (NVA) by the thermal wind is associated with ascent (descent). Recall that this forcing can be expressed in terms of the divergence of a vector field oriented parallel to the thickness isopleths ($\vec{Q}_{TR}$ discussed at the end of Chapter 6). Consequently, the initial vertical motion couplet portrayed in Figure 8.10(b) is a *shearwise* couplet. Shearwise vertical motions, by virtue of their relation to the $\vec{Q}_s$ component of the $\vec{Q}$-vector, are associated with rotation of the $\nabla\theta$-vector. Thus, the distribution of $\vec{Q}_{TR}$ shown in Figure 8.10(b) will also deform the thermal field, producing a thermal ridge in the vicinity of the developing surface low-pressure center and a thermal trough upstream. Under the influence of the cyclonic circulation associated with the developing lower tropospheric disturbance, low-level warm air advection will occur downstream of the upper-level

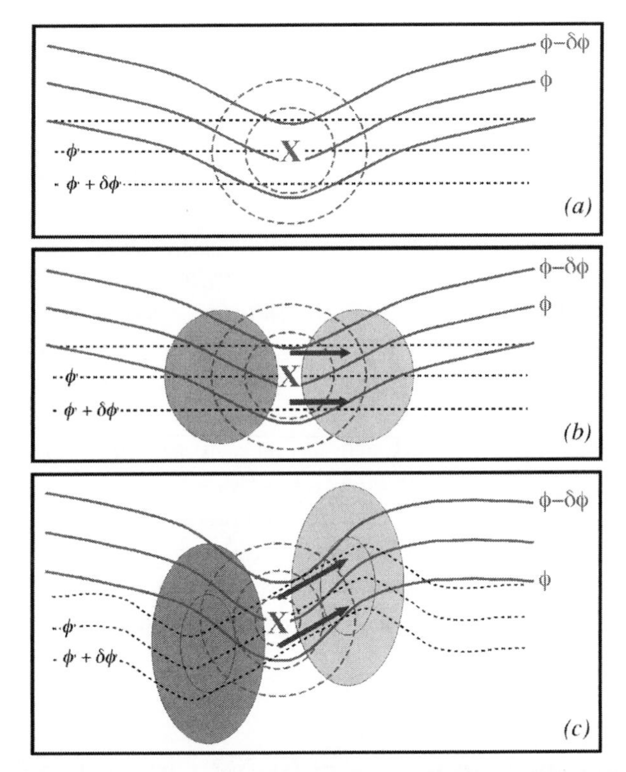

Figure 8.10 Initial thermal and mass field adjustments to cyclogenesis. (a) Upper tropospheric vorticity maxima in a zonal thermal wind. Gray solid lines are 500 hPa geopotential heights, gray dashed lines are 500 hPa geostrophic absolute vorticity, and the black dashed lines are 1000–500 hPa thickness isopleths. 'X' marks the location of the maximum absolute vorticity. (b) As for (a) with the gray arrows representing the $\vec{Q}_{TR}$ vectors. Light (dark) shaded area is a region of upward (downward) vertical motion and upper troposopheric divergence (convergence). (c) As for (b) but for a subsequent time in the cyclone's development. Note the development of the thermal ridge downstream of the upper trough axis and the thermal trough upstream of it. Larger $\vec{Q}_{TR}$ vectors and greater vertical motions are the result of intensification of the 500 hPa trough–ridge couplet

trough axis and low-level cold air advection just upstream of it. As illustrated in Figure 8.9, the tendency equation suggests that such a circumstance will serve to raise the geopotential heights in the middle troposphere to the east of the surface low and lower the heights to its west. Alternatively, the distribution of upper tropospheric convergence and divergence associated with the vertical motion couplet illustrated in Figure 8.10(b) will tend to increase the upper tropospheric vorticity in the vicinity of the trough axis while decreasing it in the vicinity of the downstream ridge. A more intense upper-level vorticity maximum leads to greater PVA by the thermal wind and attendant upward vertical motions which further intensify the surface cyclone downstream of the upper feature. The rotation of $\nabla\theta$ afforded by the shearwise vertical motions eventually orients the baroclinicity into a configuration in which the deformation fields characteristic of the cyclogenesis itself begin to increase $|\nabla\theta|$ locally. Thus, the cold and warm frontal zones begin to develop along with their frontogenetically induced transverse circulations. The combination of the shearwise and transverse couplets of vertical motion thus produced underlies the comma-shaped cloud distribution that characterizes the mid-latitude cyclone.

As the upper disturbance continues to develop and progress eastward, it begins to outrun its surface reflection. As a result, the convergence at the surface (maximized at the location of the sea-level pressure minimum) gradually becomes disconnected from its divergence valve aloft and the surface cyclone can no longer intensify. Thus, the phasing of the upper and lower disturbances is crucial to their complementary development.

The foregoing description of cyclogenesis makes no mention of the fact that clouds and precipitation (and therefore latent heat release) are involved in this process. Naturally, the interaction between the dynamic and diabatic processes is an important aspect of the overall process of cyclogenesis. Though such interaction characterizes every cyclogenesis event to some degree, these interactions are most vividly illustrated by considering cases of dramatic surface development, known as explosive cyclogenesis.

8.5 The Cyclogenetic Influence of Diabatic Processes: Explosive Cyclogenesis

Explosive cyclogenesis is the rapid development of a sea-level pressure minimum. Prior work has suggested a threshold of 24 hPa of deepening in 24 hours[5] as a reasonable distinguishing characteristic of an explosive deepener. The deepening rates for all northern hemisphere cyclones in a single year are shown in Figure 8.11. It appears that the distribution is skewed toward these rapid deepeners[6] suggesting that something may be different about these storms. In fact, there are some notable differences between the 'ordinary' cyclones that constitute the majority of all cyclones and these rarer events. One of the more significant differences between these populations is that the explosive deepeners not only deepen more rapidly but also for a longer time than the 'ordinary' cyclones. What does this mean about the contrast in physical processes that operate in these two populations? The distribution of these explosive deepeners provides a clue as to the circumstances that conspire to produce them. As illustrated in Figure 8.12, explosively deepening mid-latitude cyclones in the northern hemisphere tend to develop along the warm western boundary ocean currents such as the Kurishio and Gulf Stream. The prevailing view is that these

[5] This value was suggested by Sanders and Gyakum (1980) and is normalized for latitude according to the following formula: *Deepening Rate* = $\Delta p(\sin\phi/\sin 60^\circ)$.

[6] The reader is referred to Roebber (1984) for the complete study from which this information comes.

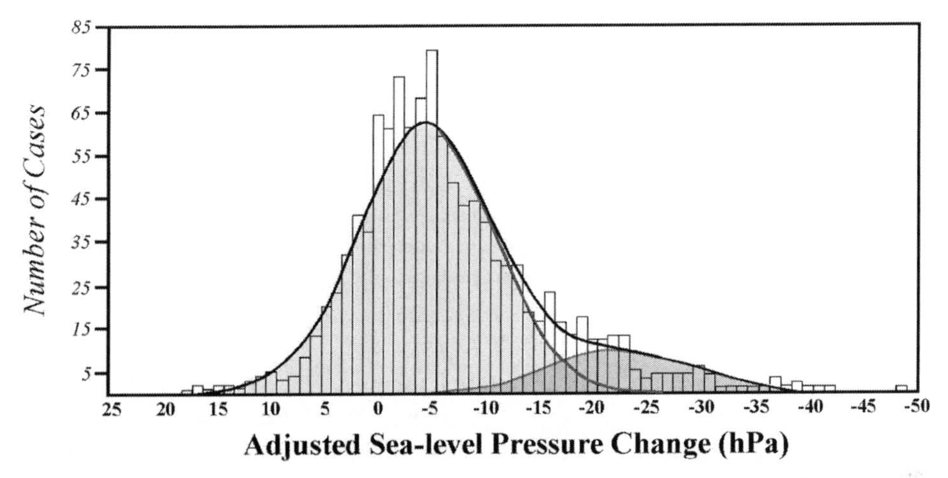

Figure 8.11 Distribution of 24 h deepening rates for all northern hemisphere surface cyclones in year. The dark solid line indicates the sum of two normal curves while the gray lines and shadings represent the separate distributions (light shading for the 'ordinary' and darker for 'explosive'cyclones). Adapted from Roebber (1984)

storms are the manifestation of physical and dynamical processes that occur to some degree in all cyclones but which are particularly vigorous in explosive deepeners. A reasonable next question, then, is: 'What makes ordinary processes so potent in these storms?' Recalling that surface development is strongly tied to upward vertical

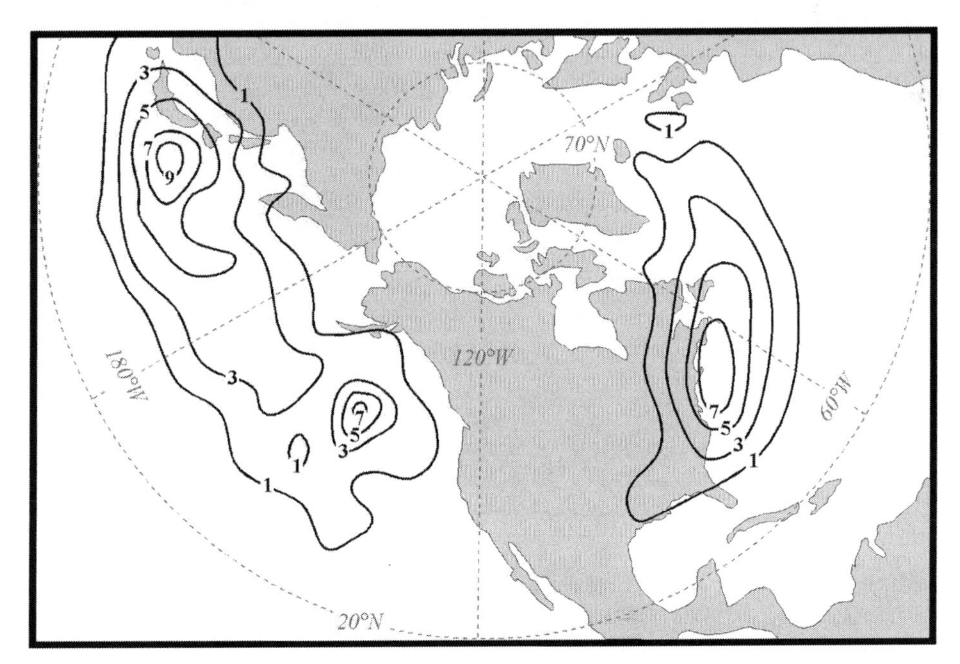

Figure 8.12 Geographical distribution of positions of maximum deepening for all northern hemisphere cyclones from 1976–1982 that intensified at a rate greater than 24 hPa $(24\ h)^{-1}$. Numbers indicate the annual frequency of such developments at the indicated locations. Adapted from Roebber (1984)

motion through the vorticity equation, let us look at the QG omega equation again,

$$\sigma\left(\nabla^2 + \frac{f_0^2}{\sigma}\frac{\partial^2}{\partial p^2}\right)\omega = -2\nabla \cdot \vec{Q}$$

and consider, hypothetically, 2 days on which the RHS forcing $(-2\nabla \cdot \vec{Q})$ in a given domain is exactly the same. Given this circumstance, the only factor that could possibly influence the production of stronger vertical motions on one day versus the other is the static stability, σ. In fact, σ acts as the amplitude modulator of the omega equation (its AM dial as it were): lower (higher) σ is associated with a greater (lesser) response to a given forcing.

Returning again to Figure 8.12, we note that the prevalence of explosively deepening cyclones over warm ocean currents is a result of the fact that these locations are characterized by consistently lower static stability and, consequently, a consistently more vigorous response to forcing for upward vertical motion. These more intense vertical motions then lead to more intense cyclogenesis. But even this physical linkage does not yet reference the effect of the characteristic cloud and precipitation distribution of cyclones.

The characteristic precipitation distribution associated with mid-latitude cyclones is asymmetric as illustrated schematically in Figure 8.13. The period of most rapid development occurs when heavy precipitation develops poleward and westward of the cyclone center. The associated latent heat release (LHR) can (1) add energy to the system, (2) focus and intensify the vertical motion pattern through a local reduction of the static stability in saturated updrafts, and, perhaps most interestingly, (3) affect the structure and dynamics of the larger-than-cyclone scale so as to intensify the cyclogenetic effect of ordinary dynamical processes. This last point bears special

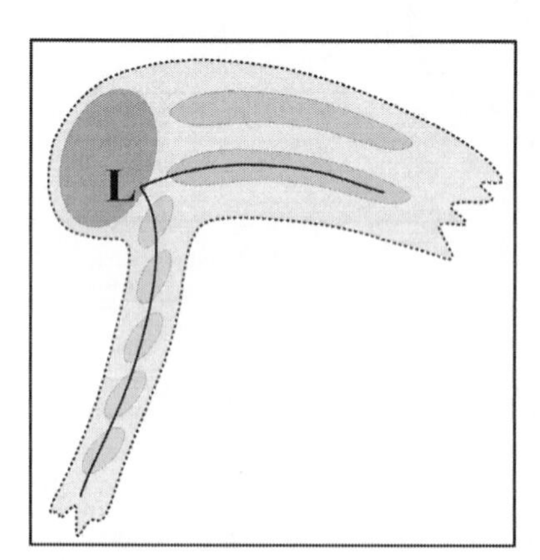

Figure 8.13 Schematic depicting the asymmetric cloud and precipitation distribution in a typical mid-latitude cyclone. The lightly shaded area enclosed by the dashed line is the cloud pattern. The solid lines within that region are the surface cold and warm fronts. The shaded subregions within the cloud mass are the precipitation elements associated with the cold front (lightest shading), warm front (darker shading), and the area to the north and northwest of the surface cyclone center (darkest shading)

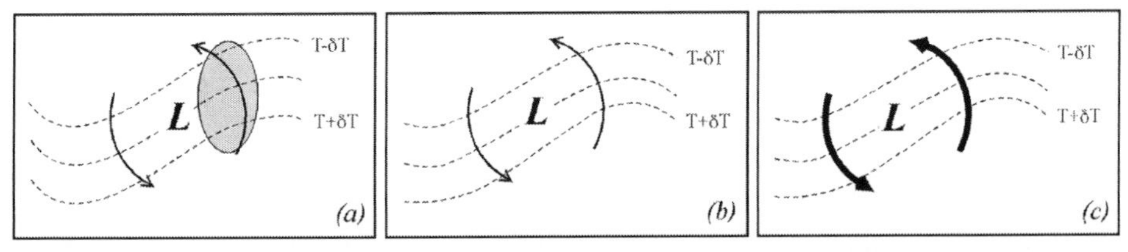

Figure 8.14 Schematic illustration of the influence that sensible and latent heat fluxes in the planetary boundary layer can have on the magnitude of the lower tropospheric temperature advection east of the surface cyclone center. (a) Prior to the influence of the heat fluxes a uniform temperature gradient exists. Dashed lines are isotherms, 'L' is the location of the sea-level pressure minimum, arrows represent the flow around the cyclone and the gray shaded area is the location where heat fluxes will warm the boundary layer. (b) Increased temperature gradient results from the heating in the boundary layer. Intensified lower tropospheric warm air advection intensifies the cyclone. (c) More intense cyclone leads to more intense lower tropospheric winds (bolder arrows) and increased warm air advection

attention as it lies at the heart of a conceptual/dynamical model of cyclogenesis known as the 'self-development' paradigm. Consider, as a first example, the feedback from sensible and latent heat fluxes in the lower troposphere to cyclogenesis. As shown schematically in Figure 8.14, poleward-directed boundary layer winds coupled with ascent on the eastern side of the cyclone warm the lower troposphere on the equatorward side of the developing warm front via sensible heating associated with the warm air advection and diabatic heating resulting from LHR in the moist, ascending air. This warming leads to an increase in the magnitude of the low-level temperature gradient and a consequent increase in the magnitude of the warm advection there. Stronger warm advection is often associated with intensified ascent in that location. Greater ascent leads to more intense baroclinic energy conversion and often to a stronger cyclone whose intensified circulation, in turn, results in a positive feedback loop. On larger scales, the LHR associated with the enhanced cloud and precipitation production produces a positive thickness anomaly just east of the upper-level short-wave trough axis (Figure 8.15). Consequently, the geopotential heights in the middle and upper troposphere increase in that region and a small-scale ridge is built up above the latent heating maxima. Positive vorticity advection to the east of the upper-level short-wave compels that feature to move eastward. In the face of the diabatic ridge building that occurs in association with the LHR in the cloud shield to the east, the wavelength between the upstream trough and downstream ridge axes shrinks. As a result of this wavelength shortening, the magnitude of the cyclonic vorticity advection by the thermal wind downstream of the trough axis greatly intensifies leading to more intense upward vertical motions. The stronger vertical motions intensify the cyclogenesis and produce more LHR just downstream of the upper-level short-wave which tends to further shorten the wavelength of the upper disturbance. In this way, a positive feedback loop is established.

A large number of numerical modeling studies investigating the influence of LHR on cyclogenesis have been undertaken in the past 30 years. The consensus

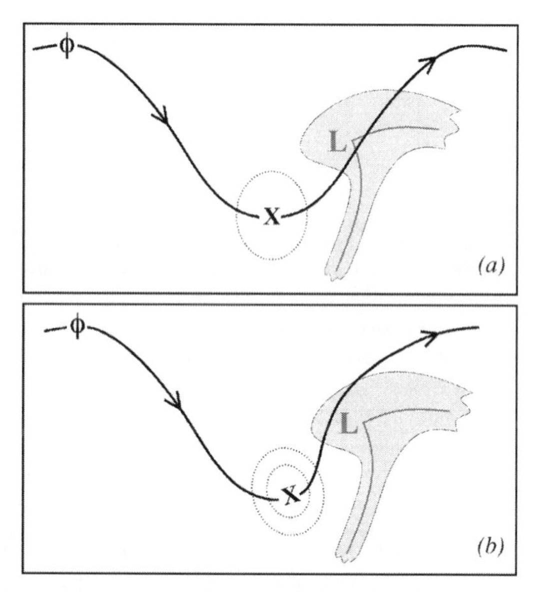

Figure 8.15 Schematic illustrating how latent heat release associated with the developing cloud and precipitation shield of a mid-latitude cyclone feeds back on the development. (a) Initial upper tropospheric wave in a geopotential height contour with the upper tropospheric vorticity maxima indicated by the 'X'. Surface low, and associated cloud and precipitation, develops downstream of 'X' (b) Latent heat release associated with the cloud and precipitation shield increases column thicknesses downstream of the upper vorticity maxima deforming the upper tropospheric geopotential as indicated. The result is a more significant ridge downstream of the upper trough and greater curvature vorticity in the vicinity of the upper trough axis as indicated

conclusion drawn from these studies is that since water vapor is not a passive scalar, its phase change tends to concentrate normal baroclinic processes onto smaller scales, which leads to feedbacks that further the scale contraction and intensification of these explosively deepening storms. From that perspective, it becomes clear that these storms do not arise as a consequence of 'special' dynamical processes but rather as a result of uncommonly intense interactions among the 'ordinary' suite of physical and dynamical processes that operate, to some degree, in all mid-latitude cyclones.

Recall that the NCM suggested that cyclones form along a pre-existing polar front that divided polar from tropical air masses throughout the depth of the troposphere. Subsequent work has proven beyond a doubt that cyclogenesis and frontogenesis are nearly concurrent processes. An idealized wave train superimposed upon a zonally oriented baroclinic zone is illustrated schematically in Figure 8.16. As the perturbations develop, regions of deformation develop as indicated. The meridional shear induced by the disturbances produces thermal ridges and troughs while the deformation (specifically diffluence) to the northeast and southwest of each cyclonic disturbance in the wave train provides an environment in which the gradient of any

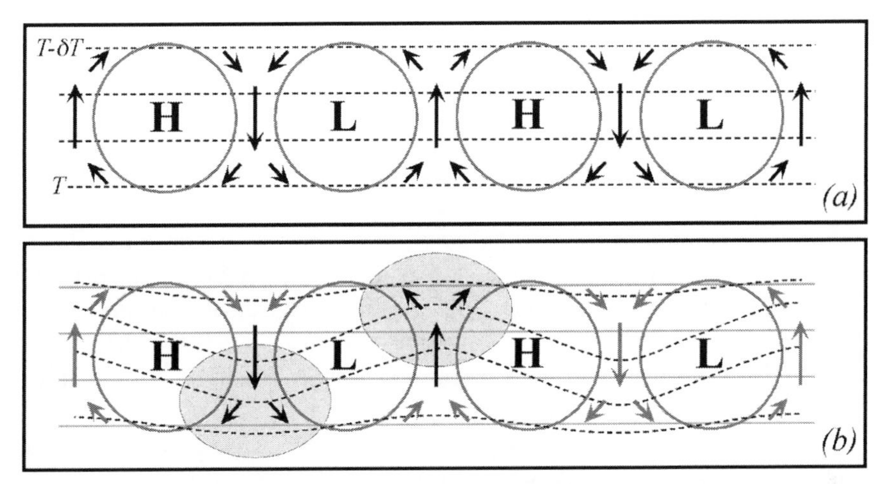

Figure 8.16 (a) Wave train superimposed upon a zonal band of isotherms in the northern hemisphere. 'H' and 'L' signify high and low sea-level pressure centers, respectively, and the arrows represent their associated geostrophic winds. (b) The effect of the deformation associated with the wave-train in (a) on the isotherms. Thin solid lines show the original orientation of the isotherms while the dashed lines represent the deformed isotherms. Light gray shaded regions identify the preferred regions of frontogenetic deformation, both of which are associated with the low-pressure center

variable in the flow may be intensified. This applies, of course, to temperature and consequently these two regions of deformation act frontogenetically to produce the warm and cold frontal zones. This simple idealized illustration demonstrates that the development of fronts is a consequence, not a cause, of cyclogenesis – a conclusion that departs radically from the ideas put forth in the NCM. In defense of the NCM, baroclinic instability theory indicates that a substantial background vertical shear, made manifest in a robust horizontal temperature contrast (through thermal wind balance), is necessary in order for cyclogenesis to occur. This temperature contrast, coupled with the presence of a discernible upper tropospheric short-wave trough, provides an environment in which cyclonic vorticity advection by the thermal wind can initiate the upward vertical motions necessary to spin-up the low-level cyclone. Since this forcing can be expressed as a portion of the along-isentrope $\vec{Q}$-vector, the lifting occurs simultaneously with the production of a thermal ridge displaced slightly downstream of the developing low-level circulation center. The circulation itself then deforms the background baroclinic zone via differential horizontal advections. Such differential horizontal advections produce regions of intensified baroclinicity (via frontogenesis) and quasi-linear vertical motion couplets which, when added to the cellular elements contributed by the shearwise forcing, result in the characteristic comma-shaped pattern of vertical motion associated with the mid-latitude cyclone. Note that this view also presents a life cycle in which the cyclogenesis and frontogenesis are nearly concurrent processes, perhaps even with cyclogenesis somewhat leading frontogenesis.

The prior discussion has implicitly assumed that surface cyclogenesis is the consequence of dynamics associated with a pre-existing upper-level disturbance. It is worth noting as we end this section that debate still rages as to whether this mechanism provides the only viable mechanism for mid-latitude surface cyclogenesis. Petterssen and Smebye (1971), recognizing that cyclones are 3-D disturbances and therefore might arise from any number of synoptic–dynamic scenarios, surveyed a large number of storms in an attempt to find some general categories that describe the possible variety of developmental scenarios. They concluded that two broad types of cyclogenesis operate in the middle latitudes: so-called Type A and Type B cyclogenesis. Type A cyclogenesis involves the amplification of lower tropospheric waves, usually along the cold front of a prior disturbance, and is thought to occur in the absence of any predecessor short-wave disturbance aloft. Typically, such storms are said to develop 'from the bottom-up' and, though relatively rare overall, were purported to be most common over the ocean basins. The manner by which such 'bottom-up' development might occur will be described in more detail in the next chapter. It is important to note that since Type A cyclogenesis was found to operate in a data-sparse region of the globe in a study undertaken in a data-sparse era, it may be that Type A cyclogenesis is a vanishingly rare occurrence.

Type B cyclogenesis is characterized by the presence of a well-defined predecessor disturbance aloft (i.e. in the form of an absolute vorticity maximum) which, upon crossing over a lower/middle tropospheric baroclinic zone, triggers surface cyclogenesis through, for instance, upward-increasing cyclonic vorticity advection. It is generally agreed that the majority of actual cyclogenesis events in the middle latitudes are examples of Type B cyclogenesis.

Thus far we have only considered the dynamics of the cyclogenesis phase of the mid-latitude cyclone life cycle. The NCM introduced the notion of occlusion as the peak of intensity and commencement of decay. In the next section we will discuss the essential distinguishing characteristics of the occluded phase of the life cycle as well as the characteristic dynamics that operate in that phase.

8.6 The Post-Mature Stage: Characteristic Thermal Structure

Since the notion of occlusion was first introduced, considerable controversy has existed concerning the nature of the occluded (post-mature) stage of the mid-latitude cyclone life cycle. Surprisingly, much of this controversy has centered around the means by which the characteristic occluded thermal structure evolves in the post-mature cyclone. In this section we therefore examine (1) the characteristic occluded thermal structure itself, and (2) an underlying dynamical mechanism (diagnosed using the QG equations) that simultaneously accounts for the development of that thermal structure and for the characteristic presence of ascent in the occluded quadrant of mid-latitude cyclones. We begin by briefly reviewing aspects of the occluded thermal structure.

As far back as the 1920s it was suggested that the process of occlusion involved the cold front encroaching upon, and subsequently overtaking and ascending, the warm frontal surface.[7] One of the main results of this process of warm occlusion was the production of a wedge of warm air aloft, displaced poleward of the surface warm and (newly created) occluded fronts. The cloudiness and precipitation associated with the development of the warm occlusion were suggested to result from lifting of warm air ahead of the upper cold front and were consequently distributed to the north and west of the sea-level pressure minimum. As a result of the gradual squeezing of warm air aloft between the two intersecting frontal surfaces, the horizontal thermal structure of a warm occlusion was characterized by a thermal ridge connecting the peak of the warm sector to the geopotential or sea-level pressure minimum. This thermal ridge is often manifested as a 1000–500 hPa thickness ridge or an axis of maximum θ or θ_e in a horizontal cross-section as shown in Figure 8.17. Note that considerable upward vertical motion is also co-located with this thermal ridge.

A vertical cross-section perpendicular to the axis of the thermal ridge in Figure 8.17 reveals the characteristic vertical structure of the warm occlusion (Figure 8.18a) consisting of a poleward-sloping axis of maximum θ_e separating two regions of concentrated baroclinicity. The surface warm occluded front is generally analyzed at the location where this axis of maximum θ_e intersects the ground, whereas the base of the warm air between the two baroclinic zones (the cold and warm fronts) sits atop their point of intersection (labeled A in Figure 8.18a). It is clear, however, from this cross-section that the upward vertical motion maximum is located at the leading edge of the cold frontal baroclinicity, significantly displaced from the position of the surface occluded front. Upon taking another vertical cross-section further along the thermal ridge toward the surface cyclone center (Figure 8.18b), we find that the same basic thermal structure and a roughly similar vertical motion distribution exist though the intersection of the warm and cold frontal baroclinic zones (labeled A in Figure 8.18b also) occurs at a higher elevation.

The observation that the cloudiness and precipitation characteristic of the occluded quadrant of cyclones often occurs in the vicinity of the thermal ridge led scientists at the Canadian Meteorological Service in the 1950s and 1960s to regard the essential feature of a warm occlusion to be the trough of warm air that is lifted aloft ahead of the upper cold front, *not* the position of the surface occluded front. The sloping line of intersection between the cold and warm frontal baroclinic zones, termed the **trough of warm air aloft (trowal)**, was found to bear a closer correspondence to the cloud and precipitation features in occluded North American cyclones than did the often weak surface warm occluded front. The trowal marks the 3-D sloping intersection of the upper cold frontal portion of the warm occlusion with the warm frontal zone and therefore represents a refined, 3-D description

[7] Although this idea was first published in Bjerknes and Solberg (1922), the notion of occlusion as described in the NCM was first devised by Tor Bergeron.

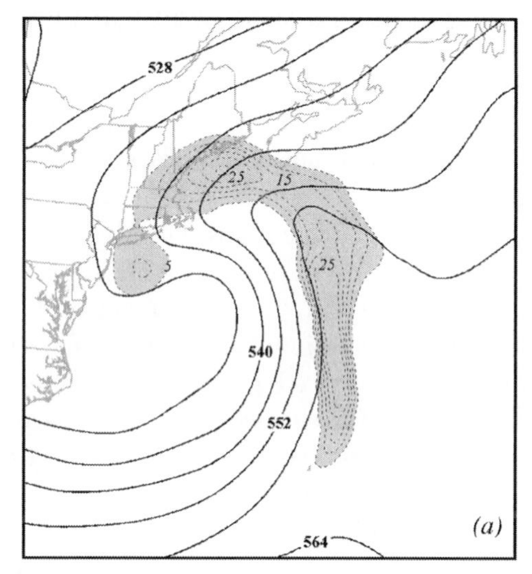
528
25
15
5
25
540
552
564
(a)

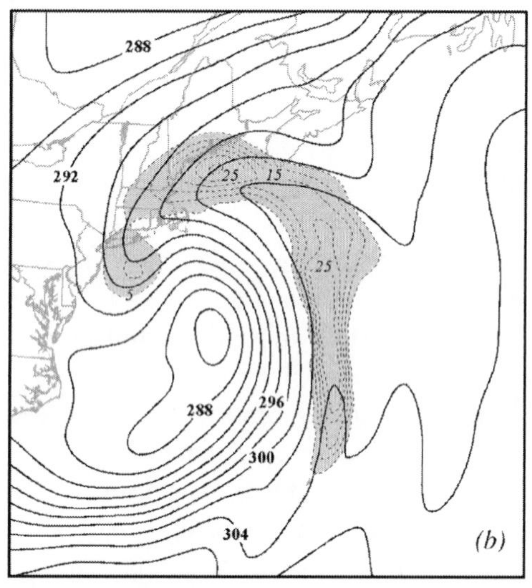
288
292
25
15
5
25
288
296
300
304
(b)

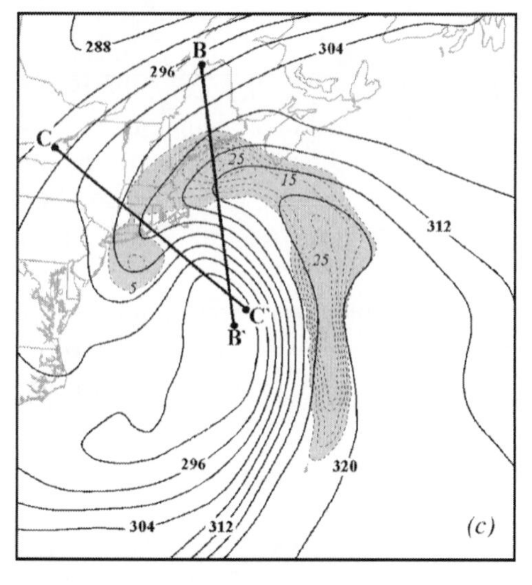
288
B
304
296
C
25
15
312
25
5
C'
B'
296
320
304
312
(c)

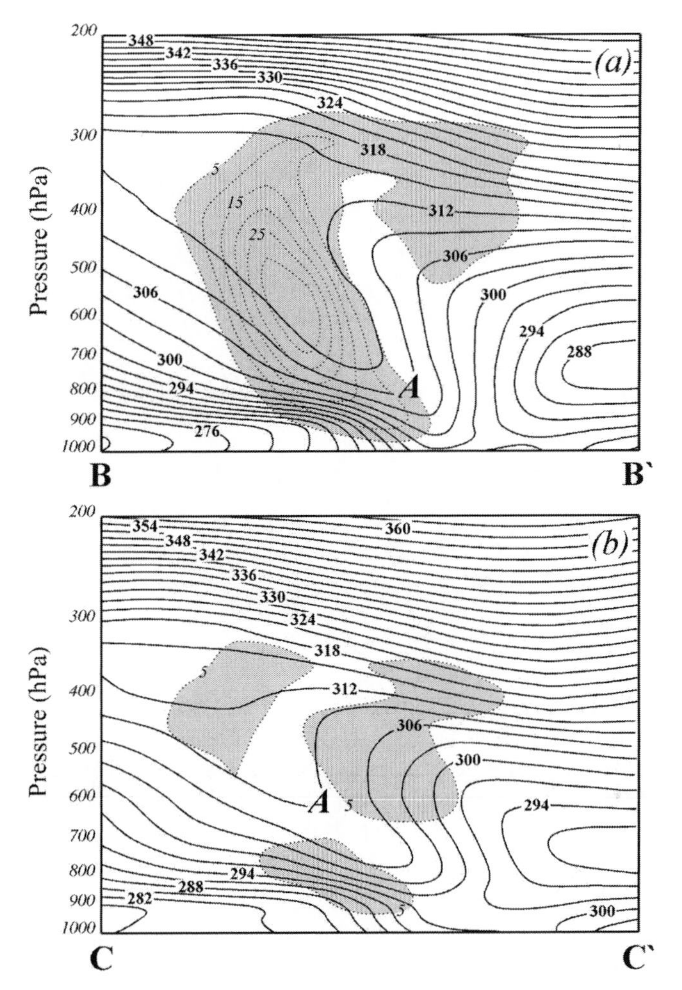

Figure 8.18 (a) Vertical cross-section of θ_e, through the occluded thermal ridge, along line B–B′ in Figure 8.17(c). Solid lines are moist isentropes labeled in K and contoured every 3 K. Shaded regions are upward vertical motions labeled in cm s^{-1} and contoured every 5 cm s^{-1}. 'A' represents the point of intersection between the cold and warm frontal zones in the warm occluded thermal structure. (b) As for (a) but for the cross-section along line C–C′ in Figure 8.17(c)

←

Figure 8.17 The characteristic occluded thermal ridge as observed at 0600 UTC 1 April 1997. (a) Solid lines are 1000–500 hPa thickness labeled in dam and contoured every 6 dam. Dashed lines with shading are 700 hPa upward vertical motion labeled in cm s^{-1} and contoured every 5 cm s^{-1}. Both variables are from an 18 h forecast of the NCEP Eta model valid at 0600 UTC 1 April 1997. (b) Solid lines are 18 h forecast of 700 hPa θ, valid at 0600 UTC 1 April, labeled in K and contoured every 2 K. Vertical motion as in (c). As for (b) but solid lines are 18 h forecast of 700 hPa θ_e, labeled in K and contoured every 4 K. Vertical motions indicated as in (a). Vertical cross-sections along lines B–B′ and C–C′ are shown in Fig. 8.18

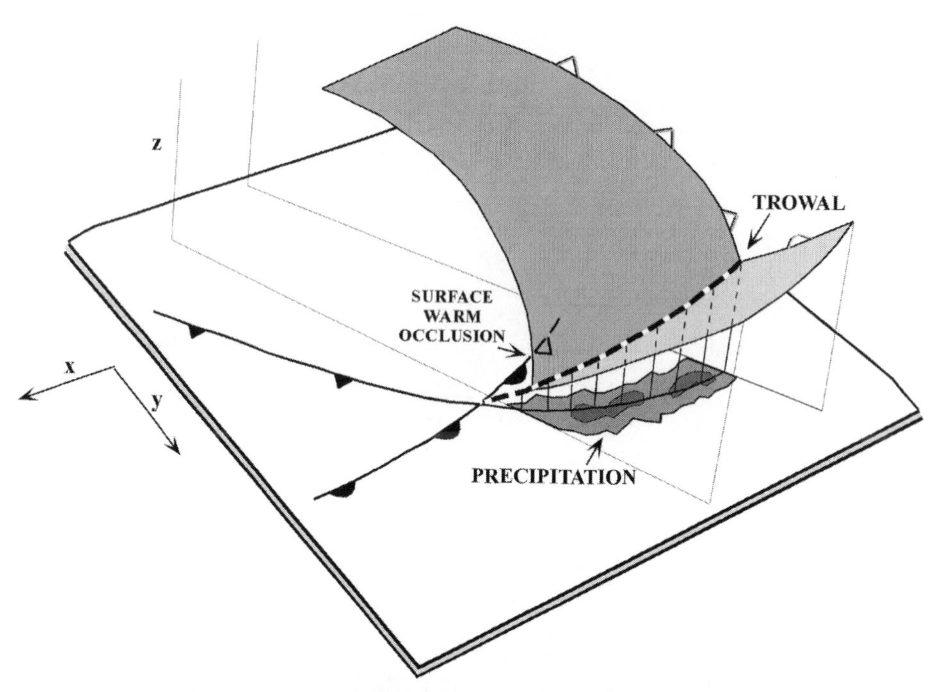

Figure 8.19 Schematic of the trowal conceptual model. The dark (light) shaded surface represents the warm edge of the cold (warm) frontal zone. The bold dashed line at the 3-D sloping intersection of those two frontal zones lies at the base of the trough of warm air aloft – the trowal. The schematic precipitation in the occluded quadrant of the cyclone lies closer to the projection of the trowal to the surface than to the position of the surface warm occluded front

of the warm occluded structure presented in the NCM. A schematic illustrating the trowal conceptual model is shown in Figure 8.19.

Given the availability of gridded output from numerical simulations of cyclones along with the graphical capability of software display packages for viewing this output, it is now relatively simple to identify the trowal structure in occluded cyclones. Referring back to Figure 8.18, notice that the 312 θ_e isentrope lies near the warm edge of both the warm and cold frontal baroclinic zones comprising the warm occluded structure. Plotting the 312 K moist isentrope every 100 hPa beginning at 1000 hPa from a gridded data set of this case reveals the isobaric topography of the 312 θ_e surface (Figure 8.20a). Clearly identifiable in this topography are (1) the steeply sloped cold frontal surface, (2) the less steeply sloped warm frontal surface, and (3) the poleward- and westward-sloping 3-D 'canyon' in the 312 K surface representing the trowal. This topography can also be viewed through inspection of the actual 312 θ_e surface produced by a different software package (Figure 8.20b).

Despite the historical controversy surrounding the nature of the occlusion process, there is fairly widespread agreement that the thermal structures just described are among the basic structural characteristics of the post-mature phase mid-latitude

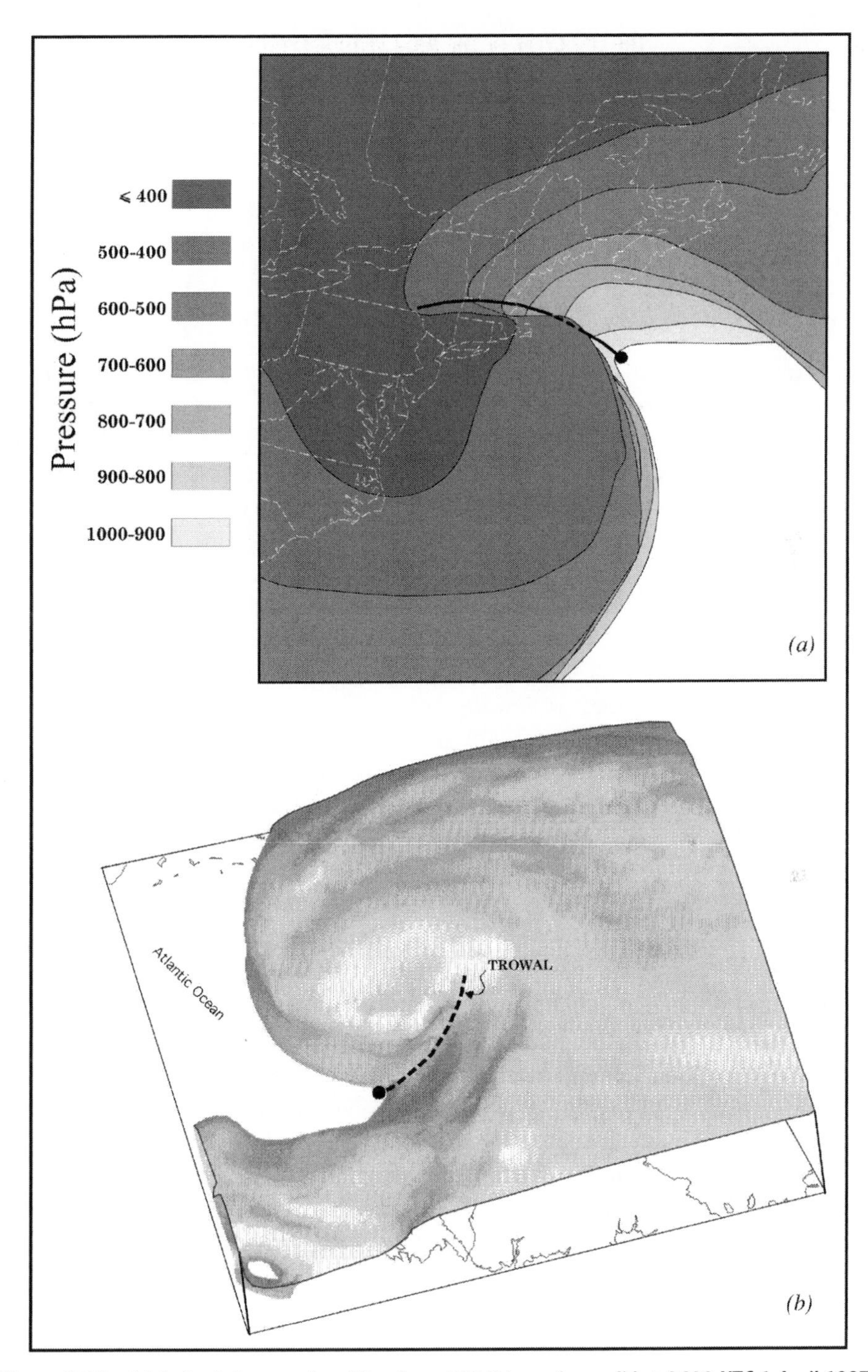

Figure 8.20 (a) Isobaric topography of the $\theta_e = 312$ K isosurface valid at 0600 UTC 1 April 1997. The thick solid line represents the position of the trowal and is clearly seen as a 3-D sloping canyon in the 312 K isosurface. (b) Elevated, northern view of the $\theta_e = 312$ K isosurface valid at 0600 UTC 1 April 1997 from the visualization software package VIS-5D. The bold dashed line represents the trowal – a sloping 3-D canyon in the 312 K isosurface

cyclone. Next we employ the QG omega equation to gain insight into a fundamental dynamical mechanism that simultaneously accounts for the presence of upward vertical motion and a thermal ridge in the occluded quadrant of such cyclones.

8.7 The Post-Mature Stage: The QG Dynamics of the Occluded Quadrant

Recall from Chapter 6 that a natural coordinate partition of the $\vec{Q}$-vector into along- and across-isentrope components was described. The across-isentrope component was later shown to be exactly equal, in magnitude, to the QG frontogenesis while we had yet to specify any similar physical meaning for the along-isentrope component. If we now consider the along-isentrope component of $\vec{Q}$ once again, with the help of Figure 8.21, we find a relation to the problem of occlusion. Figure 8.21(a) illustrates a straight baroclinic zone along which there is a region of convergence of $\vec{Q}_s$. This convergence will not only be associated with upward vertical motion, a consequence of the QG omega equation, but will also differentially rotate $\nabla\theta$ on either side of the convergence axis, as illustrated in Figure 8.21(b). Since $\vec{Q}_s$ cannot change the magnitude of $\nabla\theta$, the result is that displayed in Figure 8.21(c) – the production of a thermal ridge characterized by upward vertical motion! Figure 8.22 shows the partition of the total 500–900 hPa column averaged $\vec{Q}$-vector forcing in the occluded quadrant of a post-mature mid-latitude cyclone. It is clear that the $\vec{Q}_s$ component (Figure 8.22b) far exceeds the $\vec{Q}_n$ component (Figure 8.22c) in that region. It turns out that the $\vec{Q}_s$ component characteristically far exceeds the $\vec{Q}_n$ component in the vicinity of the occluded quadrant of mid-latitude cyclones.[8] Thus we find that rotation of $\nabla\theta$ by the geostrophic flow (which is described by $\vec{Q}_s$) is the underlying dynamical mechanism responsible for creating the occluded thermal structure and for forcing the QG ascent associated with that process in the occluded quadrant of cyclones.

Upward vertical motion is, of course, associated with adiabatic cooling. The upward vertical motion maximum that occupies the axis of the occluded thermal ridge, therefore, contributes to a local maximum in adiabatic cooling in the thermal ridge. Naturally, this cooling will tend to erode the thermal ridge. In the preceding argument, however, we have not included any of the effects of LHR. Clearly, the release of latent heat in the updraft associated with the occluded thermal ridge will mitigate against the adiabatic cooling that would otherwise erode that feature. Thus, we might suspect that LHR is an essential component of the development of occluded thermal structures in the mid-latitude atmosphere. This suspicion turns out to be correct and is best demonstrated within the framework of a potential vorticity view of the cyclone life cycle to be presented in Chapter 9. Before we develop that view, however, we first consider some dynamical aspects of the decay stage of the cyclone life cycle.

---

[8] The reader is referred to the study by Martin (1999).

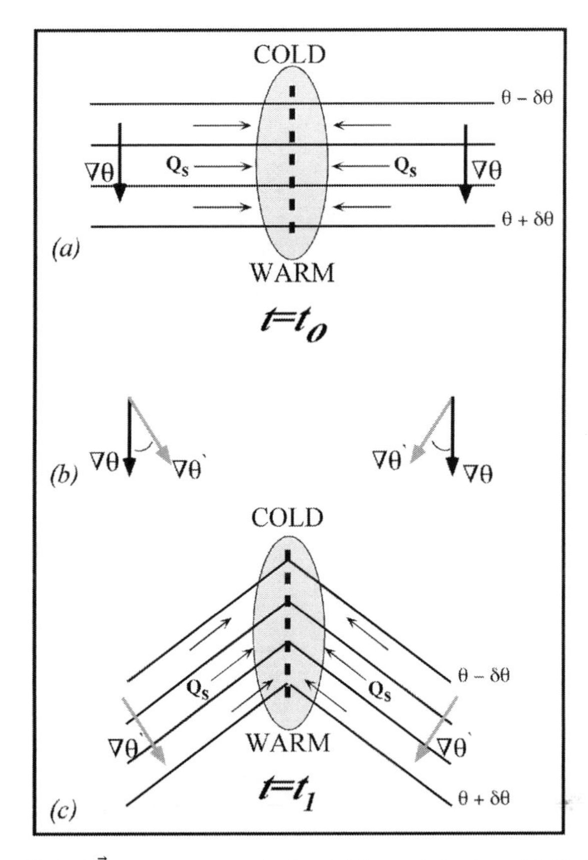

Figure 8.21 The effect of $\vec{Q}_s$ convergence on horizontal thermal structure. (a) Straight line isentropes (solid lines) in a field of $\vec{Q}_s$ convergence (shading). The thick dashed line indicates the axis of maximum $\vec{Q}_s$ convergence. The direction of $\nabla\theta$ vector on either side of the $\vec{Q}_s$ convergence maximum is indicated. (b) Rotation of $\nabla\theta$ vector implied by $\vec{Q}_s$ vectors on either side of the $\vec{Q}_s$ convergence maximum in (a). The thick black arrow denoted as $\nabla\theta$ represents the original direction of the $\nabla\theta$ vector. The thick gray arrow denoted as $\nabla\theta'$ represents the direction of $\nabla\theta$ vector after rotation implied by $\vec{Q}_s$ vectors. (c) Orientation of the baroclinic zone depicted in (a) after differential rotation of $\nabla\theta$ on either side of the $\vec{Q}_s$ convergence maximum

8.8 The Decay Stage

The decay stage of the extratropical cyclone is the least studied, and therefore the least well-understood stage of the cyclone life cycle. At a basic level, the decay stage is associated with lower tropospheric geopotential and sea-level pressure rises and, consequently, with a systematic decrease in lower tropospheric vorticity. Therefore, cyclone decay is known as cyclolysis – the opposite of cyclogenesis. We have already seen that cyclogenesis requires column stretching and upward vertical motions. It therefore seems reasonable to assume that cyclolysis requires column squashing and downward vertical motions. Certainly this set of physical circumstances represents a

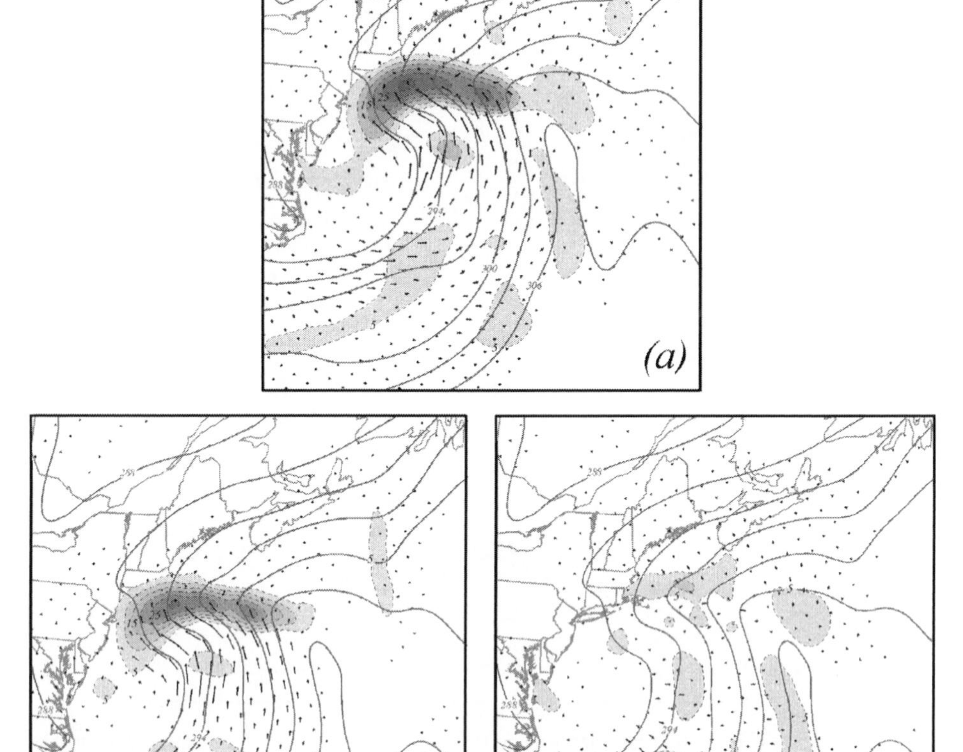

Figure 8.22 (a) The 500–900 hPa column averaged $\vec{Q}$-vectors and $\vec{Q}$-vector convergence from an 18 h forecast of the NCEP Eta model valid at 0600 UTC 1 April 1997. $\vec{Q}$ convergence is contoured and shaded in units of m kg^{-1} s^{-1} every 5×10^{16} m kg^{-1} s^{-1} beginning at 5×10^{-16} m kg^{-1} s^{-1}. Thin dashed gray lines are 500–900 hPa column-averaged isentropes labeled in K and contoured every 3 K. (b) As in (a) except for $\vec{Q}_s$. (c) As in (a) except for $\vec{Q}_n$

sufficient means of reducing the lower tropospheric vorticity; it does not, however, appear to be necessary for the occurrence of surface cyclolysis.

Recall that the vertical structure of a developing mid-latitude cyclone was such that the axis of minimum geopotential height tilted into the vertical shear (i.e. to the west as shown in Figure 8.7b). Since upward vertical motions occur downshear of upper-level vorticity maxima (i.e. minima in the upper-level geopotential height), that vertical structure ensures that an upper-level, dynamically forced divergence maximum is located directly above the sea-level pressure minimum and, via the resulting upward vertical motions, serves to evacuate the mass that accumulates into the sea-level pressure minimum as a consequence of frictionally induced surface convergence. As the cyclone matures, the vertical tilt of the geopotential minimum axis gradually becomes more vertical by the time of occlusion. A purely vertical stacking results in the displacement of the upper divergence maximum to the east of the sea-level pressure minimum. By the commencement of decay, the sea-level pressure minimum has reached its greatest intensity as has the frictionally induced surface convergence into its center. As a consequence of the eastward displacement of the upper-level divergence at this stage of the life cycle, there is no mechanism available to evacuate the accumulating mass near the center of the surface cyclone and the surface pressure rises as a consequence. This rise in surface pressure is associated with a decrease in the near surface geostrophic vorticity and therefore qualifies as a cyclolysis event. Note that this sequence of events can occur in the absence of any notable downward vertical motions over the surface cyclone center. Instead, it is the *absence of upward vertical motions sufficient to evacuate the mass accumulated near the center of the surface cyclone* that appears to be the dynamically necessary ingredient for cyclone decay. Indeed, any process that results in decreased upper-level divergence directly above the surface cyclone center leads to cyclone decay.

The results of a recently constructed synoptic climatology of surface cyclolysis in the north Pacific Ocean can be used to illustrate these characteristic elements of the decay stage. In particular, we examine the composite evolutions of the 500 hPa geopotential height and sea-level pressure distributions constructed from 180 so-called rapid cyclolysis periods (RCPs), defined as 12 h periods during which a sea-level pressure rise of at least 12 hPa occurs at the center of a mid-latitude cyclone.[9]

Twenty-four hours before the commencement of rapid cyclolysis, a fairly intense sea-level pressure minimum is located just downstream of a strongly curved, slightly negatively tilted, 500 hPa geopotential height trough (Figure 8.23a). Twelve hours later, the upper trough axis has become more negatively tilted and the more intense sea-level pressure minimum has drawn closer to the trough axis, characteristic of occluded cyclones (Figure 8.23b).

By commencement of the 12 h period of rapid cyclolysis (Figure 8.23c), the sea-level pressure minimum lies directly beneath the 500 hPa geopotential height minimum of the even more negatively tilted trough. An astounding transformation in the 500 hPa geopotential height field occurs during the 12 h RCP. The radius of curvature of the geostrophic streamlines increases dramatically while the 500 hPa geopotential height gradient weakens south of the dramatically weaker sea-level pressure minimum (Figure 8.23d). The rapid flattening of the 500 hPa trough–ridge couplet, which had been amplifying up to the commencement of cyclone decay, is associated with

[9] Such RCPs are relatively rare events, occurring in less than 7% of all mid-latitude cyclones according to the study of Martin *et al.* (2001) referenced here.

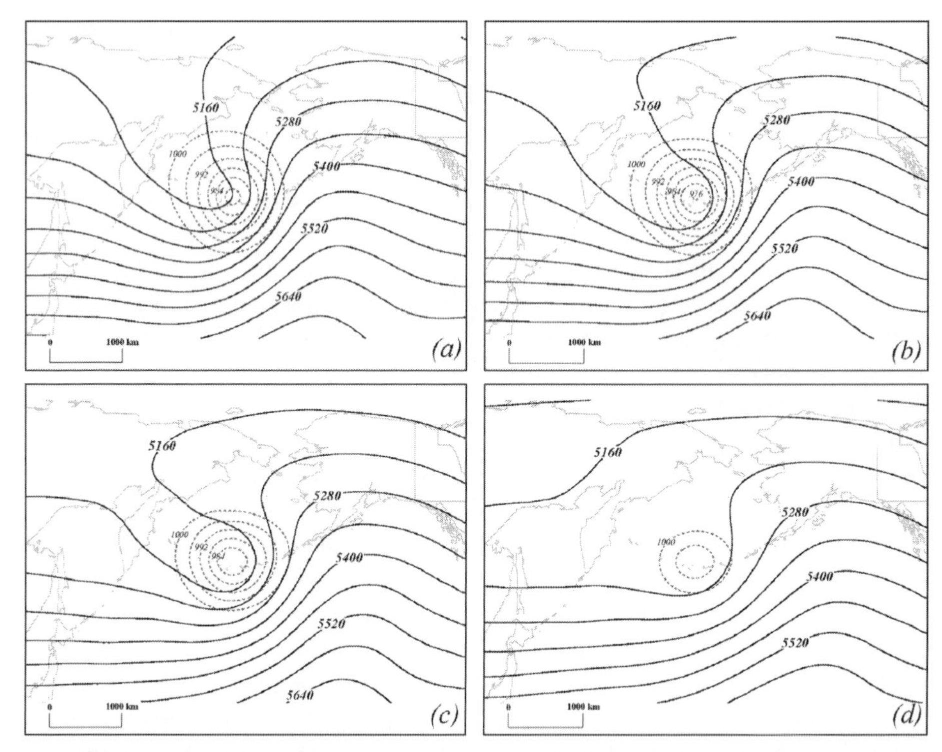

Figure 8.23 (a) Composite 500 hPa geopotential height (solid black lines) and sea-level isobars (gray dashed lines) (from 180 rapid surface cycloysis events observed in the north Pacific Ocean) valid 24 h before the commencement of surface decay. Geopotential height is labeled in m and contoured every 60 m. Sea-level pressure is labeled in hPa and contoured every 4 hPa up to 1000 hPa. (b) As for (a) but valid 12 h before the commencement of surface decay. (c) As (a) but valid at the commencement of surface decay. (d) As for (a) but valid 12 h after the commencement of surface decay. The geographical background map is given to provide scale only. Adapted from Martin *et al.* (2001)

a rapid decrease in upper tropospheric divergence downstream of the upper trough axis (i.e. to the northeast of the sea-level pressure minimum). Such a circumstance, occurring immediately after the surface cyclone reaches its maximum intensity, provides the key ingredient for the subsequent rapid cyclolysis at the surface. As the surface cyclone reaches its greatest intensity, presumably so does the lower tropospheric mass convergence into it, forced by friction in the lower troposphere. With the abrupt reduction in cyclonic curvature and, consequently, in the mass divergence aloft, the accumulating mass in the lower troposphere is less efficiently evacuated from the column and the sea-level pressure rises rapidly as a result. Through subsequent study of these events,[10] it appears that rapid surface cyclolysis, though influenced by

[10] The reader is referred to McLay and Martin (2002) for the complete study from which this conclusion is drawn.

friction in the boundary layer, is initiated and largely controlled by synoptic-scale dynamical processes. Less intense 'garden variety' cyclolysis events most likely proceed in a similar fashion relying more on the gradual acquisition of a downshear tilted structure, characteristic of all cyclones in the post-mature phase, than on the eradication of upper tropospheric flow curvature.

Examination of individual examples of particularly rapid surface cyclolysis has revealed that the surface decay in such cases is associated with a rapid erosion of the associated upper tropospheric short-wave disturbance. There are many different means of accomplishing this end in the atmosphere. Use of the potential vorticity perspective, which we will develop more fully in the next chapter, will provide us with an additional, powerful tool for examining the nature of both ordinary and rapid cyclone decay.

Selected References

Bjerknes and Solberg (1922) is the paper that introduces the NCM.

Bluestein, *Synoptic-Dynamic Meteorology in Midlatitudes, Volume I*, contains a thorough discussion of the QG height tendency equation.

Holton, *An Introduction to Dynamic Meteorology*, discusses the QG height tendency equation as well.

Martin (2006) examines the roles of the shearwise and transverse QG vertical motions in the mid-latitude cyclone life cycle.

Sutcliffe and Forsdyke (1950) introduce the concept of 'self-development'.

Palmén and Newton, *Atmospheric Circulation Systems*, contains a number of illustrative examples of cyclone life cycles.

Martin (1999) explores the QG forcing for ascent in the occluded quadrant of mid-latitude cyclones.

Schultz and Mass (1993) provide a comprehensive list of references on the occlusion process.

Posselt and Martin (2004) examine the effect of LHR on the development of warm occluded thermal structures in mid-latitude cyclones.

Martin and Marsili (2002) undertake a synoptic case study of rapid surface cyclolysis.

Problems

8.1. Briefly describe the major differences between modern understanding of the frontal cyclone and the early ideas presented by Bjerknes and Solberg (1922) with respect to the following:

(a) The 'polar front' and its role in cyclogenesis.
(b) The nature of fronts themselves.
(c) The nature of the precipitation distribution in frontal regions.
(d) The production of vertical motions in frontal regions.
(e) The relationship between cyclogenesis and frontogenesis.

8.2. The typical cloud distribution associated with a mid-latitude cyclone is illustrated in Figure 8.1A. Explain how the latent heat release associated with this distribution reinforces the differential geostrophic temperature advection forcing in the quasi-geostrophic height tendency equation.

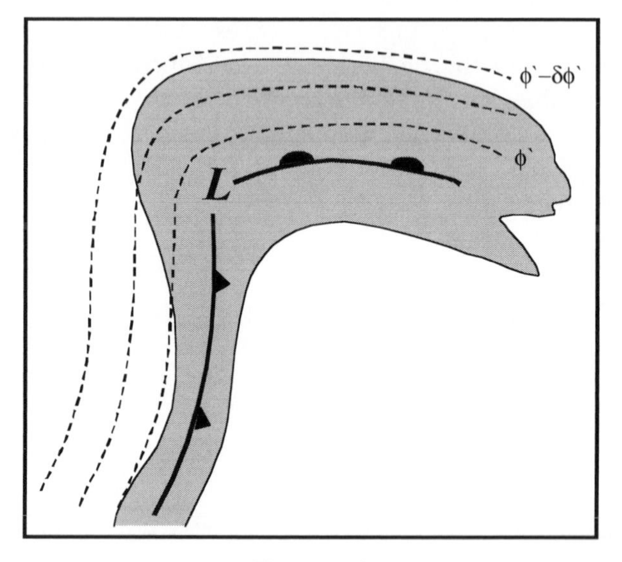

Figure 8.1 A

8.3. A portion of Figure 8.7(b) describes the vertical structure of a *developing* mid-latitude cyclone in terms of the locations of the SLP minimum and the corresponding upper tropospheric geopotential minimum.

(a) Draw a similar idealized picture for a cyclone that is occluded.
(b) Why is it reasonable to suggest that the point of occlusion represents the commencement of decay?
(c) Are strong downward vertical motions necessary for surface cyclone decay? Explain your answer.
(d) Draw a similar idealized picture of a decaying cyclone.
(e) Explain why the tilt of the geopotential minimum with height is so crucial to diagnosing the life cycle stage of a cyclone.

8.4. According to the classical definition of warm and cold occlusions, the type of occlusion that is expected to form in a given cyclone depends upon whether the air mass poleward of the warm front or upstream of the cold front is colder. Use the result from Problem 7.13, along with a basic analysis characteristic of the surface occluded front, to determine what physical parameter actually controls the slope of an occluded front. How does this answer compare to the classical view? Is there a physical reason underlying the observation that warm occlusions account for nearly all occluded structures? Explain.

8.5. Figure 8.2A shows a surface cyclone, its associated precipitation and cloud shield, and a 500 hPa geopotential height contour at $t = 0$.

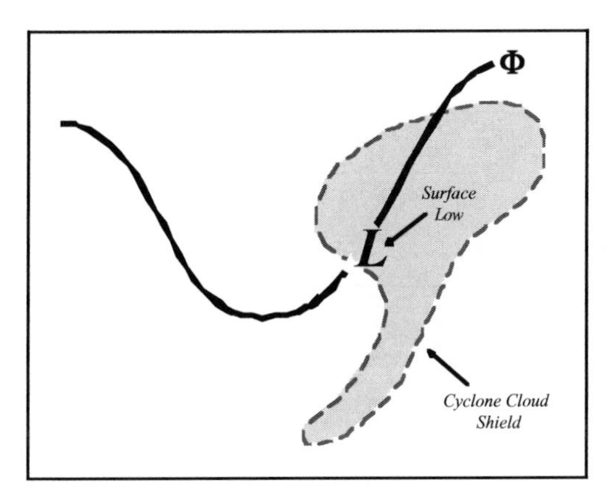

Figure 8.2A

(a) Qualitatively sketch the 500 mb geopotential height line at some later time, demonstrating how it is altered by latent heat release. Explain your reasoning.
(b) How does the diabatically altered 500 hPa trough affect the magnitude of the vorticity maximum at that level? Explain your answer.
(c) Will the diabatically altered 500 mb trough increase or decrease the intensity of cyclogenesis at the surface? Why?

8.6. If the Earth were completely dry (i.e. no water substance in any form existed in its atmosphere) would explosive cyclogenesis still be possible at middle latitudes? Defend your answer.

8.7. The Norwegian Cyclone Model hypothesized that the polar front was a necessary condition for cyclogenesis suggesting that frontogenesis must precede cyclogenesis. Later research suggested that frontogenesis and cyclogenesis are concurrent processes. With reference to Figure 8.16, describe why it might be even more accurate to say that cyclogenesis slightly precedes frontogenesis. Explain your reasoning.

8.8. Figure 8.3A is a subjectively analyzed vertical cross-section of θ_e through a warm occlusion in the central United States at 0000 UTC 20 January 1995. What θ_e surface lies at the warm edge of both the warm and cold frontal zones constituting this occluded thermal structure? Describe how you would construct the isobaric topography of this θ_e surface given gridded numerical model output for this case. What feature of interest would clearly emerge from such an exercise?

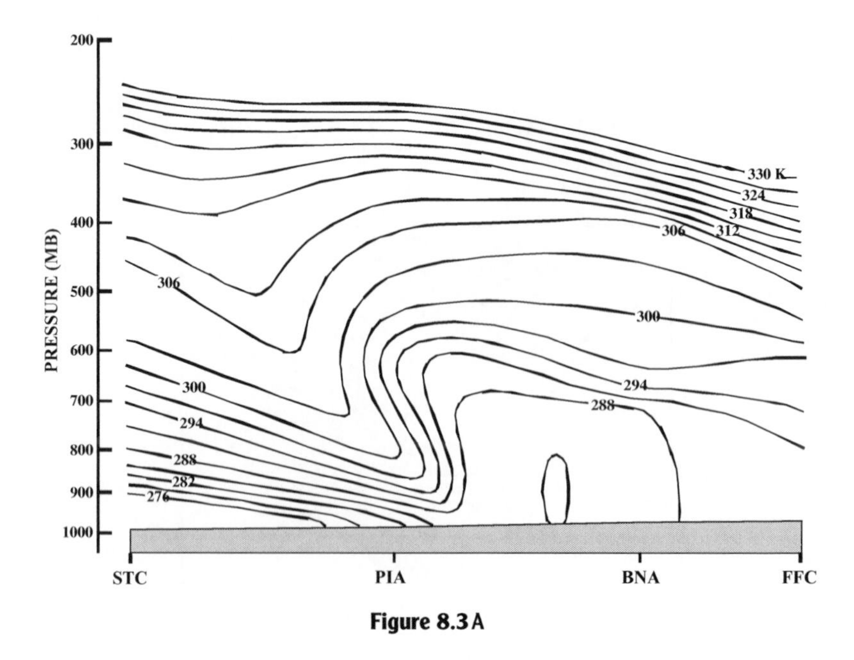

Figure 8.3 A

8.9. Describe how the process of cyclolysis is physically distinct from that of anti-cyclogenesis. Specifically consider the nature of the dynamical and diabatic processes that characterize each phenomenon.

8.10. In the middle latitudes, surface cyclones propagate poleward while surface anticyclones propagate equatorward. Give a physical explanation of this basic observational characteristic of mid-latitude weather systems.

8.11. On 21 January 2000 an intense surface cyclone developed south of Nova Scotia. Listed below are surface observations from the LeHave Bank buoy at 11 a.m. and 1 p.m. Local Standard Time (LST). Assuming that the wind makes a 20° cross-isobar angle at all times and that the distance from the sea-level pressure minimum to the buoy is the same at 11 a.m. as it is at 1 p.m., answer the following questions:

(a) What path did the sea-level pressure minimum take relative to the buoy over the 2 hour interval? Explain your answer.

(b) What was the minimum sea-level pressure of this cyclone during the same time interval? Explain your answer.

Time (LST)	Temperature	Wind direction	Wind speed	SLP
11 a.m.	8.4°C	100°	10 m s^{-1}	950.4 hPa
1 p.m.	6.5°C	270°	15 m s^{-1}	951.0 hPa

8.12. Show that

$$\frac{\vec{Q}\cdot(\hat{k}\times\nabla\theta)}{|\nabla\theta|}=\frac{f_0\gamma\,|\nabla\theta|}{2}\left[\frac{2F_{1_g}\dfrac{\partial\theta}{\partial x}\dfrac{\partial\theta}{\partial y}+F_{2_g}\left(\dfrac{\partial\theta}{\partial y}^2-\dfrac{\partial\theta}{\partial x}^2\right)}{|\nabla\theta|^2}\right]+\frac{f_0\gamma\,|\nabla\theta|\,\zeta_g}{2}$$

where $\vec{Q}$ is the $\vec{Q}$-vector, and F_{1_g} and F_{2_g} are the geostrophic stretching and shearing deformations, respectively.

Solutions

8.1. (a) From 205° to 25°, roughly from SSW to NNE (b) $p_{min} = 949.2\,\text{hPa}$

9

Potential Vorticity and Applications to Mid-Latitude Weather Systems

Objectives

Thus far we have considered the dynamics of the middle latitudes from what might be termed the 'basic state variables' perspective in which a number of separate variables (pressure/geopotential, temperature, omega) are considered simultaneously in the context of the physical relationships and mathematical expressions that relate them. In the quasi-geostrophic system which has formed the basis of this book, considerable diagnostic power is available by simply keeping track of the geopotential at a variety of levels as we have seen in the development of the QG omega and height tendency equations. In this chapter we will discover that knowledge of the distribution of a single variable, the so-called potential vorticity, enables us to develop alternative, but equivalent, understanding of the dynamical processes operating in the mid-latitude atmosphere.

Our investigation begins by exploring the curious relationship between vorticity and static stability in the isentropic coordinate system. We will find that the definition and diagnostic properties of potential vorticity are a straightforward extension of this physical connection. Next, we consider the characteristic kinematic and thermodynamic structure of the environments associated with positive and negative anomalies in the potential vorticity distribution. This discussion leads us to a conceptualization of the process of cyclogenesis from the potential vorticity perspective which will include consideration of the influence of diabatic processes, particularly those associated with latent heat release. Finally, we will consider some additional applications of the potential vorticity perspective. We begin this pursuit with an investigation of the effect of horizontal divergence in isentropic coordinates.

Mid-Latitude Atmospheric Dynamics Jonathan E. Martin

9.1 Potential Vorticity and Isentropic Divergence

Assume that a given flow is adiabatic and that we will describe that flow in isentropic coordinates. One might reasonably wonder what will be the effects of horizontal divergence on the fluid when viewed in these isentropic coordinates. Based on the invariance of both the divergence and vorticity, and their physical connection as manifest in the vorticity equation, one effect of horizontal divergence in isentropic coordinates is to change the relative vorticity. This effect can be expressed in the isentropic coordinate form of the vorticity equation

$$\frac{d(\zeta_\theta + f)}{dt} = -(\zeta_\theta + f)(\nabla \cdot \vec{V}_\theta). \tag{9.1}$$

Next, consider the hypothetical isentropic column of air portrayed in Figure 9.1(a). If horizontal convergence into this column occurs, there must be an increase in the mass of the column. The mass of the column is directly related to the pressure interval between the bounding isentropic surfaces ($M = -\delta p/g$). Thus, the pressure interval δp between the isentropes θ and $\theta + \delta\theta$ must be increased in the face of horizontal convergence as illustrated in Figure 9.1(b). As a consequence, the ratio $-(1/g)\delta p/\delta\theta$ becomes larger in the indicated column. This relationship underlies the expression for the continuity equation in isentropic coordinates

$$\frac{d}{dt}\left(-\frac{1}{g}\frac{\partial p}{\partial \theta}\right) = -\left(-\frac{1}{g}\frac{\partial p}{\partial \theta}\right)(\nabla \cdot \vec{V}_\theta). \tag{9.2}$$

Now, letting

$$\sigma = -\frac{1}{g}\frac{\partial p}{\partial \theta}$$

for simplicity of notation, we can isolate the expressions for $\nabla \cdot \vec{V}_\theta$ from both (9.1) and (9.2) yielding

$$-\nabla \cdot \vec{V}_\theta = \frac{d \ln(\zeta_\theta + f)}{dt} \quad \text{and} \quad -\nabla \cdot \vec{V}_\theta = \frac{d \ln \sigma}{dt},$$

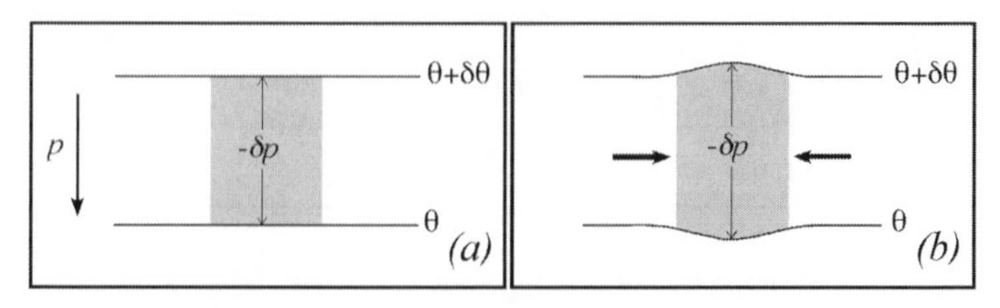

Figure 9.1 (a) Column of air (shaded) confined between two isentropes. The pressure interval between the bounding isentropic surfaces is $-\delta p$. (b) Horizontal convergence of air (represented by the heavy arrows) increases the pressure interval between the two isentropes

respectively. Equating these two expressions we get

$$\frac{d\ln(\zeta_\theta + f)}{dt} = \frac{d\ln\sigma}{dt} \tag{9.3}$$

which can be solved by first multiplying by dt to get

$$d\ln(\zeta_\theta + f) = d\ln\sigma \quad \text{or} \quad \frac{d(\zeta_\theta + f)}{(\zeta_\theta + f)} = \frac{d\sigma}{\sigma}. \tag{9.4}$$

Integration of (9.4) proceeds from

$$\int_{(\zeta_\theta+f)_0}^{(\zeta_\theta+f)} \frac{d(\zeta_\theta + f)}{(\zeta_\theta + f)} = \int_{\sigma_0}^{\sigma} \frac{d\sigma}{\sigma}$$

where the subscript indicates an initial value of the indicated variable. The integration yields

$$\ln\frac{(\zeta_\theta + f)}{(\zeta_\theta + f)_0} = \ln\frac{\sigma}{\sigma_0} \quad \text{or} \quad \frac{(\zeta_\theta + f)}{(\zeta_\theta + f)_0} = \frac{\sigma}{\sigma_0}$$

which can be rearranged into

$$\frac{(\zeta_\theta + f)}{\sigma} = \frac{(\zeta_\theta + f)_0}{\sigma_0}.$$

The above expression demonstrates that, for adiabatic flow, the quantity

$$(\zeta_\theta + f) \Big/ -\frac{1}{g}\frac{\partial p}{\partial \theta}$$

is a constant. We will call the related expression

$$-g(\zeta_\theta + f)\left(\frac{\partial \theta}{\partial p}\right) \tag{9.5}$$

the **isentropic potential vorticity** (IPV). It is clear from (9.5) that the IPV (or PV for short) is a product of the absolute vorticity and the static stability. Potential vorticity derives its name from the fact that there is a potential for creating relative vorticity by changing latitude (through manipulation of f) and by adiabatically changing the separation between isentropic layers (through modification of $-\partial\theta/\partial p$). Why such a product should be conserved is rather a mystery at first glance. Since it is derived by combining the vorticity and continuity equations, PV describes a mass-weighted circulation and conservation of PV suggests that a parcel may exchange stratification for circulation or vice versa, but that the stratification times the circulation will not change so long as no flow is permitted across isentropic surfaces (i.e. the flow remains adiabatic).

There are two important, exploitable characteristics of PV that bear mentioning here. The first of these we have already considered – the conservative nature of PV.

For adiabatic, frictionless flow a parcel of air will retain its value of PV forever. Thus, if the PV distribution in a given domain is known at some initial time, and it is known that the flow in that domain is adiabatic, any subsequent changes in the distribution of PV in that domain must have occurred as a result of advection of PV. Conversely, if the flow is not adiabatic, as is the case in the real atmosphere, some component of the change in the PV distribution in the domain must have resulted from frictional generation/dissipation or diabatic heating of some kind. We will exploit this property of PV later in this chapter.

The second important property of PV is that it is *invertible*. This means that a lot of information about the characteristics of a given flow exists in the PV distribution of the flow. This is a consequence of the fact that knowledge of the vorticity field proceeds from information about the horizontal winds, u and v. Similarly, knowledge of the static stability proceeds from information about the vertical distribution of temperature. The hydrostatic relationship allows vertical temperature information to be converted into knowledge about the geopotential height field, ϕ. Knowledge of the ϕ and (u, v) fields provides information concerning the ageostrophic wind distribution, and, consequently, ω itself. With appropriate specification of the conditions on the boundary of a domain, those characteristics can be retrieved from knowledge of the PV distribution so long as independent knowledge of a relationship between the mass and momentum fields is known as well. Thus, the information contained in PV can be recovered given (1) a knowledge of the distribution of PV in a given domain, (2) knowledge of the boundary conditions on that domain, and (3) a balance condition within the domain that relates the mass to the momentum field. For mid-latitude flows on Earth, the primary example of a balance condition is the geostrophic balance but there are others (i.e. gradient wind balance). The concept of invertibility, and the importance of boundary conditions, is best demonstrated with a simple example.

Let us consider the barotropic vorticity equation expressed as

$$\frac{d\eta}{dt} = \frac{\partial \eta}{\partial t} + \vec{V}_\psi \cdot \nabla \eta = 0 \tag{9.6}$$

where $\vec{V}_\psi = \hat{k} \times \nabla \psi$, η is the absolute vorticity ($\eta = \nabla^2 \psi$), and ψ is a non-divergent streamfunction describing the flow. In such a case the balance condition is geostrophy. Imagine a limited domain in which only η is known within that domain. Since $\eta = \nabla^2 \psi$, it is possible to solve for ψ under these circumstances. However, in the absence of knowledge of the values of ψ along the boundaries of the limited domain, there are many solutions that will satisfy the condition that $\eta = \nabla^2 \psi$ and there is therefore no unique solution for ψ. A simple way to demonstrate this is to consider the case in which $\eta = 0$ everywhere in the limited domain. In such a case, there can be no curvature or horizontal shear in the flow, meaning that the streamlines must be everywhere parallel and equally spaced. As suggested in Figure 9.2, there are an infinite number of solutions that satisfy these characteristics. Given knowledge of the boundary conditions on the limited domain, however, there is only one solution

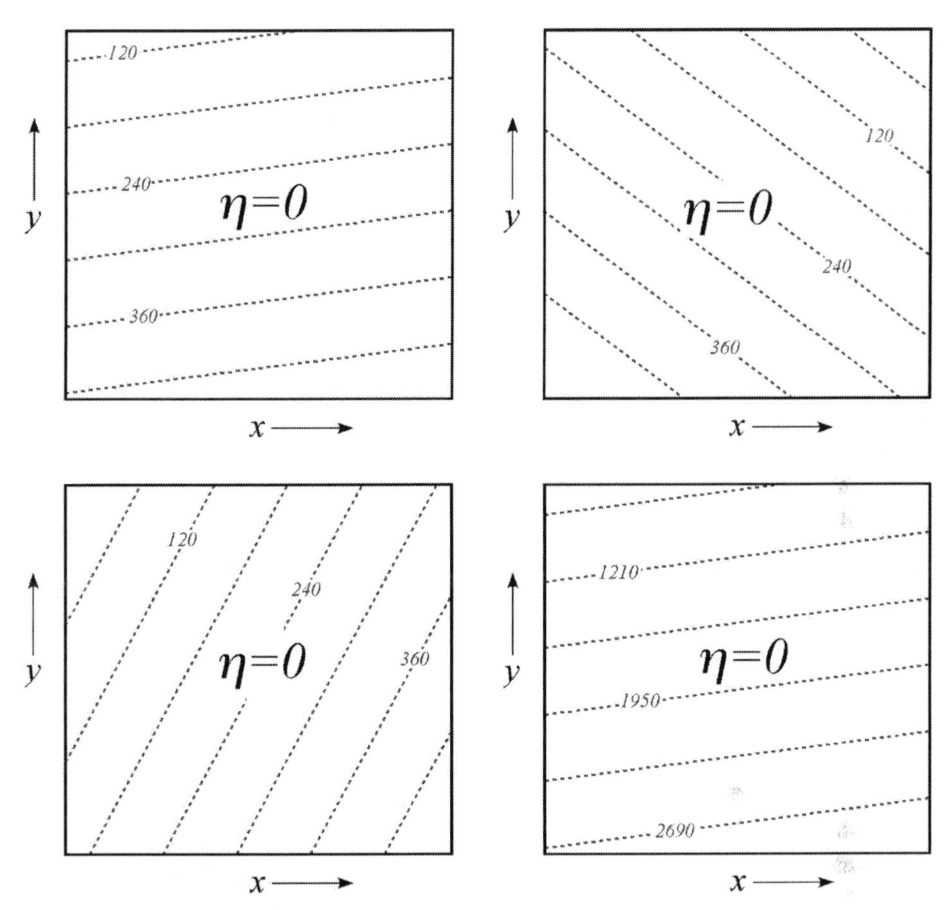

Figure 9.2 A series of four different solutions to the barotropic vorticity equation ($\eta = 0$) in a limited domain with unspecified boundary conditions

that simultaneously satisfies the balance condition (i.e. $\eta = \nabla^2\psi$) and the boundary conditions. In this one case, a unique solution for ψ exists.

Once a solution for the ψ field is found in this way, it can be used to calculate the balanced wind in the domain which can then be used to advect η to the next time step. Once the new η field is produced by this advection, it can be inverted for the new ψ field (provided one has knowledge of the *updated* conditions on the boundary of the domain), and so on. In this manner, the η field is invertible given (1) global (or domain-wide) knowledge of η, (2) a balance condition relating η to the mass field (i.e. $\eta = \nabla^2\psi$, geostrophy), and (3) knowledge of boundary conditions for ψ. The exact same thing is true about PV, though the inversion is much more complicated than for the barotropic vorticity equation. If the global distribution of PV is known, along with a balance condition and boundary conditions, it is possible to invert PV to determine values of ϕ, u, v, T, ω, and static stability within the domain.

Local anomalies in the distribution of PV (i.e. departures from a long time/large-scale average) are features of greatest interest since they have associated with them identifiable and discrete circulations. Consequently, in exploiting the property of invertibility we want to focus on these *PV anomalies.* Before examining the nature of PV inversion and the insight that it provides concerning mid-latitude weather systems, we must investigate the characteristic kinematic and thermodynamic structure of the environments associated with these PV anomalies; specifically, that associated with a **positive PV anomaly**.

9.2 Characteristics of a Positive PV Anomaly

A schematic of an upper-level, positive PV anomaly is shown in Figure 9.3. The anomaly itself is drawn with a + sign at 300 hPa and represents a local region in which PV is larger than the local spatial or temporal average. In more formal terms, this PV anomaly represents a location at which the product $-(\zeta_\theta + f)\partial\theta/\partial p$ exceeds the local average. This could mean one of three things: (1) the vorticity is larger than average, (2) the static stability is larger than average, or (3) both the vorticity and static stability are larger than average. In order to determine which of these three possibilities is correct, we will perform the following thought experiment. Let us try to construct a positive PV anomaly that is manifest entirely as a vorticity anomaly in an atmosphere in thermal wind balance. Using a copy of the schematic from Figure 9.3, we see that, since the PV anomaly is maximized at 300 hPa, the winds must also be at their maximum at that level (Figure 9.4a). Thus, the winds in the column straddling the positive PV anomaly in our example must be increasing with height beneath 300 hPa. Given the assumed thermal wind balance, this implies that

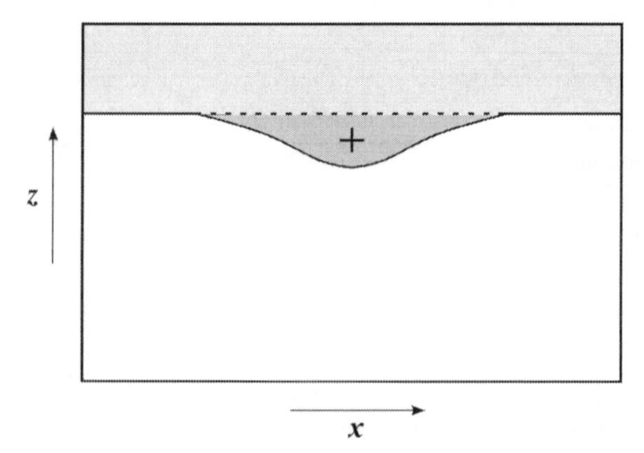

Figure 9.3 Schematic of an upper tropospheric positive PV anomaly. The darker shaded region with the + sign indicates the anomaly. The lighter shaded region represents the lower stratosphere and the unshaded region the troposphere

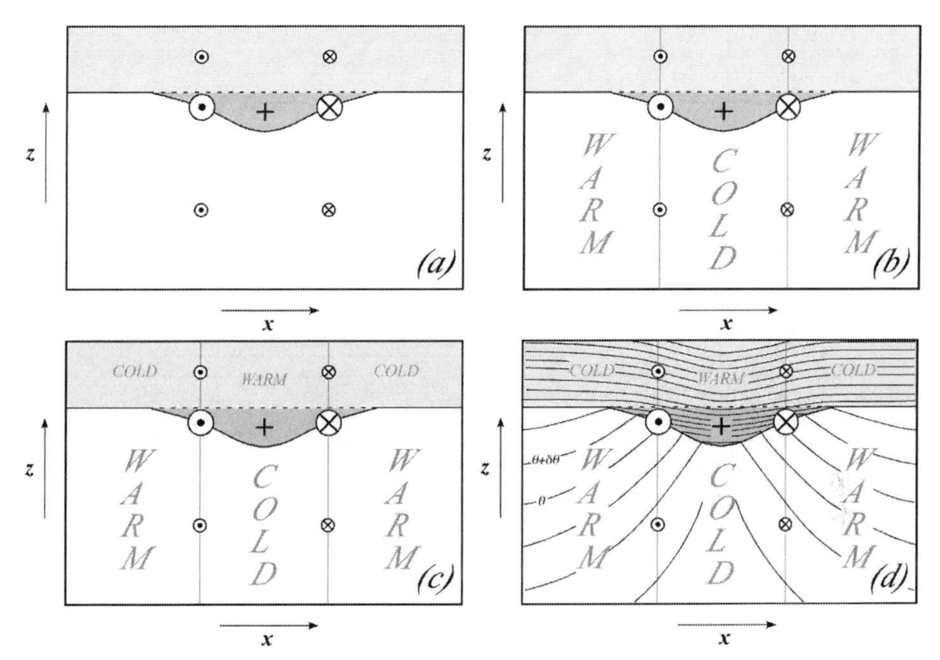

Figure 9.4 (a) Positive upper tropospheric PV anomaly characterized by a positive vorticity anomaly near the tropopause. Large circle with a cross (a dot) in it indicates wind into (out of) the page. Smaller such circles indicate lesser wind speeds in the indicated directions. (b) The resulting relative temperatures in the troposphere assuming thermal wind balance. (c) The resulting relative temperatures in the lower stratosphere assuming thermal wind balance. (d) The overall isentropic distribution in the cross-section. Note that the positive PV anomaly is a positive static stability anomaly as well as a positive vorticity anomaly

a relatively cold column of air must lie directly beneath the positive PV anomaly with a relatively warm ring surrounding it as shown in Figure 9.4(b). Conversely, since the winds must decrease with increasing height above 300 hPa, there must be a relatively warm column of air above the 300 hPa level with a ring of relatively colder air surrounding it as shown in Figure 9.4(c). Now, constructing schematic isentropes that conform to this relative distribution of temperatures shows that the positive PV anomaly must be characterized by both positive vorticity and positive static stability anomalies (Figure 9.4d). By extension, there is no ambiguity regarding the structure of any PV anomaly; it must be characterized by vorticity and static stability anomalies of the same sign!

Thus, we can identify characteristic structures and circulations associated with PV anomalies. A negative PV anomaly is characterized by anticyclonic flow, whose magnitude is maximized at the level of the anomaly but which extends throughout some depth of the atmosphere above and below the anomaly. The vertical extent of the circulatory influence of a PV anomaly is known as the penetration depth of the

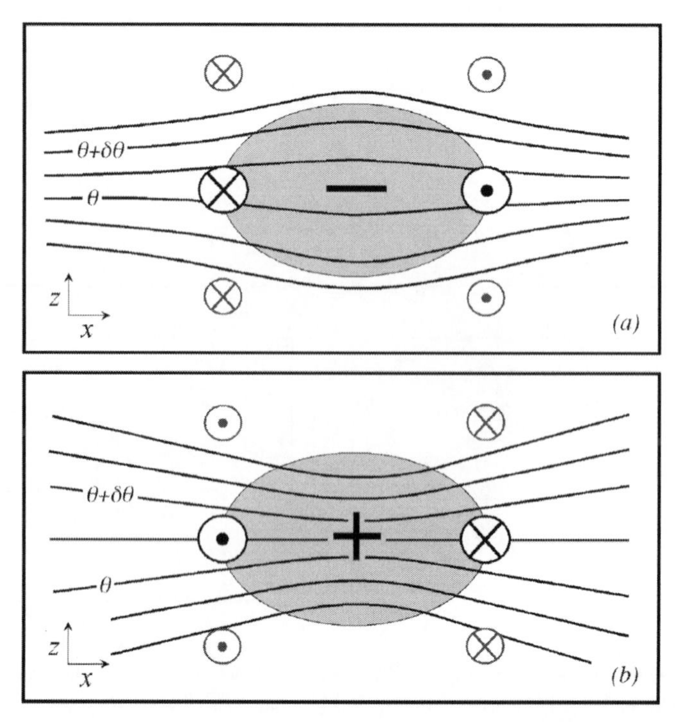

Figure 9.5 (a) Characteristic structure of a negative PV anomaly. Gray shaded area delineates the negative anomaly and thin solid lines are isentropes. Wind into the page is indicated by a cross while wind out of the page is signified by a dot. (b) Characteristic structure of a positive PV anomaly

anomaly and is given by

$$H = \frac{fL}{N} \tag{9.7}$$

where L is the characteristic length scale of the anomaly and N is the Brünt–Vaisala frequency. Thus, the penetration depth of a given PV anomaly varies as the scale of the anomaly and inversely as the ambient static stability. The negative anomaly will also be characterized by isentropes that bow *around* the anomaly, indicating the presence of a negative static stability anomaly in its vicinity (Figure 9.5a). A positive PV anomaly is characterized by cyclonic flow, whose magnitude is maximized at the level of the anomaly but which extends throughout some depth above and below the anomaly according to the constraints imposed by (9.7). The positive anomaly will be characterized by isentropes that bow *toward* the anomaly, indicating the presence of a positive static stability anomaly in its vicinity (Figure 9.5b).

The structure of an upper-level, positive PV anomaly can be exploited to gain insight into the relationship between positive PV advection and cyclogenesis. Imagine a situation in which there is initially barotropic flow (i.e. there is no $\nabla_p\theta$) and the isentropes are unique to individual isobars as shown in Figure 9.6(a). Now, if a

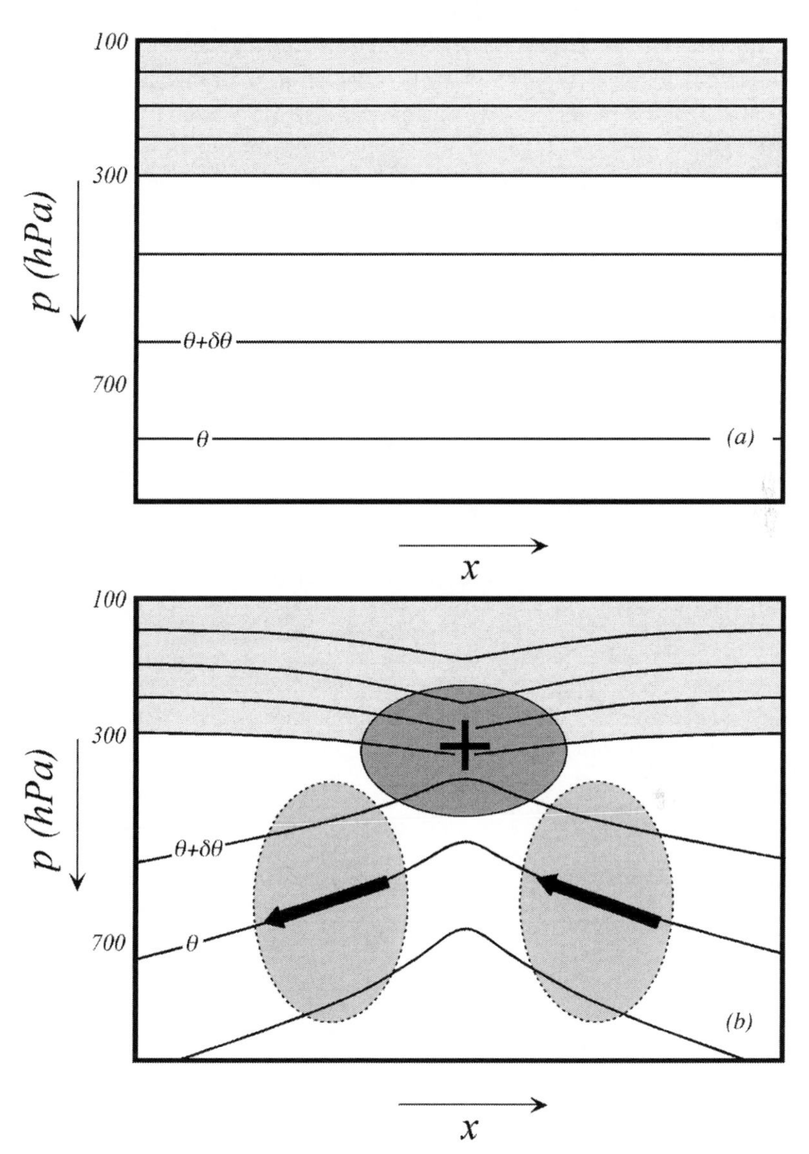

Figure 9.6 Illustration of the 'vaccuum cleaner' effect associated with a mobile positive PV anomaly near the tropopause. (a) Vertical cross-section of isentropes in a barotropic state. Light gray shading represents the stratosphere. (b) A westward-moving, positive anomaly intrudes into this environment and deforms the thermal structure. Heavy black arrows represent adiabatic flow relative to the moving PV anomaly which forces ascent downstream and descent upstream of the anomaly

positive, upper-level PV anomaly enters this domain from the west, the isentropes must deform so as to take on the characteristic structure of a positive PV anomaly. As the PV anomaly migrates eastward, adiabatic flow *relative to the PV anomaly* will head westward along the newly sloping isentropes. Thus, there will be upward vertical motion to the east of the anomaly and downward vertical motions to the west as shown schematically in Figure 9.6(b). This distribution of vertical motions is identically that which characterizes any mid-latitude synoptic short-wave disturbance: ascent (descent) downstream (upstream) of the maximum cyclonic vorticity!

Confirmation of this simple physical picture comes from extending our previous examination of the QG height tendency equation. Recall that when we left it (as (8.8)) we had rewritten the tendency equation as

$$\left(\nabla^2 + \frac{f_0^2}{\sigma}\frac{\partial^2}{\partial p^2}\right)\chi = -f_0\vec{V}_g \cdot \nabla\left(\frac{1}{f_0}\nabla^2\phi + f\right) - \frac{f_0^2}{\sigma}\left(\vec{V}_g \cdot \nabla\left(\frac{\partial^2\phi}{\partial p^2}\right)\right). \tag{9.8}$$

Recalling that $\chi = \partial\phi/\partial t$ and noting that both terms on the RHS of (9.8) involve geostrophic advection, (9.8) can be rewritten as

$$\frac{\partial}{\partial t}\left(\nabla^2\phi + \frac{f_0^2}{\sigma}\frac{\partial^2\phi}{\partial p^2}\right) = -\vec{V}_g \cdot \nabla\left(\nabla^2\phi + ff_0 + \frac{f_0^2}{\sigma}\frac{\partial^2\phi}{\partial p^2}\right). \tag{9.9a}$$

Note that by adding $\partial(ff_0)/\partial t$ (which is equal to zero) to the LHS of (9.9a) we arrive at the equivalent expression

$$\frac{\partial}{\partial t}\left(\nabla^2\phi + ff_0 + \frac{f_0^2}{\sigma}\frac{\partial^2\phi}{\partial p^2}\right) = -\vec{V}_g \cdot \nabla\left(\nabla^2\phi + ff_0 + \frac{f_0^2}{\sigma}\frac{\partial^2\phi}{\partial p^2}\right). \tag{9.9b}$$

If we now divide both sides of (9.9b) by f_0, we get

$$\frac{\partial}{\partial t}\left(\frac{1}{f_0}\nabla^2\phi + f + \frac{f_0}{\sigma}\frac{\partial^2\phi}{\partial p^2}\right) = -\vec{V}_g \cdot \nabla\left(\frac{1}{f_0}\nabla^2\phi + f + \frac{f_0}{\sigma}\frac{\partial^2\phi}{\partial p^2}\right). \tag{9.9c}$$

Next we define the QG potential vorticity (PV_g) as

$$PV_g = \left(\frac{1}{f_0}\nabla^2\phi + f + \frac{f_0}{\sigma}\frac{\partial^2\phi}{\partial p^2}\right) \tag{9.9d}$$

allowing us to rewrite (9.9c) as

$$\frac{\partial}{\partial t}(PV_g) = -\vec{V}_g \cdot \nabla(PV_g) \tag{9.10a}$$

or

$$\frac{d}{dt_g}(PV_g) = 0 \tag{9.10b}$$

where

$$\frac{d}{dt_g} = \frac{\partial}{\partial t} + \vec{V}_g \cdot \nabla.$$

Thus, the QG tendency equation can be interpreted as a statement that PV_g is conserved following adiabatic, geostrophic flow. Returning to the schematic in Figure 9.6(b), the region of upward vertical motion depicted there is clearly a region of positive PV advection. In the QG system, (9.10a) requires that such a region be characterized by a local increase in PV_g (i.e. that the LHS of the traditional QG tendency equation be positive). Whenever

$$\left(\nabla^2 + \frac{f_0^2}{\sigma}\frac{\partial^2}{\partial p^2}\right)\chi$$

is greater than zero, χ itself must be negative and so the heights must fall in the region of upward vertical motion shown in Figure 9.6(b). This is consistent with the effect of adiabatic cooling in the ascending air which reduces the column thickness and consequently lowers heights throughout the troposphere.

Thus far our discussion has been concerned only with *upper-level* PV anomalies. In nature, PV anomalies can occur at a variety of levels including at the surface. In order to develop a reasonable first description of cyclogenesis from the PV perspective, it is necessary to consider the structure of PV anomalies at the surface of the Earth. Figure 9.7(a) illustrates a warm potential temperature (θ) anomaly at the surface, such as might be observed ahead of a cold front. At the top of the atmosphere there

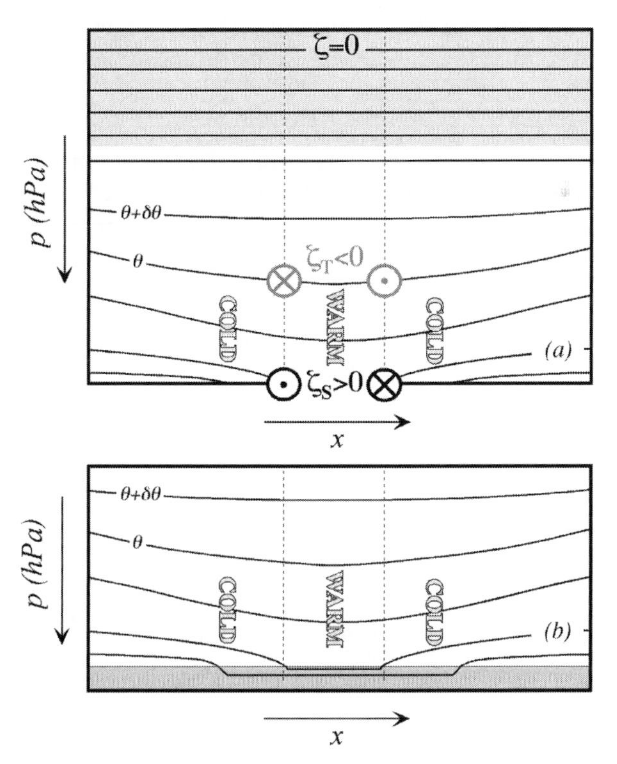

Figure 9.7 Illustration of the equivalence of a surface warm anomaly to a positive PV anomaly. (a) Surface warm anomaly produces a lower tropospheric warm column within which there is anticyclonic thermal vorticity (ζ_T, indicated by light gray circles). Since there is no vorticity at the top of the atmosphere, there must be positive vorticity at the surface (ζ_S, indicated by black circles). The stratosphere is shaded gray. (b) Positive static stability anomaly is created by connecting (underground) the isentropes that straddle the warm anomaly. The surface is shaded gray

is no vorticity since there is no wind. The contribution to the horizontal shear made by the warm core above the surface warm anomaly, however, is clearly anticyclonic. If there is no vorticity at the top of the atmosphere and there is anticyclonic *thermal* vorticity (i.e. anticyclonic thermal wind vorticity), then there must be a cyclone at the surface associated with the positive θ anomaly! There is also a maximum in static stability at the surface in association with the warm anomaly if one considers artificially connecting the isentropes that straddle the warm anomaly at the surface under the ground (Figure 9.7b). This mathematical trick, invented by F. P. Bretherton, allows us to consider positive (negative) low-level θ anomalies precisely as positive (negative) low-level PV anomalies. Now that we have some knowledge of the nature of PV anomalies near the surface and at upper levels (near the tropopause) we can offer a description of cyclogenesis from the PV perspective.

9.3 Cyclogenesis from the PV Perspective

The first step in developing a PV-based description of cyclogenesis is to consider separately the behavior of upper- and lower-level PV anomalies. A schematic upper-level PV anomaly is illustrated at three different times in Figure 9.8. At the initial time ($T = 0$), the anomaly is represented by an equatorward protuberance of high-PV air indicated by the + sign. This positive PV anomaly will be associated with a cyclonic circulation as indicated by the heavy arrows straddling the anomaly. The indicated circulation will have the effect of advecting high PV southward to the west of the anomaly and low-PV air northward to its east. Such advective tendencies have

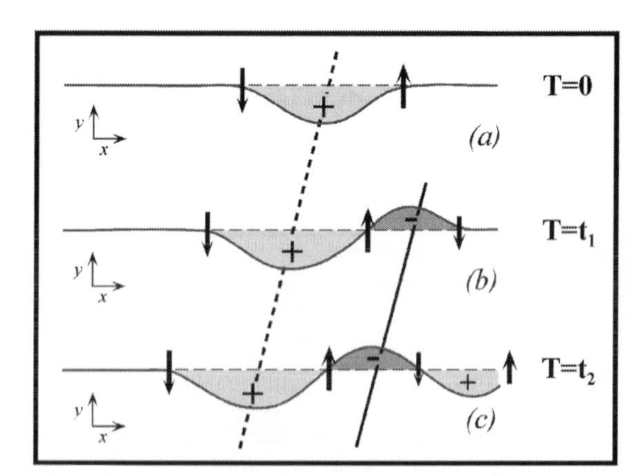

Figure 9.8 (a) Positive upper tropospheric PV anomaly in the northern hemisphere indicated by the + sign and gray shading. Solid gray line is a single PV contour and the arrows indicate the cyclonic circulation associated with the PV anomaly. (b) The same PV contour at some later time. Darker shaded area with a – sign indicates a negative upper tropospheric PV anomaly forced by northward advection of low PV to the east of the original positive anomaly. (c) The same PV contour at an even later time. The long dashed (solid) line is the phase line of the original positive (newly formed negative) PV anomaly

two obvious effects: (1) to propagate the initial anomaly *upstream* (i.e. to the west), and (2) to produce a negative PV anomaly to the east of the original feature, as indicated. The negative PV anomaly that develops downstream of the original PV anomaly also has a circulation associated with it indicated by the solid arrows in Figure 9.8(b). At a still later time, the original PV anomaly continues to propagate westward while the circulation associated with the negative PV anomaly spawns a secondary positive PV anomaly even further to the east (Figure 9.8c). Notice that the phase lines of the upper-level PV anomalies all suggest that the anomalies will propagate upstream if left to their own devices. The same is true of large-scale waves (i.e. Rossby waves) which propagate westward by virtue of the fact that the meridional gradient of the Coriolis parameter compels positive (negative) vorticity tendency via cyclonic (anticyclonic) planetary vorticity advection to the west (east) of cyclonic disturbance in the westerlies.

Turning our attention now to a low-level potential temperature anomaly, the surrogate for a low-level PV anomaly, we find a different behavior. Any low-level warm anomaly can be considered a positive PV anomaly as previously shown in Figure 9.7. As such, a low-level warm anomaly has a cyclonic circulation associated with it indicated by the solid arrows straddling the + sign in Figure 9.9(a). The southerly (northerly) winds downstream (upstream) of the anomaly center are associated with

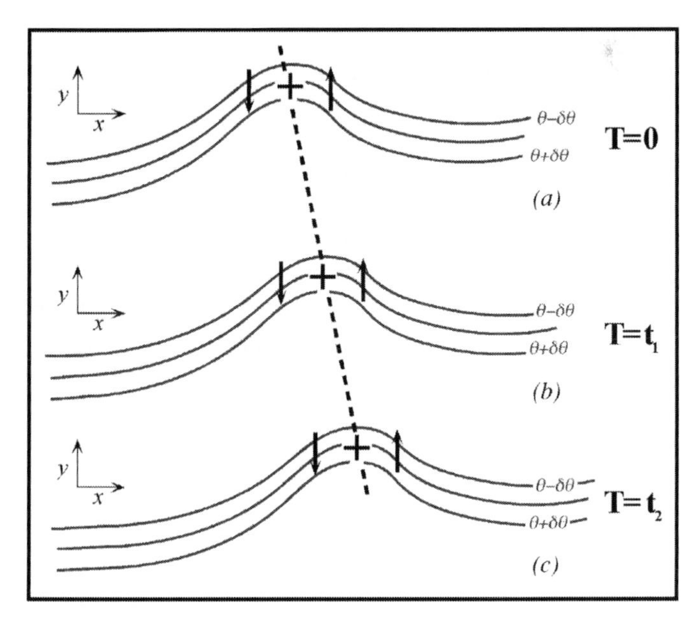

Figure 9.9 (a) Schematic of a surface warm anomaly in the northern hemisphere. Solid gray lines are surface isentropes, + sign indicates the surrogate positive PV anomaly center associated with the warm anomaly, and the arrows represent the cyclonic circulation associated with the feature. (b) The same warm anomaly at some later time. (c) The same warm anomaly at an even later time. The thick dashed line indicates the phase line of the warm anomaly through time

horizontal warm (cold) air advection. The net effect of warm advection downstream and cold advection upstream of a warm anomaly is to propagate the warm anomaly downstream, toward the warm advection and away from the cold advection (Figure 9.9b). Still later in time, the circulation continues to propagate the warm anomaly downstream (Figure 9.9c). Notice that there is very little upstream development in the behavior of the low-level warm anomaly in Figure 9.9; only the original anomaly persists through time.

Recalling that any PV anomaly will have associated with it a circulation that penetrates the atmosphere through a certain depth, the possibility exists that the lower tropospheric portion of the circulation associated with an upper-level PV anomaly may be able to penetrate far enough downward to influence the development of a low-level warm anomaly through horizontal advection. Likewise, the upper-level portion of the circulation associated with a low-level warm anomaly may penetrate far enough upward through the troposphere to affect the amplitude of an upper-level positive PV anomaly through horizontal PV advection at upper levels. In short, the upper- and lower-level PV anomalies might be able to amplify one another if they are properly phased in space. Notable surface cyclones are the product of persistent, significant development. Viewed from the PV perspective, such development depends critically upon a prolonged period of the mutual amplification of upper- and lower-level anomalies just described. However, as analysis of Figures 9.8 and 9.9 suggests, the upper- and lower-level PV anomalies act as counter-propagating Rossby waves – each headed in opposite directions when left to their own devices. Therefore, the likelihood that prolonged mutual amplification of the separate anomalies will occur depends upon the upper- and lower-level anomalies being in sufficiently close proximity to one another for an extended period of time. Given the different propagation tendencies of the two anomalies, this would seem a difficult proposition. On the other hand, the fact that cyclones are a ubiquitous feature of the mid-latitude atmosphere suggests, of course, that it is not. Now we finally turn to the basic description of cyclogenesis from the PV perspective to see why this is the case.

Imagine an upper-level PV anomaly migrating over a lower tropospheric baroclinic zone as shown schematically in Figure 9.10(a). The upper anomaly has a cyclonic circulation associated with it that, while maximized at the level of the anomaly, extends with some vigor throughout the depth of the troposphere. The influence of the upper PV anomaly is felt at sea level, to a degree determined by the penetration depth, in the form of a weak cyclonic circulation that acts to deform the sea-level isotherms through horizontal temperature advection (Figure 9.10a). The warm air advection portion of that sea-level circulation produces a low-level warm anomaly (signified by the + sign in Figure 9.10b) that acts as a positive PV anomaly near sea level. This anomaly has its own circulation that, though strongest at the surface, penetrates upward through the troposphere to a degree determined by the penetration depth. Accordingly, the influence of the low-level PV anomaly is felt near the tropopause in the form a weak cyclonic circulation that acts to intensify the upper-level PV anomaly by inducing positive PV advection into the eastern half of the anomaly. The

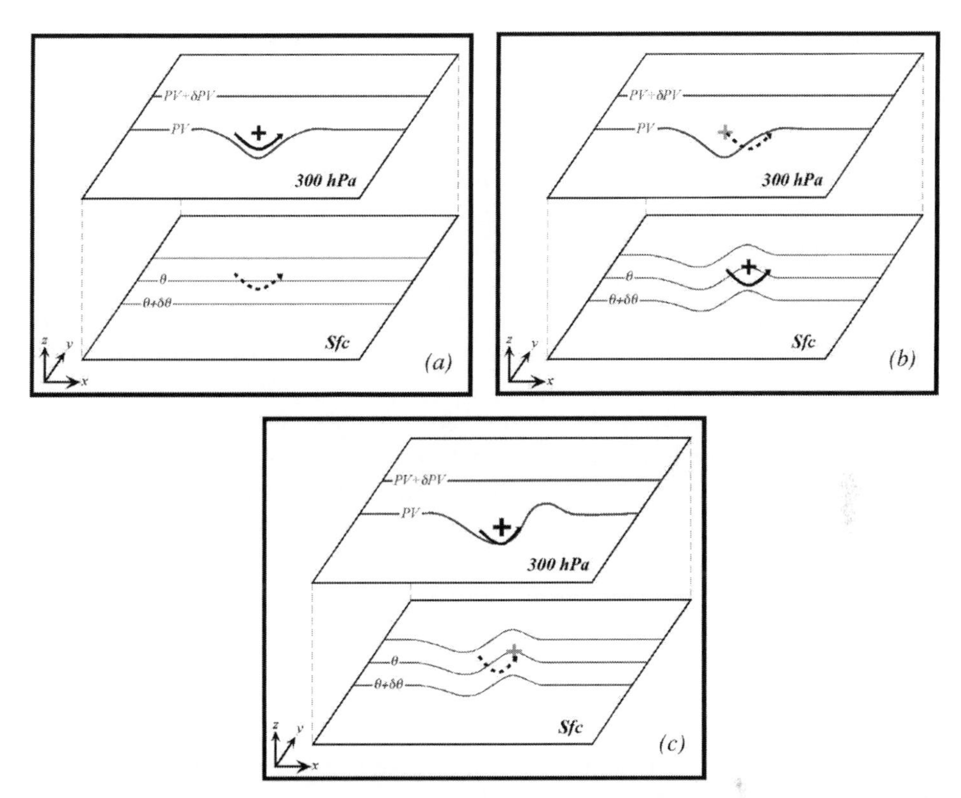

Figure 9.10 Upper tropospheric positive PV anomaly moving over a low-level baroclinic zone. (a) Circulation associated with the upper PV anomaly ('+' sign) is indicated by the bold arrow. The surface reflection of that circulation is given by the dashed arrow at the surface. (b) Low-level thermal advections produce a surface warm anomaly (dark '+' sign) whose circulation, indicated by the bold arrow, has an upper tropospheric reflection (indicated by the dashed arrow). (c) Upper tropospheric PV advections intensify the upper PV anomaly (dark '+' sign) and its circulation, indicated by the bold arrow, intensifies. The surface reflection of that circulation (dashed arrow) results in thermal advections that serve to intensify the surface warm anomaly

concurrent negative PV advection to the east of the original upper-level anomaly, coupled with this positive PV advection in the eastern half of the original anomaly, produces a tendency for the upper PV anomaly to propagate downstream, quite the opposite of its natural tendency to propagate westward! The invigorated upper-level PV anomaly then exerts an invigorated cyclonic influence on the low-level thermal field as shown in Figure 9.10(c). Since the upper-level PV anomaly lies upstream of the surface warm anomaly, this influence produces maximum warm air advection into the center of the warm anomaly and maximum cold air advection to its west. This distribution of upper-level PV-induced warm and cold air advection not only acts to intensify the low-level warm anomaly, but also tends to promote a westward propagation of that feature, quite the opposite of its natural tendency to propagate eastward! Thus, when upper- and lower-level PV anomalies come into sufficiently close proximity to one another, their influences on one another promote not only mutual amplification but also a beneficial 'phase locking' whereby the natural tendency for each anomaly to run away from the other is countermanded through the interaction of their respective circulations and prolonged interaction is promoted.

Notice that, since the PV perspective is most correctly viewed as a complementary perspective on cyclogenesis, many elements of the cyclogenesis process with which we were already familiar are evident in the PV view. Primary among these is that *development cannot occur unless there is an upshear tilt between lower and upper disturbances.* In the basic state variable (or omega-centric) view of cyclogenesis, this requirement was manifest in the upshear tilt of the geopotential minima with height that ensured that the region downstream of the upper-level geopotential minimum (i.e. the region characterized by PVA by the thermal wind) would lie directly above the sea-level pressure minimum and thereby contribute to its subsequent intensification. Another physical similarity is the ubiquity of a lower tropospheric warm anomaly in the development of a cyclone. Yet another is the fact that the basic elements of cyclogenesis appear to be quite physically separate from the attendant process of frontogenesis, though this result seems more obvious from the PV perspective.

Once again, we have thus far considered the process of cyclogenesis from a purely adiabatic perspective. In order to appreciate more fully the PV view of cyclogenesis and to further understand the connections between this view and the classical view of the process, we must also consider the effect of latent heat release from the PV perspective. Before we can effectively consider such effects, however, we must first understand the effect of diabatic heating on PV itself.

9.4 The Influence of Diabatic Heating on PV

Recall from (9.5) that the isentropic PV is defined as

$$PV = -g(\zeta_\theta + f)\left(\frac{\partial\theta}{\partial p}\right).$$

We seek to derive an expression for the Lagrangian rate of change of this PV. This process is made considerably easier if we first derive an equivalent isobaric form of the expression for PV.

Since the relative vorticity in isentropic coordinates is $\zeta_\theta = (\partial v/\partial x)_\theta - (\partial u/\partial y)_\theta$, we need expressions for $(\partial v/\partial x)_\theta$ and $(\partial u/\partial y)_\theta$. Recall that the differentials of u and v, evaluated on surfaces of constant potential temperature, can be written as

$$du_\theta = \left(\frac{\partial u}{\partial x}\right)_{y,p} dx_\theta + \left(\frac{\partial u}{\partial y}\right)_{x,p} dy_\theta + \left(\frac{\partial u}{\partial p}\right)_{x,y} dp_\theta \tag{9.11a}$$

$$dv_\theta = \left(\frac{\partial v}{\partial x}\right)_{y,p} dx_\theta + \left(\frac{\partial v}{\partial y}\right)_{x,p} dy_\theta + \left(\frac{\partial v}{\partial p}\right)_{x,y} dp_\theta. \tag{9.11b}$$

Rearranging (9.11a) to solve for $(du/dy)_\theta$ (which is equal to $(\partial u/\partial y)_\theta$) we get

$$\left(\frac{du}{dy}\right)_\theta = \left(\frac{\partial u}{\partial y}\right)_\theta = \left(\frac{\partial u}{\partial y}\right)_{x,p} + \left(\frac{\partial u}{\partial p}\right)_{x,y} \left(\frac{dp}{dy}\right)_\theta \tag{9.12a}$$

with a similar result for $(dv/dx)_\theta$ upon rearranging (9.11b) into

$$\left(\frac{dv}{dx}\right)_\theta = \left(\frac{\partial v}{\partial x}\right)_\theta = \left(\frac{\partial v}{\partial x}\right)_{y,p} + \left(\frac{\partial v}{\partial p}\right)_{x,y} \left(\frac{dp}{dx}\right)_\theta . \tag{9.12b}$$

A simple rearrangement of the Poisson equation leads to the expression

$$p = 1000 \left(\frac{T}{\theta}\right)^{c_p/R}$$

from which we find that

$$\left(\frac{dp}{dy}\right)_\theta = c_p \rho \left(\frac{dT}{dy}\right)_\theta \quad \text{and} \quad \left(\frac{dp}{dx}\right)_\theta = c_p \rho \left(\frac{dT}{dx}\right)_\theta . \tag{9.13}$$

Similar to (9.11), the differential of Ton an isentropic surface can be written as

$$dT_\theta = \left(\frac{\partial T}{\partial x}\right)_{y,p} dx_\theta + \left(\frac{\partial T}{\partial y}\right)_{x,p} dy_\theta + \left(\frac{\partial T}{\partial p}\right)_{x,y} dp_\theta \tag{9.14}$$

so that

$$\left(\frac{dT}{dx}\right)_\theta = \left(\frac{\partial T}{\partial x}\right)_\theta = \left(\frac{\partial T}{\partial x}\right)_{y,p} + \left(\frac{\partial T}{\partial p}\right)_{x,y} \left(\frac{dp}{dx}\right)_\theta \tag{9.15a}$$

and

$$\left(\frac{dT}{dy}\right)_\theta = \left(\frac{\partial T}{\partial y}\right)_\theta = \left(\frac{\partial T}{\partial y}\right)_{x,p} + \left(\frac{\partial T}{\partial p}\right)_{x,y} \left(\frac{dp}{dy}\right)_\theta . \tag{9.15b}$$

Substituting the expressions for $(dp/dx)_\theta$ and $(dp/dy)_\theta$ from (9.13) into (9.15) yields

$$\frac{1}{c_p \rho} \left(\frac{dp}{dx}\right)_\theta = \left(\frac{\partial T}{\partial x}\right)_{y,p} + \left(\frac{\partial T}{\partial p}\right)_{x,y} \left(\frac{dp}{dx}\right)_\theta \tag{9.16a}$$

$$\frac{1}{c_p \rho} \left(\frac{dp}{dy}\right)_\theta = \left(\frac{\partial T}{\partial y}\right)_{x,p} + \left(\frac{\partial T}{\partial p}\right)_{x,y} \left(\frac{dp}{dy}\right)_\theta \tag{9.16b}$$

which can be solved to render

$$\left(\frac{dp}{dx}\right)_\theta = \left(\frac{\partial T}{\partial x}\right)_{y,p} \Bigg/ \left[\frac{1}{c_p \rho} - \left(\frac{\partial T}{\partial p}\right)_{x,y}\right] \tag{9.17a}$$

and

$$\left(\frac{dp}{dy}\right)_\theta = \left(\frac{\partial T}{\partial y}\right)_{x,p} \Big/ \left[\frac{1}{c_p\rho} - \left(\frac{\partial T}{\partial p}\right)_{x,y}\right]. \tag{9.17b}$$

Now, taking the vertical ($-\partial/\partial p$) derivative of the Poisson equation yields

$$-\frac{T}{\theta}\frac{\partial\theta}{\partial p} = \frac{1}{c_p\rho} - \left(\frac{\partial T}{\partial p}\right)_{x,y}.$$

Similarly, the x and y derivatives of potential temperature, evaluated on isobaric surfaces, are

$$\frac{T}{\theta}\left(\frac{\partial\theta}{\partial x}\right)_p = \left(\frac{\partial T}{\partial x}\right)_p \quad \text{and} \quad \frac{T}{\theta}\left(\frac{\partial\theta}{\partial y}\right)_p = \left(\frac{\partial T}{\partial y}\right)_p,$$

respectively. Substituting these expressions into (9.17) yields

$$\left(\frac{dp}{dx}\right)_\theta = -\frac{\partial\theta}{\partial x}\Big/\frac{\partial\theta}{\partial p} \tag{9.18a}$$

and

$$\left(\frac{dp}{dy}\right)_\theta = -\frac{\partial\theta}{\partial y}\Big/\frac{\partial\theta}{\partial p}. \tag{9.18b}$$

We can now rewrite (9.12) as

$$\left(\frac{\partial v}{\partial x}\right)_\theta = \left(\frac{\partial v}{\partial x}\right)_{y,p} + \left(\frac{\partial v}{\partial p}\right)_{x,y}\left[-\frac{\partial\theta}{\partial x}\Big/\frac{\partial\theta}{\partial p}\right] \tag{9.19a}$$

$$\left(\frac{\partial u}{\partial y}\right)_\theta = \left(\frac{\partial u}{\partial y}\right)_{x,p} + \left(\frac{\partial u}{\partial p}\right)_{x,y}\left[-\frac{\partial\theta}{\partial y}\Big/\frac{\partial\theta}{\partial p}\right]. \tag{9.19b}$$

Subtracting (9.19b) from (9.19a) yields an expression for the isentropic relative vorticity

$$\begin{aligned}\zeta_\theta = \left(\frac{\partial v}{\partial x}\right)_\theta - \left(\frac{\partial u}{\partial y}\right)_\theta &= \left(\frac{\partial v}{\partial x}\right)_{y,p} - \left(\frac{\partial u}{\partial y}\right)_{x,p} \\ &+ \left(\frac{\partial v}{\partial p}\right)_{x,y}\left[-\frac{\partial\theta}{\partial x}\Big/\frac{\partial\theta}{\partial p}\right] - \left(\frac{\partial u}{\partial p}\right)_{x,y}\left[-\frac{\partial\theta}{\partial y}\Big/\frac{\partial\theta}{\partial p}\right]\end{aligned}$$

which can be rewritten, by noting that $\zeta_p = (\partial v/\partial x)_{y,p} - (\partial u/\partial y)_{x,p}$ and by multiplying the entire expression by $-g\partial\theta/\partial p$, as

$$-g\zeta_\theta\frac{\partial\theta}{\partial p} = -g\frac{\partial\theta}{\partial p}\zeta_p - g\frac{\partial v}{\partial p}\frac{\partial\theta}{\partial x} + g\frac{\partial u}{\partial p}\frac{\partial\theta}{\partial y}. \tag{9.20}$$

Every term on the RHS of (9.20) is evaluated on an isobaric surface and so that collection of terms nearly represents PV in isobaric coordinates. By including the planetary vorticity on both sides of (9.20) the resulting expression can be rearranged into the vector expression

$$PV = -g(\zeta_\theta + f)\frac{\partial\theta}{\partial p} = -g(f\hat{k} + \nabla \times \vec{V}_h) \cdot \nabla\theta. \tag{9.21}$$

Note that if the horizontal wind is assumed to be geostrophic, then (9.21) reduces to the QG potential vorticity (9.9d). In fact, a more general derivation of the full PV exists which extends to all three dimensions (i.e. includes gradients of ω as well) and yet retains the same conservation property as the simple isentropic expression.[1] We are now interested in discovering how diabatic heating alters PV and so must derive the Lagrangian derivative of (9.21) next.

This derivation is a long exercise that begins by expanding (9.21) into all of its components. Then one operates upon that component expression with the isobaric Lagrangian operator

$$\frac{d}{dt} = \frac{\partial}{\partial t} + u\frac{\partial}{\partial x} + v\frac{\partial}{\partial y} + \omega\frac{\partial}{\partial p}.$$

After considerable algebraic manipulation and careful attention to the chain rule of differentiation, the resulting expression is

$$\frac{d(PV)}{dt} = -g(\vec{\eta}_a \cdot \nabla\dot{\theta}) \tag{9.22}$$

where $\vec{\eta}_a$ is the 3-D absolute vorticity vector and $\dot{\theta}$ is the diabatic heating rate. Keeping only the vertical component of (9.22) we have

$$\frac{d}{dt}(PV) \approx -g(\zeta + f)\frac{\partial\dot{\theta}}{\partial p} \tag{9.23}$$

so that PV is increased (decreased) where the vertical gradient of diabatic heating is positive (negative). This result is illustrated schematically in Figure 9.11 in which the reasonable assumption is made that the diabatic heating maximum in a typical mid-latitude cyclone is located in the middle troposphere (between 400 and 600 hPa). In such a case, it is clear that PV 'production' occurs in the lower troposphere while PV 'destruction' occurs near the tropopause. The low-level, positive PV anomaly thus created has an associated cyclonic circulation just like any other positive PV feature and so can contribute to the intensification of the low-level circulation associated with a surface cyclone. In fact, if we consider the foregoing schematic within the context of a developing cyclone, as in Figure 9.12, a more comprehensive view of the effect of latent heat release on the PV structure arises.

The heating maximum occurs slightly downstream of an upper-level positive PV anomaly since that is where the air is rising most vigorously. As just described, the

[1] This more complete version of PV was first derived by Ertel (1942) and is thus often referred to as the 'Ertel PV'.

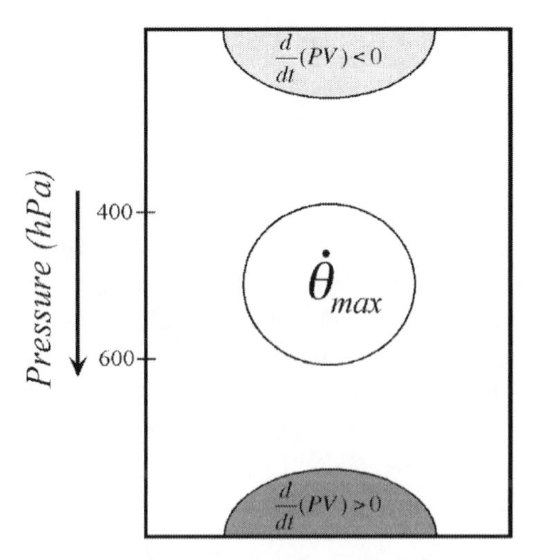

Figure 9.11 Lagrangian PV tendencies associated with diabatic heating. The circle labeled $\dot{\theta}_{max}$ is the diabatic heating maximum. The light (dark) shading above (below) it indicates the region of PV destruction (production)

effects of that heating are to create the low-level positive PV anomaly as well as to erode upper tropospheric PV. This erosion of the upper tropospheric PV serves to steepen the slope of the PV isopleth downstream of the upper-level positive PV anomaly. Such steepening is the PV equivalent of shortening the wavelength between the upper-level trough and the downstream ridge that was emphasized in the description of self-development. From the PV perspective, such an increase in slope also contributes to making the upper PV feature more anomalous. Simultaneously, the cyclonic circulation associated with the low-level positive PV feature enhances both

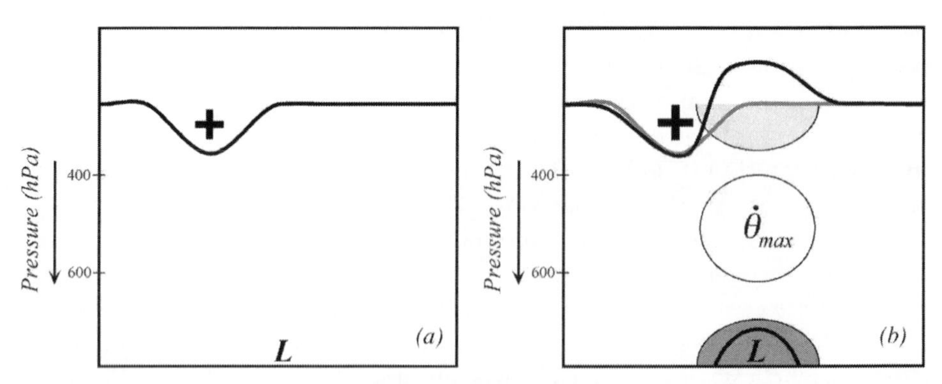

Figure 9.12 (a) Relationship between an upper tropospheric positive PV anomaly (+ sign) and a surface low-pressure center ('L'). (b) Ascent downstream of the PV anomaly produces latent heat release manifest as a $\dot{\theta}_{max}$. PV erosion aloft deforms the bold PV contour to the east of the original anomaly, making that anomaly even more anomalous (larger + sign). PV production in the lower troposphere intensifies the surface cyclone with high values of PV developing near the center indicated by the bold black line surrounding the 'L'

the mutual amplification and phase-locking effects described earlier, leading to continued intensification of the cyclone. Finally, the penetration depth of each anomaly is increased in the presence of latent heating which generally serves to reduce the static stability in such regions.

9.5 Additional Applications of the PV Perspective

The diagnostic power of the PV perspective can be extended by exploiting the invertibility principle in what is known as piecewise PV inversion. Use of this diagnostic tool affords considerable insight into a number of canonical mid-latitude developments. In this section, we first examine the nature of piecewise PV inversion and then consider a number of ubiquitous mid-latitude phenomena from the PV perspective.

9.5.1 Piecewise PV inversion and some applications

Perhaps chief among the applications of the PV perspective to understanding mid-latitude cyclones is the use of piecewise inversion of PV to elucidate the contributions to the atmospheric flow corresponding to discrete pieces of the PV distribution. Inversion of the PV allows one to recover the portion of the total flow that is directly attributable to each of several discrete pieces of the perturbation PV. At any given time the total perturbation PV at a certain location in a domain can be defined as

$$P' = P - \overline{P} \tag{9.24}$$

where $\overline{P}$ is the time-averaged (or spatially averaged) PV at that location and P is the instantaneous PV at that location and time. There are a number of ways to subsequently partition the total perturbation PV (P') into pieces. For instance, one could simply divide the atmospheric column in half and consider the P' in the layer above 500 hPa as one piece and all the P' below 500 hPa as another piece. Inverting each piece of the P' would then yield the portions of the total circulation, geopotential height field, temperature, vertical motions, etc., associated with each piece.

More sophisticated partitioning of the total perturbation PV distribution allows for more physical insight to be gained through the subsequent inversion. For instance, one might consider one piece of the P' to be that associated with the upper tropospheric PV. Another piece might be that portion of the P' associated with the low-level potential temperature anomaly and still another piece might be the diabatically generated PV which, as we have just seen, is relatively easy to isolate in principle. One can employ criteria relating to isobaric level and ambient relative humidity to make these suggested partitions.[2]

[2] A functioning scheme for inversion of the full (Ertel) PV has been developed by Davis and Emanuel (1991). A set of such criteria is employed, for example, by Korner and Martin (2000) in their piecewise inversion of Ertel PV, following Davis and Emanuel (1991).

Inversion of the full PV is a non-trivial numerical operation. The details of the method, and its companion piecewise PV inversion, are described in the original sources and will not be derived here. Nonetheless, the nature of the inversion process can be illustrated by outlining the considerably easier inversion of the QG PV. Recalling the definition of the QG PV (9.9d), it follows that the perturbation QG PV is given by

$$P'_g = \frac{1}{f_0}\nabla^2\phi' + f + f_0\frac{\partial}{\partial p}\left(\frac{1}{\sigma}\frac{\partial \phi'}{\partial p}\right) \tag{9.25}$$

where ϕ' is the perturbation geopotential and

$$\sigma = -\frac{\alpha}{\theta}\frac{d\Theta}{dp},$$

Θ being the domain-wide average of θ at each isobaric level. By subtracting f from P'_g we get $P^*_g = P'_g - f$ and P^*_g can be partitioned into any number of pieces such that

$$P^*_g = \sum_{i=1}^{n} P^*_{g_i} \tag{9.26}$$

where each piece of the partitioned QG PV is denoted by $P^*_{g_i}$. Associated with each $P^*_{g_i}$ is its own piece of perturbation geopotential height, ϕ'_i. From (9.25) we have

$$P^*_{g_i} = \ell(\phi'_i) \tag{9.27a}$$

where ℓ is the linear operator

$$\ell = \left[\frac{1}{f_0}\nabla^2 + f_0\frac{\partial}{\partial p}\left(\frac{1}{\sigma}\frac{\partial}{\partial p}\right)\right]. \tag{9.27b}$$

It is clear, then, that each ϕ'_i can be obtained through inversion of (9.27a) as

$$\phi'_i = \ell^{-1}(P^*_{g_i}). \tag{9.27c}$$

As was the case with the QG omega equation, the inversion of QG PV through (9.27c) can be achieved through successive overrelaxation with appropriate boundary conditions. Inverting the full PV is not as simple because the operator relating the perturbation PV to the perturbation geopotential is not as simple but the nature of the inversion is the same.

Since it is the explicit accounting for diabatic processes that distinguishes the PV perspective from the basic state variables perspective on mid-latitude dynamics, most of the useful applications of the PV view of dynamics seek to exploit that advantage. As we have already discussed, the cloud and precipitation distribution associated with a typical mid-latitude cyclone is composed of contributions from the transverse updrafts in the vicinity of the frontal zones as well as the larger-scale shearwise updrafts which produce the cloud head. Taken as a whole, the cloud and precipitation production that accompanies these upward vertical motion regions results in considerable lower tropospheric PV production. A natural question is: 'What portion of the circulation of the cyclone is directly attributable to this diabatically generated lower tropospheric PV?' A number of studies, employing piecewise PV inversion, have considered this question. Figure 9.13 illustrates a result from just one of these many studies. The total perturbation geopotential height at 950 hPa at a certain time in the evolution of a north central Pacific Ocean cyclone is shown in Figure 9.13(a). The contributions from the upper tropospheric PV (Figure 9.13b), diabatically generated PV (Figure 9.13c), and lower-level warm anomaly (Figure 9.13d) all contribute nearly equally to the geopotential minimum associated with this storm at this time. Estimates from other similar studies place the influence of the diabatically generated PV as high as 50% of the total circulation in intense mid-latitude cyclones.

The frontal zones themselves are also associated with diabatically generated lower tropospheric PV as shown schematically in Figure 9.14.[3] Cold frontal precipitation is often distributed in a narrow band oriented parallel to the front itself (Figure 9.14a). The latent heat released produces a similarly oriented strip of high PV in the lower troposphere (Figure 9.14b). The circulation associated with this lower tropospheric PV enhances the cyclonic shear across the frontal zone and contributes substantially to the strength of the cold frontal low-level jet (LLJ). The consequence of this intensified LLJ is an increase in the moisture transport into the warm sector of the cyclone which can serve to enhance the overall diabatic contribution to the intensification of the cyclone itself through promoting greater cloud and precipitation production.

9.5.2 A PV perspective on occlusion

Among the many applications of the PV perspective on the cyclone life cycle, one of the most enlightening is related to the evolution of occluded cyclones. It has been noted that the development of some occluded cyclones is accompanied by the development of a characteristic upper tropospheric PV distribution termed the 'treble clef.'[4] The treble clef PV distribution is characterized by an isolated, low-latitude, high-PV feature that is connected to a high-latitude reservoir of high PV by a thin filament of high PV as shown in Figure 9.15(a). As we have already seen, in an atmosphere in approximate thermal wind (or gradient wind) balance, regions of high PV in the upper troposphere sit atop relatively cold columns of air, while relative minima of upper tropospheric PV sit atop relatively warm columns of air. Thus, the characteristic tropospheric thermodynamic structure associated with the horizontal juxtaposition of two positive upper-level PV anomalies of unequal magnitude,

[3] Examination of the diabatically generated PV associated with a cold front is carried out by Lackmann (2002).

[4] A full description of this structural connection between the tropopause-level PV and the tropospheric thermal structure is given in Martin (1998a).

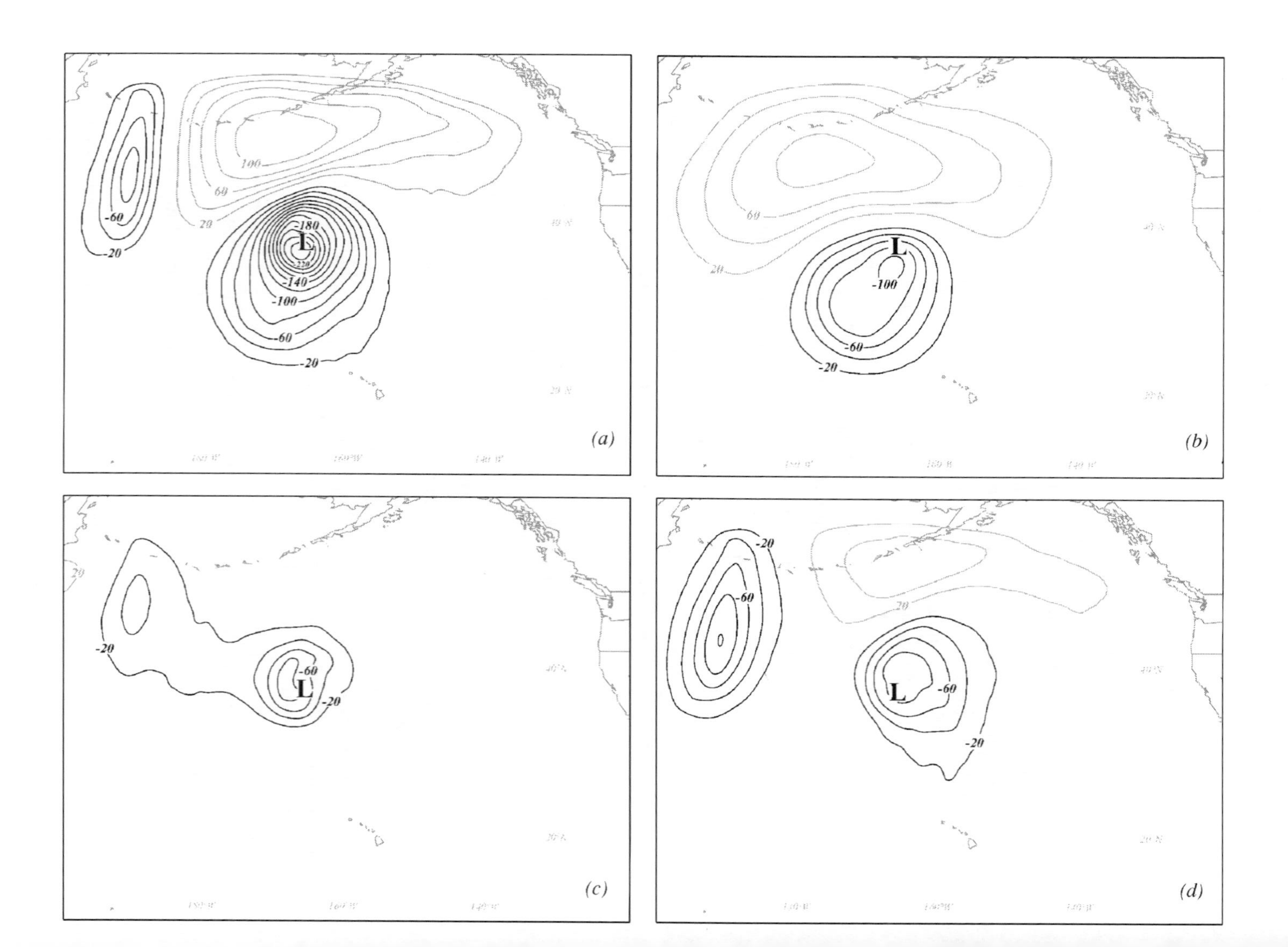
(a)
-20
-60
100
60
20
-180
L
-220
-140
-100
-60
-20
(b)
60
20
L
-100
-60
-20
(c)
20
-20
-60
L
-20
(d)
-20
-60
20
L
-60
-20

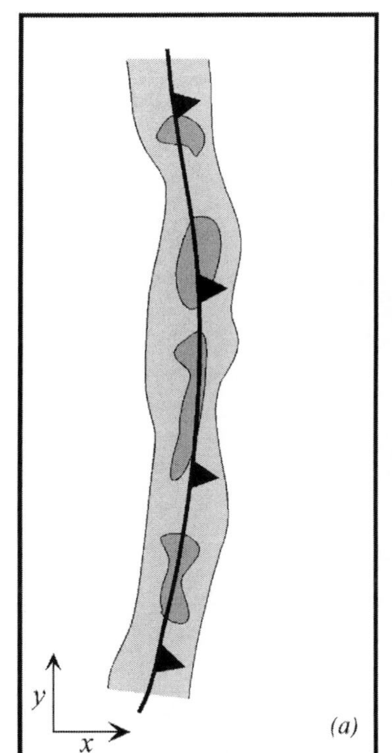

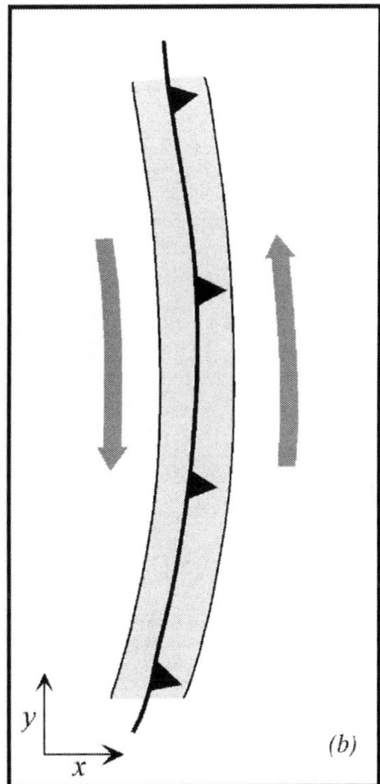

Figure 9.14 (a) Schematic cold frontal precipitation distribution with shading representing radar echoes in the precipitation along the front. (b) Gray shading indicates the thin, lower tropospheric positive PV anomaly created via diabatic heating associated with the cold frontal precipitation. Bold arrows represent the low-level cyclonic flow associated with the diabatically generated PV

separated by a relative minimum in PV (such as along the cross-section line in Figure 9.15a), precisely depicts the canonical warm occluded thermal structure (Figure 9.15b). Consequently, the presence of an upper tropospheric treble clef PV signature serves as a sufficient condition for asserting the presence of a warm occluded thermal structure in the underlying troposphere. The production of this treble clef PV structure depends upon development of the 'notch' of low PV highlighted in

←

Figure 9.13 Perturbation geopotential height at 950 hPa at 0000 UTC 6 November 1986. Black (gray) lines are negative (positive) perturbation heights labeled in m and contoured every 20 m. 'L' marks the location of the 950 hPa geopotential height minimum at that time. (b) Contribution to the total perturbation geopotential height perturbation made by the upper tropospheric PV anomalies. Heights contoured and labeled as in (a). (c) Contribution to the total geopotential height perturbation made by the diabatically generated PV anomalies. Height perturbations contoured and labeled as in (a). (d) Contribution to the total geopotential height perturbation made by the near surface PV anomalies. Height perturbations contoured and labeled as in (a)

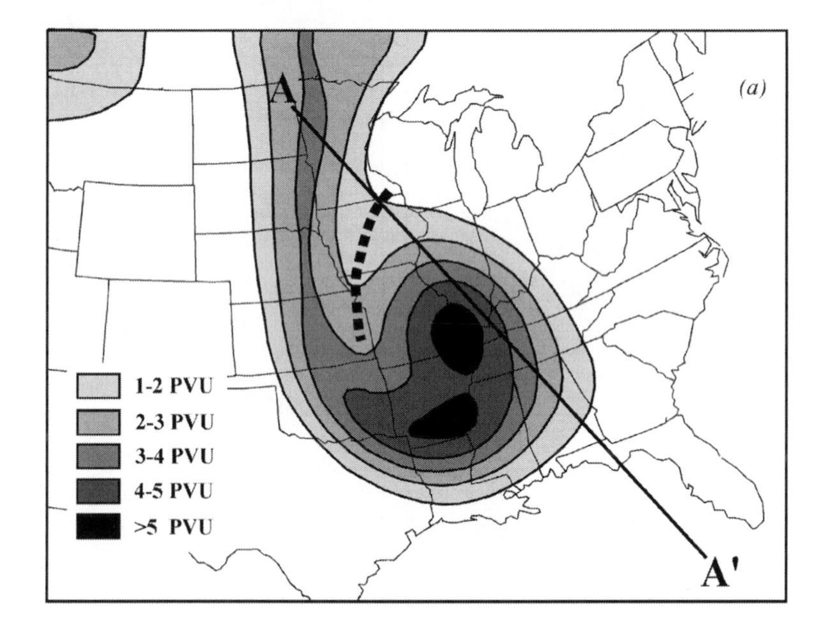

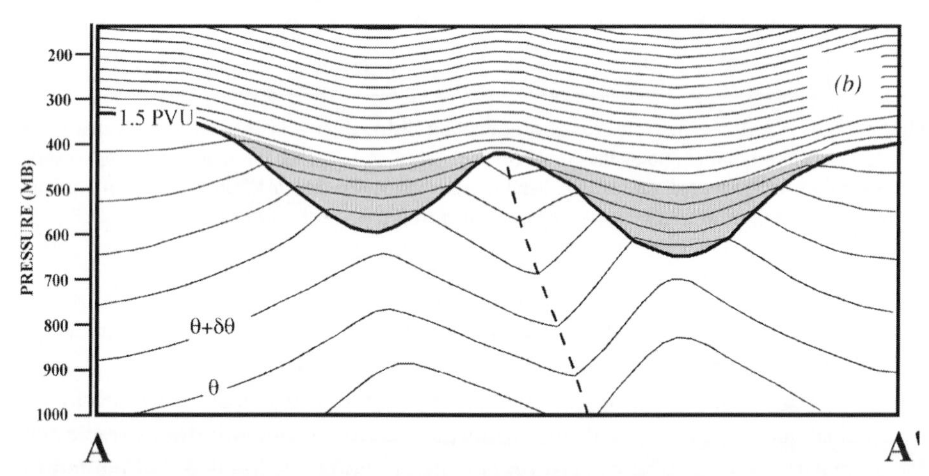

Figure 9.15 (a) Schematic of treble-clef-shaped upper tropospheric PV structure described in the text. Solid lines are isopleths of PV on an isobaric surface contoured and shaded in PVU (1 PVU = 10^{-6} m^2 K kg^{-1}s^{-1}). The thick dashed line identifies the PV 'notch' described in the text. (b) Schematic cross-section of potential temperature (θ) in the vicinity of a treble-clef-shaped upper tropospheric PV signature. The dashed axis denotes the sloping axis of warm air in the troposphere characteristic of an occluded cyclone

Figure 9.15(a). This notch development, in turn, depends upon tropopause-level PV erosion via diabatic heating (in the form of latent heat release) in the occluded quadrant of the cyclone.[5] This dependence is illustrated schematically in Figure 9.16.

[5] The reader is referred to a recent study by Posselt and Martin (2004) for the details of this argument.

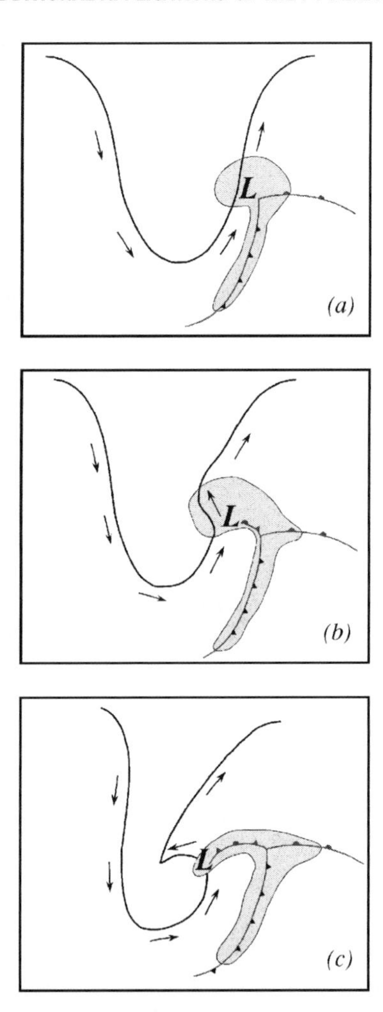

Figure 9.16 Schematic illustrating the synergy between diabatic erosion of PV and negative advection of PV at the tropopause during occlusion. Gray shading represents the erosion of tropopause PV by diabatic heating associated with the cyclone, the surface position of which is marked by the 'L'. Traditional surface frontal symbols indicate surface frontal locations. The thick soild line represents the PV = 2 PVU isopleth at the tropopause. Arrows represent the tropopause-level flow associated with the upper tropospheric PV feature. (a) The open wave stage. Heating is concentrated along the cold front and in the vicinity of the developing surface cyclone. (b) Commencement of occlusion. Persistent diabatic erosion in the northwest quadrant of the cyclone deforms the upper tropospheric PV contour northwest of the surface cyclone. Tropopause-level flow is also deformed in that vicinity. (c) Fully occluded stage. Cyclone is far removed from the peak of the surface warm sector. Heating is no longer proximate to the 'notch' in the upper tropospheric PV. Tropopause-level flow controls intensification of the notch through negative PV advection in the upper troposphere

In the open wave stage of the cyclone life cycle, significant ascent occurs in the near vicinity of the SLP minimum, which is located just downstream of an upper tropospheric positive PV anomaly as shown schematically in Figure 9.16(a). Given sufficient moisture, this ascent produces clouds and precipitation and release of latent heat which, in turn, serves to erode the upper tropospheric PV in accord with (9.23). Persistent diabatic erosion of upper tropospheric PV forms a 'notch' in the upper tropospheric PV structure (Figure 9.16b) which, coupled with the eastward progression of the upstream ridge, initiates the isolation of a low-latitude, upper tropospheric PV maximum. The circulation associated with this feature then begins to contribute to negative PV tendencies in the developing notch via negative PV advection, further isolating the low-latitude PV maximum and accelerating the cutoff process (Figure 9.16c). In the underlying troposphere the response to the development of the local upper tropospheric PV minimum in the 'notch' is the simultaneous development of an isolated, warm, weakly stratified column of air. Based along the near surface thermal ridge, this column slopes poleward and westward and its axis is identically the trowal, the essential structural characteristic of the warm occlusion.

9.5.3 A PV perspective on leeside cyclogenesis

Another application of the PV perspective applies to orographic cyclogenesis such as occurs in the lee of the Rocky Mountains in North America. One of the favored regions for development of cyclones in North America is in the lee of the Colorado and Alberta Rockies (Figure 9.17). Westerly flow over a mountain barrier results in subsidence in its lee as illustrated in Figure 9.18(a). This leeside subsidence advects middle tropospheric θ downward toward the ground. Consequently, lower tropospheric air in the lee is warmed up leading to the characteristic warm axis of a leeside trough. The downward advection of high θ is associated with an increased separation between adjacent isentropes just above mountain height (Figure 9.18b) so that $-\partial\theta/\partial p$ decreases. Thus, in order for the product $-g(\zeta_\theta + f)\partial\theta/\partial p$ to be conserved, there must be an increase in ζ_θ. As a result, the characteristic warm axis of a leeside pressure trough is accompanied by a cyclonic vorticity maximum at lower tropospheric levels. The presence of cyclogenesis maxima in the lee of meridionally oriented mountain ranges proceeds from the fact that the mid-latitude flow is predominantly westerly and the subsidence in the lee compels increases in ζ_θ in order to conserve PV.

9.5.4 The effects of PV superposition and attenuation

The circulation associated with a given PV anomaly is greatly influenced by the morphology of that anomaly. For instance, as shown in Figure 9.19(a), a linear, positive PV filament can be thought of as a 'string' of positive PV anomalies, each with its own cyclonic circulation. When these circulations are arranged in a linear geometry, considerable cancellation between adjacent features occurs, resulting in the creation of a cyclonic shear line. If one imagines this linear PV feature located in the upper troposphere, its associated lower tropospheric circulation will take the

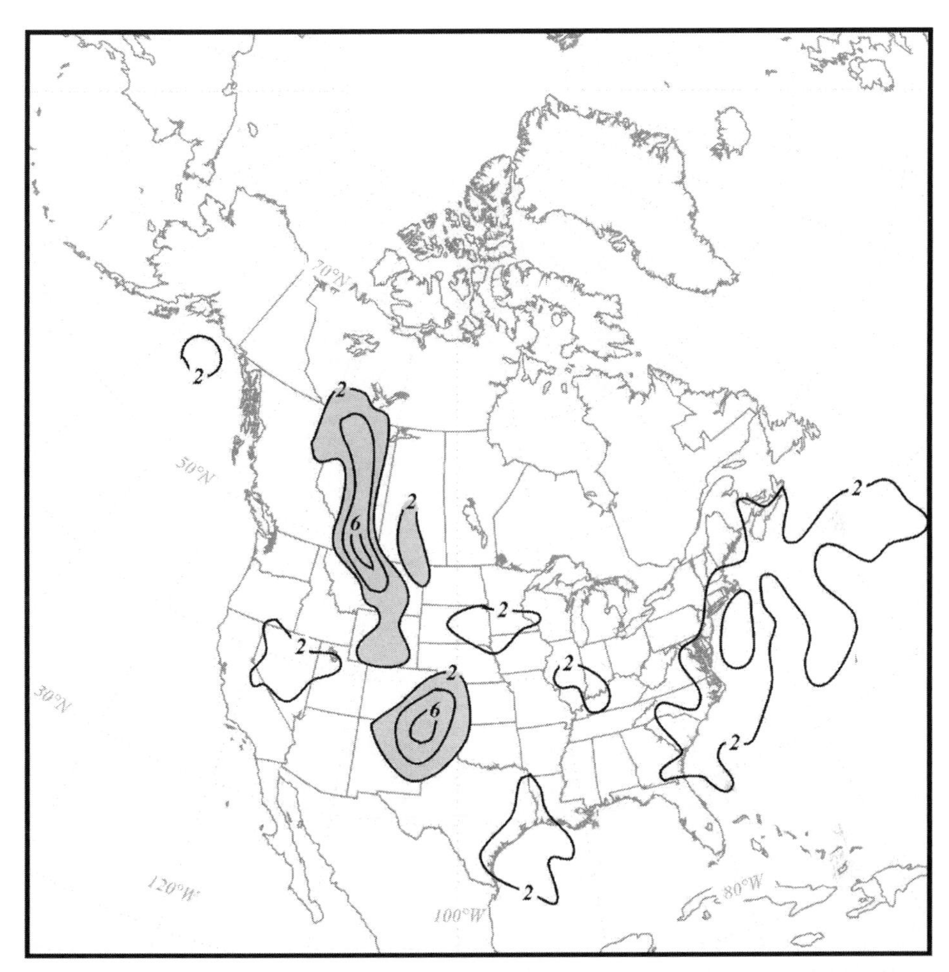

Figure 9.17 Areal distributions of cyclogenesis frequencies for the month of January from 1950 to 1977. Gray shaded regions represent lee cyclogenesis areas east of the Rocky Mountains. Adapted from Zishka and Smith (1980)

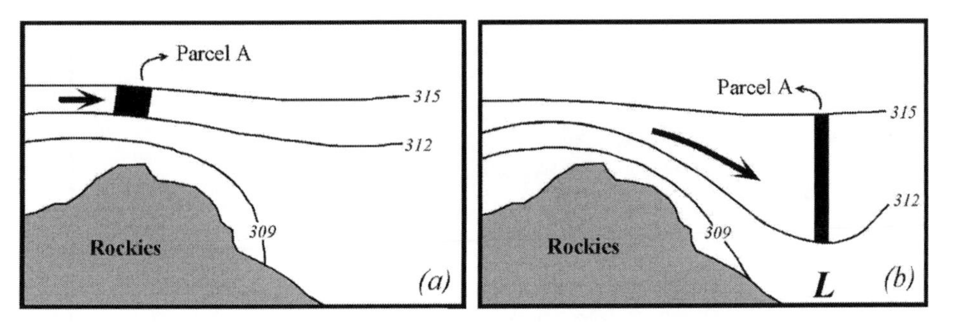

Figure 9.18 (a) Westerly flow (bold arrow) impinging on the Rocky Mountains. Parcel A is confined between the 312 K and 315 K isentropes. (b) As the flow pushes Parcel A over the ridge of the Rockies, the 312 K isentrope is forced toward the surface and the parcel is stretched in the vertical. A surface low-pressure center ('L') develops in response to the conservation of PV

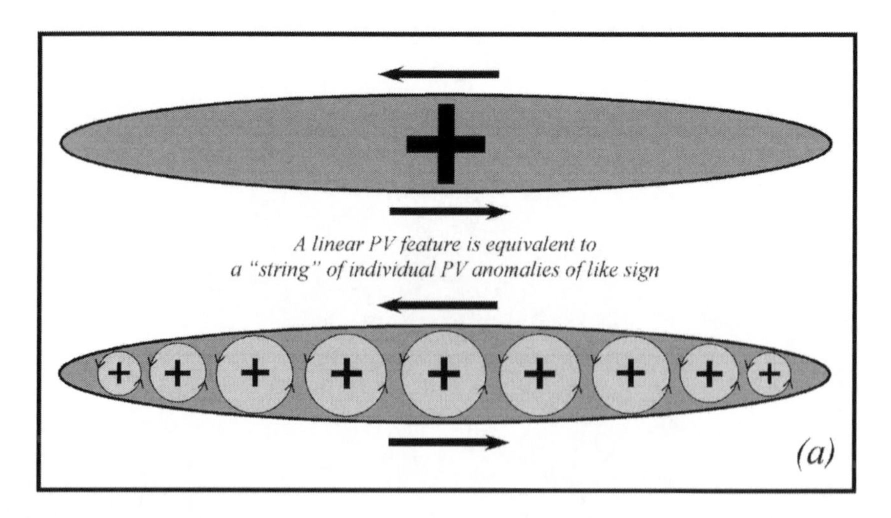
A linear PV feature is equivalent to
a "string" of individual PV anomalies of like sign
(a)

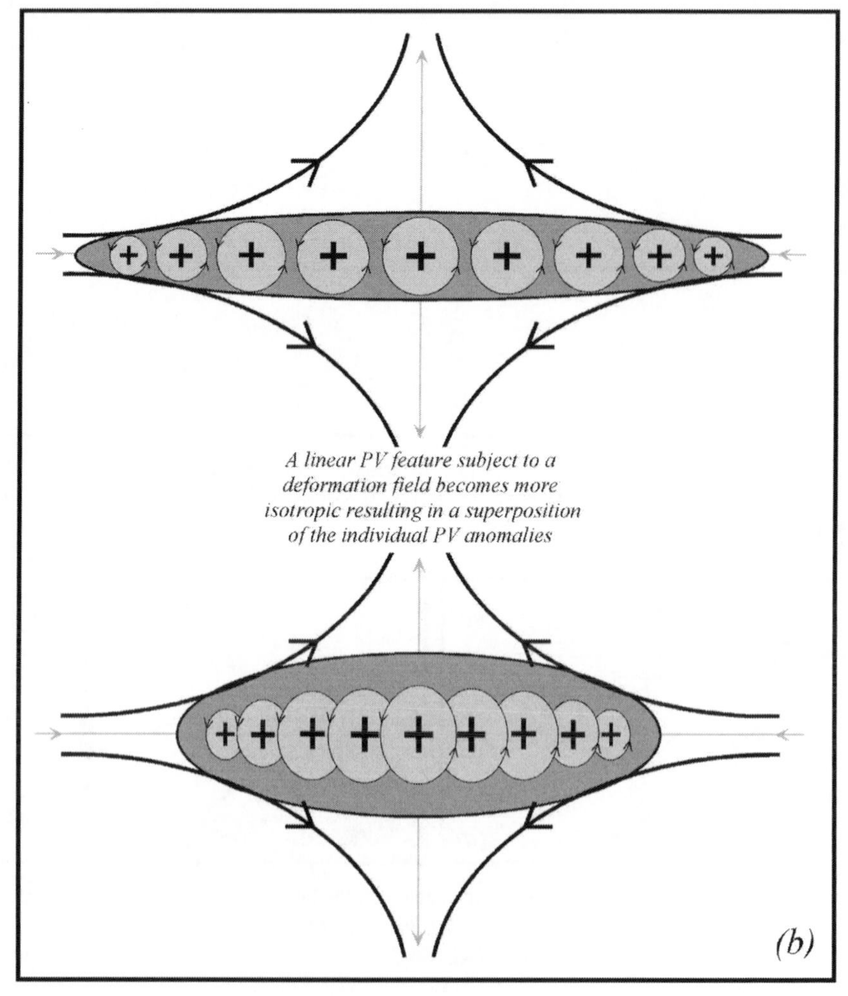
A linear PV feature subject to a
deformation field becomes more
isotropic resulting in a superposition
of the individual PV anomalies
(b)

form of a shear line as well. Such a feature is not the most likely progenitor of a surface cyclone. Quite a number of studies have demonstrated the influence of environmental deformation in promoting an increased isotropy in the perturbation PV field. As shown in Figure 9.19(b), if the axis of dilatation of the environmental deformation is at a sufficiently large angle to the long axis of a PV anomaly (90° is optimal) then the PV anomaly tends to become more circular with time. The resulting circular geometry means that the collection of positive PV anomalies originally aligned in the 'string' in Figure 9.19(a) have become superposed on one another. Since each of these PV elements retains its own circulation and anomalous geopotential height field, the superposition of these discrete anomalies additively produces a more substantial, more isotropic circulation maximum and geopotential height minimum. The intensification of the perturbation circulation and geopotential height anomalies associated with a PV anomaly when the anomaly is made more isotropic is a result of what is known as the **superposition principle.**

Conversely, environmental deformation might just as well promote an increased anisotropy in the perturbation PV (Figure 9.20). When a PV anomaly becomes more anisotropic, its associated circulation and geopotential height anomalies weaken. This process, the opposite of PV superposition, has been termed **PV attenuation** and it has been demonstrated to exert an important influence on surface cyclolysis – even when that cyclolysis is particularly rapid. To some extent, therefore, surface cyclolysis is likely to occur in large-scale environments that promote the attenuation (i.e. the thinning and stretching) of upper tropospheric PV anomalies associated with synoptic-scale short waves.

This brief survey may suggest to you that the PV perspective on the cyclone life cycle offers an alternative perspective more than it offers brand new insight into the processes. It is still necessary, for instance, that an upper-level disturbance migrate over a surface baroclinic zone in order to initiate cyclogenesis. All of the vorticity advection considerations from the QG dynamical perspective on cyclogenesis can be reinterpreted in terms of the PV perspective and vice versa. One rather obvious and important benefit of the PV perspective is that one can take explicit accounting of the effect of latent heat release on the development of the cyclone and its characteristic thermal structure. The omega-centric view of the cyclone problem does not as readily yield this specific information. Clearly, not every problem in mid-latitude dynamic meteorology is amenable to analysis using one perspective. It is most fruitful to view the QG and PV views as different, but complementary, tools by which investigation of nature might proceed and, as anyone who has had to fix a household item at one

←

Figure 9.19 (a) A linear positive PV anomaly can be thought of as a 'string' of positive PV anomalies. The dark gray shading represents the linear PV feature while the lighter gray circles represent the individual PV anomalies that constitute the larger linear feature. (b) Effect of a deformation field on the linear PV feature. When the long axis of the linear PV feature is nearly the same as the axis of contraction of the deformation field, the individual PV anomalies become superposed, enhancing the overall cyclonic circulation

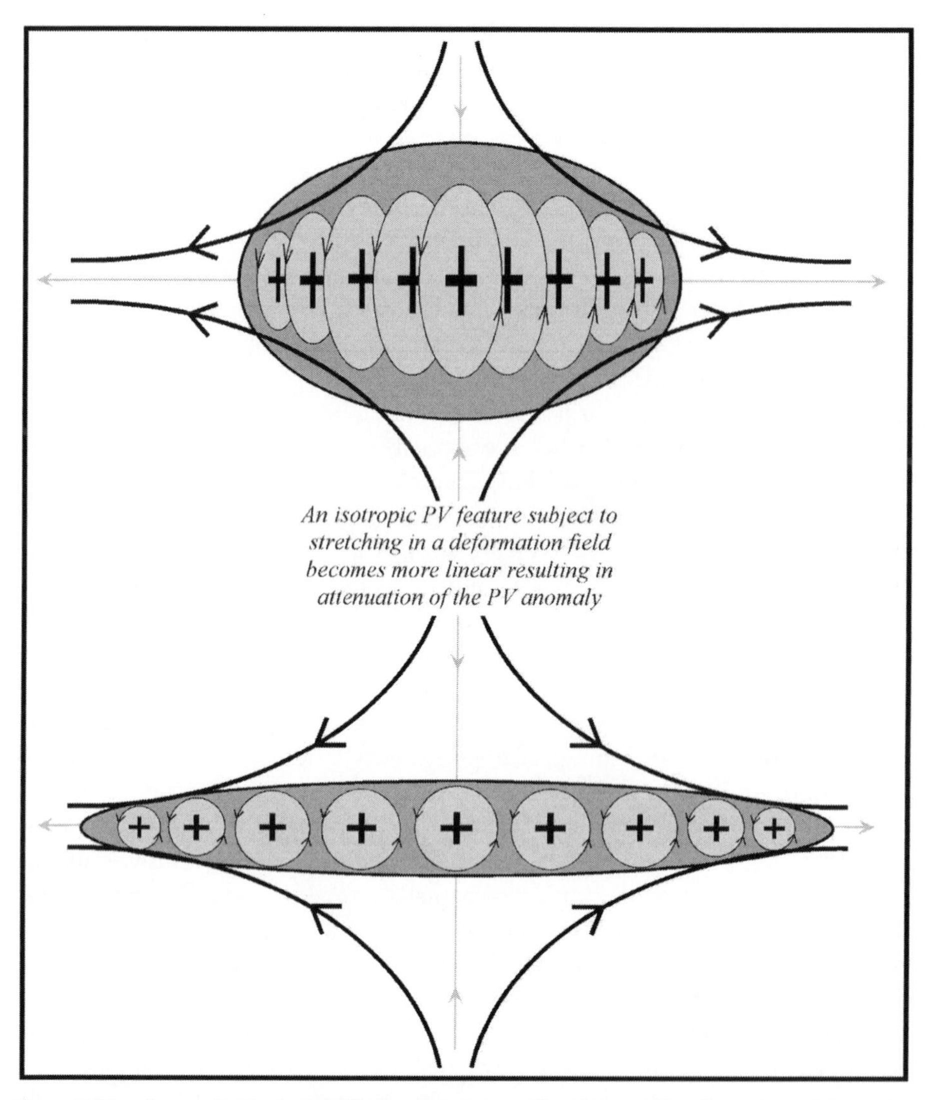

Figure 9.20 Schematic illustrating PV attenuation. A positive PV anomaly subjected to a deformation field can be stretched into a thin line of individual PV anomalies thus reducing the potency of the circulation associated with the anomaly. Such a process is known as PV attenuation

point in his/her life knows, different jobs require different tools. We humans have developed a host of different languages to describe our common experience with the world, yet none of those languages affords any more insight into the mysteries of our existence than any other. Analogously, neither of the two 'languages' of mid-latitude dynamics explored in this book – the basic state variables and PV 'languages' – provides exclusive insight into any aspect of the manner in which mid-latitude cyclones are created and how they live their relatively short, profoundly important, and fascinating life cycles.

Selected References

Hoskins *et al.* (1985) provide a comprehensive overview of the theory and use of Ertel potential vorticity in the diagnosis and prediction of mid-latitude cyclones.

Ertel (1942) is the seminal work on potential vorticity (the original is in German, but translations are available).

Eliassen and Kleinschmidt, *Dynamic Meteorology*, discuss cyclone life cycles from the potential vorticity perspective.

Davis and Emanuel (1991) develop a scheme for the piecewise inversion of potential vorticity. This scheme has been used extensively in diagnostic studies of mid-latitude dynamics.

Nielsen-Gammon and Lefevre (1996) develop a scheme for discerning height tendencies associated with a variety of physical processes through inversion of the QG potential vorticity.

Stoelinga (1996) examines the sources and sinks of potential vorticity in the life cycle of a robust mid-latitude cyclone.

Hoskins and Berrisford (1988) diagnose aspects of the Great October Storm of 1987 from a potential vorticity perspective.

Morgan and Nielsen-Gammon (1998) discuss the use of maps of potential temperature on the dynamic tropopause (what they term *tropopause maps*) in the diagnosis of mid-latitude cyclones.

Problems

9.1. (a) Imagine an upper tropospheric negative PV anomaly is introduced into a zonal, geostrophic vertical shear in the northern hemisphere as shown in Figure 9.1A. Based upon the characteristic thermodynamic structure of such a PV anomaly, what must be the distribution of vertical motion associated with this setting?

(b) Demonstrate that a consistent distribution of vertical motion can be diagnosed from the quasi-geostrophic omega-equation perspective.

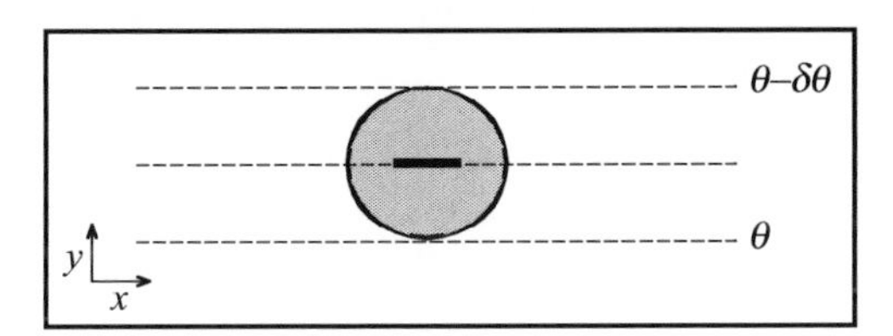

Figure 9.1 A

9.2. Show that for quasi-geostrophic flow on an f plane, geopotential height changes are governed by the flux divergence of quasi-geostrophic potential vorticity.

9.3. Construct a diagram analogous to Figure 9.10 that describes the process of anticyclogenesis, including the distribution of vertical motions associated with the process.

(a) Does the concept of mutual amplification apply in this case?

(b) Is there something specific to anti-cyclogenesis that conspires to limit rather than enhance the mutual amplification?

(c) Based upon your answer, suggest a physical reason why extreme negative departures from standard sea-level pressure (1013 hPa) exceed extreme positive departures.

9.4. Figure 9.2A shows a vertical cross-section through an upper tropospheric PV anomaly. How does the horizontal scale of the PV anomaly affect the magnitude of the vorticity anomaly? You may assume that the anomaly is axisymmetric.

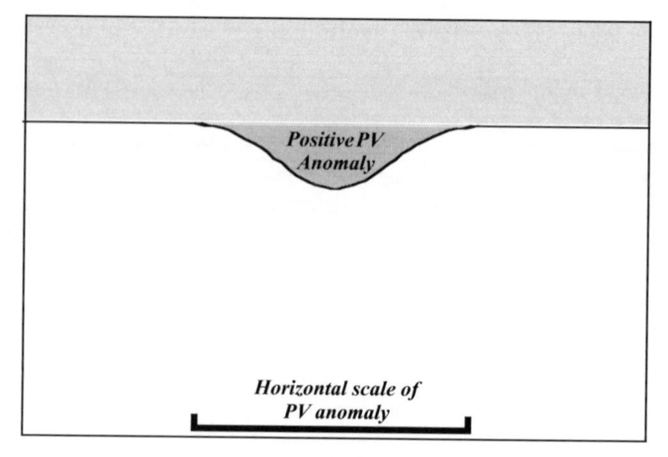

Figure 9.2A

9.5. Beginning with the isobaric expression for PV

$$PV = g(-f\hat{k} + \nabla \times \vec{V}_h) \cdot \nabla\theta,$$

develop an expression for PV in an atmosphere with a statically stable, purely zonal, baroclinic flow.

(a) Describe the contribution of the baroclinic basic state to PV in such an atmosphere.
(b) Under what conditions would this atmosphere exhibit dry symmetric instability?
(c) What does this suggest about the conditions necessary for large-scale slantwise motions such as those associated with baroclinic instability?

9.6. Beginning with (9.21), prove that

$$\frac{d(PV)}{dt} = -g(\eta_a \cdot \nabla\dot{\theta})$$

where

$$\frac{d}{dt} = \frac{\partial}{\partial t} + u\frac{\partial}{\partial x} + v\frac{\partial}{\partial y} + \omega\frac{\partial}{\partial p}.$$

9.7. (a) Figure 9.3A shows a vertical cross-section through a developing cyclone. If the maximum latent heat release occurs at about 500 hPa, sketch the diabatically altered 2 PVU surface (at some later time) in this cross-section. Explain your sketch.

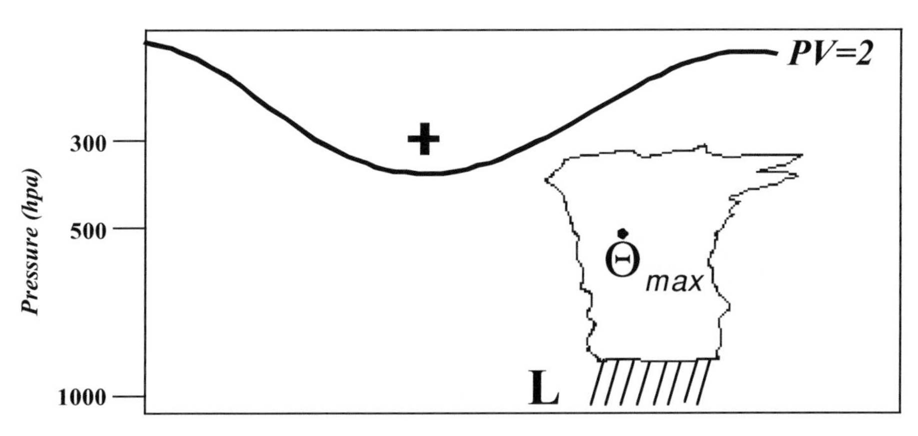

Figure 9.3 A

(b) Based upon your answer in (a), speculate as to the effect that the migration of tropical cyclones into the middle latitudes might have on the subsequent development of extratropical cyclones. Consider both the dynamic and thermodynamic impacts that such migration can have on mid-latitude weather systems.

9.8. For an atmosphere in thermal wind balance, demonstrate that a positive PV anomaly cannot be manifest solely as a positive static stability anomaly.

9.9. Perturbation PV anomalies are often defined as deviations from a time mean.

(a) Show that in such a case the local rate of change of the perturbation PV is the result of four distinct physical processes.
(b) Describe each of these processes in words.
(c) Estimate which two of these four are likely to be the largest in magnitude. Explain your answer.

9.10. By considering the characteristic distribution of clouds and precipitation in a developing cyclone, describe the effect of latent heat release (LHR) on

(a) the distribution of upper and lower tropospheric PV,
(b) the lower tropospheric vorticity, and
(c) the static stability.
(d) Describe in detail how the mutual interaction and amplification of upper and lower PV anomalies, depicted in Figure 9.10 (i.e. the PV view of cyclogenesis), is influenced by LHR.
(e) Explain the equivalence between the elements of the view outlined in (d) and the physical elements constituting the classical self-development theory of Sutcliffe and Forsdyke (1950).

9.11. Figure 9.4A depicts a lower tropospheric positive PV anomaly associated with the diabatic residue of precipitation along a frontogenetically active surface cold front.

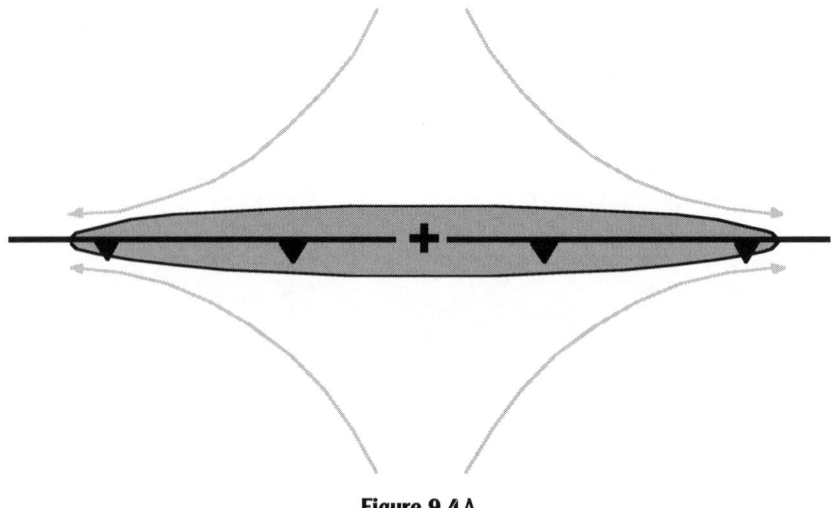

Figure 9.4 A

Assume that the surface front itself is the axis of dilatation of the total deformation field.

(a) Describe an alternative way to represent this elongated positive PV anomaly.
(b) With respect to the concept of PV superposition, describe what might occur if the frontogenetic deformation is systematically relaxed over time.
(c) Could the situation you've described in (b) be associated with a Petterssen Type A cyclogenesis event? Explain.

Solutions

9.4. As the horizontal scale increases, the magnitude of the vorticity decreases.

9.5. $PV = -g\frac{\partial u}{\partial p}\frac{\partial \theta}{\partial y} - g\left(f - \frac{\partial u}{\partial y}\right)\frac{\partial \theta}{\partial p}$

Appendix A
Virtual Temperature

The air has varying amounts of water vapor mixed into it. This highly variable constituent will serve to reduce the density of the air by virtue of the fact that its molecular weight (18 g mol^{-1}) is lower than that of 'dry' air whose apparent molecular weight is 28.97 g mol^{-1}. Consequently, the gas constant for 1 kg of moist air is larger than that for dry air. Application of the ideal gas law under this circumstance would necessitate employment of a variable gas constant whose value depends upon the exact water vapor content of the air. Alternatively, we can employ the gas constant for dry air in conjunction with an adjusted temperature known as the virtual temperature.

Imagine that a volume V of moist air is at temperature T and exerts a total pressure P. Further imagine that this moist air mixture contains a mass m_d of dry air and a mass m_v of water vapor. In such a case, the density of the moist air is given by

$$\rho = \frac{m_d + m_v}{V} = \rho_d' + \rho_v' \tag{A1}$$

where ρ_d' and ρ_v' are the fractional densities of the dry air and water vapor, respectively. We can apply the ideal gas law to both the water vapor and the dry air individually to yield

$$e = R_v \rho_v' T \tag{A2}$$

and

$$p_d' = R_d \rho_d' T \tag{A3}$$

where e and p_d' are the partial pressures exerted by the water vapor and dry air, while R_v and R_d are the gas constants for water vapor and dry air, respectively.

We know from Dalton's law of partial pressures that

$$P = p_d' + e. \tag{A4}$$

Mid-Latitude Atmospheric Dynamics Jonathan E. Martin

Combining (A2), (A3), and (A4) results in an alternative expression for density:

$$\rho = \frac{P - e}{R_d T} + \frac{e}{R_v T} \tag{A5}$$

or

$$\rho = \frac{P}{R_d T}\left[1 - \frac{e}{P}(1 - \varepsilon)\right] \tag{A6}$$

where ε, the ratio of the gas constants, is equivalently the ratio of the molecular weight of water vapor to that of dry air ($\varepsilon = M_w/M_d = 18/28.97 = 0.622$).

Equation (A6) can be rewritten as

$$P = R_d \rho T_v \tag{A7}$$

where

$$T_v = \frac{T}{1 - (e/P)(1 - \varepsilon)} \tag{A8}$$

and is known as the **virtual temperature**. Physically the virtual temperature is the temperature that dry air would have to have in order that its pressure and density be equal to those of a sample of moist air at temperature T.

The expression for virtual temperature can be simplified by again considering Dalton's law. The partial pressure exerted by any constituent in a mixture of gases is equal to the proportion of kilomoles of the constituent in the mixture. Thus, the vapor pressure, e, is given by

$$e = \left(\frac{m_v/M_w}{m_d/M_d + m_v/M_w}\right) P$$

or

$$e = \left(\frac{m_v M_w M_d}{M_w^2 m_d + m_v M_w M_d}\right) P = \left[\frac{(m_v/m_d) M_w M_d}{M_w^2 + (m_v/m_d) M_w M_d}\right] P. \tag{A9}$$

By definition, m_v/m_d is the mixing ratio (w) of water vapor for the gaseous mixture. Thus, (A9) can be expressed as

$$e = \left(\frac{w}{\varepsilon + w}\right) P. \tag{A10}$$

Substituting (A10) into (A8) yields

$$T_v = \frac{T}{1 - [w/(\varepsilon + w)](1 - \varepsilon)} = T\left(\frac{w + \varepsilon}{\varepsilon w + \varepsilon}\right) \tag{A11}$$

which can be further simplified by performing the indicated division and neglecting terms of order w^2 and higher. The final expression then becomes

$$T_v = T(1 + 0.61w) \tag{A12}$$

where w is expressed in units of kg kg^{-1}. Even for large values of w, the virtual temperature is usually only about 1% larger than the actual air temperature.

Bibliography

Acheson, D. J., 1990: *Elementary Fluid Dynamics*, Oxford University Press, New York.

Bennetts, D. A., and B. J. Hoskins, 1979: Conditional symmetric instability – a possible explanation for frontal rainbands. *Q. J. R. Meteorol. Soc.*, **105**, 945–962.

Bergeron, T., 1928: Über die dreidimensional verknüpfende Wetteranalyse I. *Geofys. Publ.*, **5**, 1–111.

Bjerknes, J., and H. Solberg, 1922: Life cycle of cyclones and the polar front theory of atmospheric circulation. *Geofys. Publ.*, **3**(1), 1–18.

Bleck, R., 1973: Numerical forecasting experiments based on conservation of potential vorticity on isentropic surfaces. *J. Appl. Meteorol.*, **12**, 737–752.

Bluestein, H., 1993: *Synoptic-Dynamic Meteorology in Midlatitudes, Volume I*, Oxford University Press, New York.

Bluestein, H., 1993: *Synoptic-Dynamic Meteorology in Midlatitudes, Volume II*, Oxford University Press, New York.

Bretherton, F. P., 1966: Baroclinic instability and the short wavelength cutoff in terms of potential vorticity. *Q. J. R. Meteorol. Soc.*, **92**, 335–345.

Brown, R. A., 1991: *Fluid Mechanics of the Atmosphere*, Academic Press, Orlando, FL.

Cammas, J.-P., D. Keyser, G. M. Lackmann, and J. Molinari, 1994: Diabatic redistribution of potential vorticity accompanying the development of an outflow jet within a strong extratropical cyclone. Preprints, *Int. Symp. On the Life Cycles of Extratropical Cyclones, Volume II*, Bergen, Norway, Geophysical Institute, University of Bergen, 403–409.

Carlson, T. N., 1991: *Mid-Latitude Weather Systems*, Routledge, New York.

Crocker, A. M., W. L. Godson, and C. M. Penner, 1947: Frontal contour charts. *J. Meteorol.*, **4**, 95–99.

Danielsen, E. F., 1964: Project Springfield report. DASA 1517, Defense Atomic Support Agency, Washington, DC [NTIS AD-607980].

Davis, C. A., 1997: The modification of baroclinic waves by the Rocky Mountains. *J. Atmos. Sci.*, **54**, 848–868.

Davis, C., and K. A. Emanuel, 1991: Potential vorticity diagnostics of cyclogenesis. *Mon. Weather Rev.*, **119**, 1929–1953.

Eliassen, A., 1962: On the vertical circulation in frontal zones. *Geofys. Publ.*, **24**, 147–160.

Eliassen, A., 1984: Geostrophy. *Q. J. R. Meteorol. Soc.*, **110**, 1–12.

Eliassen, A., and E. Kleinschmidt, 1957: Dynamic meteorology, in *Handbuch der Physik*, Vol. 48, 1–154. Springer-Verlag, Berlin.

Mid-Latitude Atmospheric Dynamics Jonathan E. Martin

Emanuel, K. A., 1979: Lagrangian parcel dynamics of moist symmetric stability. *J. Atmos. Sci.*, **36**, 2368–2376.

Ertel, H., 1942: Ein neuer hydrodynamischer Wirbelsatz. *Meteor. Z.*, **59**, 271–281.

Galloway, J. L., 1958: The three-front model: its philosophy, nature, construction and use. *Weather*, **13**, 3–10.

Galloway, J. L., 1960: The three-front model, the developing depression and the occluding process. *Weather*, **15**, 293–301.

Godson, W. L., 1951: Synoptic properties of frontal surfaces. *Q. J. R. Meteorol. Soc.*, **77**, 633–653.

Halliday, D., and R. Resnick, 1981: *Fundamentals of Physics* (2nd Edn), John Wiley & Sons, Inc., New York.

Hess, S. L., 1959. *Introduction to Theoretical Meteorology*, Holt, New York.

Holton, J. R., 1992: *An Introduction to Dynamic Meteorology* (3rd Edn), Academic Press, New York.

Hoskins, B. J., and P. Berrisford, 1988: A potential vorticity perspective of the storm of 15-16 October 1987. *Weather*, **43**, 122–129.

Hoskins, B. J., and F. P. Bretherton, 1972: Atmospheric frontogenesis models: mathematical formulation and solution. *J. Atmos. Sci.*, **29**, 11–37.

Hoskins, B. J., and M. A. Pedder, 1980: The diagnosis of mid-latitude synoptic development. *Q. J. R. Meteorol. Soc.*, **106**, 707–719.

Hoskins, B. J., I. Draghici, and H. C. Davies, 1978: A new look at the ω-equation. *Q. J. R. Meteorol. Soc.*, **104**, 31–38.

Hoskins, B. J., M. E. McIntyre, and A. W. Robertson, 1985: On the use and significance of isentropic potential vorticity maps. *Q. J. R. Meteorol. Soc.*, **111**, 877–946.

Keyser, D., and M. A. Shapiro, 1986: A review of the structure and dynamics of upper-level frontal zones. *Mon. Weather Rev.*, **114**, 452–496.

Keyser, D., B. D. Schmidt, and D. G. Duffy, 1992: Quasigeostrophic vertical motions diagnosed from along- and cross-isentrope components of the Q vector. *Mon. Weather Rev.*, **120**, 731–741.

Korner, S. O., and J. E. Martin, 2000: Piecewise frontogenesis from a potential vorticity perspective: Methodology and a case study. *Mon. Weather Rev.*, **128**, 1266–1288.

Lackmann, G. M., 2002: Cold-frontal potential vorticity maxima, the low level jet, and moisture transport in extratropical cyclones. *Mon. Weather Rev.*, **130**, 59–74.

Margules, M., 1906: Uber temperaturschichtung in stationar bewegter und ruhender luft Hann-Band. *Meteorol. Z.*, 243–254.

Martin, J. E., 1998a: The structure and evolution of a continental winter cyclone. Part I: Frontal structure and the classical occlusion process. *Mon. Weather Rev.*, **126**, 303–328.

Martin, J. E., 1998b: On the deformation term in the quasi-geostrophic omega equation. *Mon. Weather Rev.*, **126**, 2000–2007.

Martin, J. E., 1999: Quasi-geostrophic forcing of ascent in the occluded sector of cyclones and the trowal airstream. *Mon. Weather Rev.*, **127**, 70–88.

Martin, J. E., 2006: The role of shearwise and transverse quasi-geostrophic vertical motions in the mid-latitude cyclone life cycle. *Mon. Weather Rev.*, **134** (in press).

Martin, J. E., and N. Marsili, 2002: Surface cyclolysis in the north Pacific Ocean. Part II: Piecewise potential vorticity analysis of a rapid cyclolysis event. *Mon. Weather Rev.*, **130**, 1264–1281.

Martin, J. E., and J. A. Otkin, 2004: The rapid growth and decay of an extratropical cyclone over the central Pacific Ocean. *Weather and Forecasting*, **19**, 358–376.

Martin, J. E., J. P. Locatelli, and P. V. Hobbs, 1992: Organization and structure of clouds and precipitation on the mid-Atlantic coast of the United States. Part V: The role of an upper-level front in the generation of a rainband. *J. Atmos. Sci.*, **49**, 1293–1303.

Martin, J. E., R. A. Grauman, and N. Marsili, 2001: Surface cyclolysis in the north Pacific Ocean. Part I: A synoptic-climatology. *Mon. Weather Rev.*, **129**, 748–765.

McLay, J. G., and J. E. Martin, 2002: Surface cyclolysis in the North Pacific Ocean. Part III: Composite local energetics of tropospheric-deep cyclone decay associated with rapid surface cyclolysis. *Mon. Weather Rev.*, **130**, 2507–2529.

Miller, J. E., 1948: On the concept of frontogenesis. *J. Meteorol.*, **5**, 169–171.

Montgomery, R. B., 1937: A suggested method for representing gradient flow in isentropic surfaces. *Bull. Am. Meteorol. Soc.*, **18**, 210–212.

Moore, J. T., and T. E. Lambert, 1993: The use of equivalent potential vorticity to diagnose regions of conditional symmetric instability. *Weather and Forecasting*, **8**, 301–308.

Morgan, M. C., and J. W. Nielsen-Gammon, 1998: Using tropopause maps to diagnose midlatitude weather systems. *Mon. Weather Rev.*, **126**, 2555–2579.

Nielsen-Gammon, J. W., and R. J. Lefevre, 1996: Piecewise tendency diagnosis of dynamical processes governing the development of an upper-tropospheric mobile trough. *J. Atmos. Sci.*, **53**, 3120–3142.

Palmén, E., and C. W. Newton, 1969: *Atmospheric Circulation Systems*, Academic Press, New York.

Penner, C. M., 1955: A three-front model for synoptic analyses. *Q. J. R. Meteorol. Soc.*, **81**, 89–91.

Petterssen, S., 1936: Contribution to the theory of frontogenesis. *Geofys. Publ.*, **11**, 1–27.

Petterssen, S., 1957: *Weather Analysis and Forecasting*, McGraw-Hill, New York.

Petterssen, S., and S. J. Smebye, 1971: On the development of extratropical cyclones. *Q. J. R. Meteorol. Soc.*, **97**, 457–482.

Petterssen, S., D. L. Bradbury, and K. Pedersen, 1962: The Norwegian cyclone models in relation to heat and cold sources. *Geofys. Publ.*, **24**, 243–280.

Posselt, D. J., and J. E. Martin, 2004: The effect of latent heat release on the evolution of a warm occluded thermal structure. *Mon. Weather Rev.*, **132**, 578–599.

Reed, R. J., 1955: A study of a characteristic type of upper-level frontogenesis. *J. Meteorol.*, **12**, 226–237.

Reed, R. J., and E. F. Danielsen, 1959: Fronts in the vicinity of the tropopause. *Arch. Meteorol. Geophys. Bioklimatol.*, **A11**, 1–17.

Reed, R. J., and F. Sanders, 1953: An investigation of the development of a mid-tropospheric frontal zone and its associated vorticity field. *J. Meteorol.*, **10**, 338–349.

Roebber, P., 1984: Statistical analysis and updated climatology of explosive cyclones. *Mon. Weather Rev.*, **112**, 1577–1589.

Sanders, F., 1955: An investigation of the structure and dynamics of an intense surface frontal zone. *J. Meteorol.*, **12**, 542–552.

Sanders, F., and L. F. Bosart, 1985: Mesoscale structure in the Megalopolitan snowstorm, 11-12 February 1983. Part I: Frontogenetical forcing and symmetric instability. *J. Atmos. Sci.*, **42**, 1050–1061.

Sanders, F., and J. R. Gyakum, 1980: Synoptic-dynamic climatology of the "bomb". *Mon. Weather Rev.*, **108**, 1589–1606.

Sanders, F., and B. J. Hoskins, 1990: An easy method for estimating *Q*-vectors from weather maps. *Weather and Forecasting*, **5**, 346–353.

Saucier, W. J., 1955: *Principles of Meteorological Analysis*, University of Chicago Press, Chicago.

Sawyer, J. S., 1956: The vertical circulation at meteorological fronts and its relation to frontogenesis. *Proc. R. Soc.*, **A234**, 246–262.

Schultz, D. M., and C. F. Mass, 1993: The occlusion process in a midlatitude cyclone over land. *Mon. Weather Rev.*, **121**, 918–940.

Schultz, D. M., and P. N. Schumacher, 1999: The use and misuse of conditional symmetric instability. *Mon. Weather Rev.*, **127**, 2709–2732.

Spiegel, M. R., 1959: *Vector Analysis and an Introduction to Tensor Analysis*, McGraw-Hill, New York.

Stoelinga, M., 1996: A potential vorticity-based study of the role of diabatic heating and friction in a numerically simulated baroclinic cyclone. *Mon. Weather Rev.*, **124**, 849–874.

Sutcliffe, R. C., 1938: On development in the field of barometric pressure. *Q. J. R. Meteorol. Soc.*, **64**, 495–504.

Sutcliffe, R. C., 1939: Cyclonic and anticyclonic development. *Q. J. R. Meteorol. Soc.*, **65**, 518–524.

Sutcliffe, R. C., 1947: A contribution to the problem of development. *Q. J. R. Meteorol. Soc.*, **73**, 370–383.

Sutcliffe, R. C., and A. G. Forsdyke, 1950: The theory and use of upper air thickness patterns in forecasting. *Q. J. R. Meteorol. Soc.*, **76**, 189–217.

Sutcliffe, R. C., and O. H. Godart, 1942: Isobaric analysis. *SDTM 50*, Meteorological Office, London.

Thomas, G. B. Jr, and R. L. Finney, 1980: *Calculus and Analytic Geometry*, Addison-Wesley, London.

Thorpe, A. J., 1985: Diagnosis of balanced vortex structure using potential vorticity. *J. Atmos. Sci.*, **42**, 397–406.

Trenberth, K. E., 1978: On the interpretation of the diagnostic quasi-geostrophic omega equation. *Mon. Weather Rev.*, **106**, 131–137.

Uccellini, L. W., 1990: Processes contributing to the rapid development of extratropical cyclones, in *Extratropical Cyclones: The Erik Palmen Memorial Volume*, C. W. Newton and E. O. Holopainen, Eds, American Meteorological Society, Boston, MA, 81–105.

Warsh, K. L., K. L. Echternacht, and M. Garstang, 1971: Structure of near-surface currents east of Barbados. *J. Phys. Oceanogr.*, **1**, 123–129.

Williams, J., and S. A. Elder, 1989: *Fluid Physics for Oceanographers and Physicists*, Pergamon Press, New York.

Winn-Nielsen, A., 1959: On a graphical method for an approximate determination of the vertical velocity in the mid-troposphere. *Tellus*, **11**, 432–440.

Zishka, K. M., and P. J. Smith, 1980: The climatology of cyclones and anticyclones over North America and surrounding ocean environs for January and July, 1950–1977. *Mon. Weather Rev.*, **108**, 387–401.

Index

Q

R

Printed and bound in the UK by
CPI Antony Rowe, Eastbourne